Ionic Liquids in Catalysis

Ionic Liquids in Catalysis

Editor

Hieronim Maciejewski

MDPI • Basel • Beijing • Wuhan • Barcelona • Belgrade • Manchester • Tokyo • Cluj • Tianjin

Editor
Hieronim Maciejewski
Adam Mickiewicz University in Poznan
Poland

Editorial Office
MDPI
St. Alban-Anlage 66
4052 Basel, Switzerland

This is a reprint of articles from the Special Issue published online in the open access journal *Catalysts* (ISSN 2073-4344) (available at: https://www.mdpi.com/journal/catalysts/special_issues/ion_liquid).

For citation purposes, cite each article independently as indicated on the article page online and as indicated below:

LastName, A.A.; LastName, B.B.; LastName, C.C. Article Title. *Journal Name* **Year**, *Volume Number*, Page Range.

ISBN 978-3-0365-1246-4 (Hbk)
ISBN 978-3-0365-1247-1 (PDF)

© 2021 by the authors. Articles in this book are Open Access and distributed under the Creative Commons Attribution (CC BY) license, which allows users to download, copy and build upon published articles, as long as the author and publisher are properly credited, which ensures maximum dissemination and a wider impact of our publications.

The book as a whole is distributed by MDPI under the terms and conditions of the Creative Commons license CC BY-NC-ND.

Contents

About the Editor vii

Hieronim Maciejewski
Ionic Liquids in Catalysis
Reprinted from: *Catalysts* **2021**, *11*, 367, doi:10.3390/catal11030367 1

Mar López, Sandra Rivas, Carlos Vila, Valentín Santos and Juan Carlos Parajó
Performance of 1-(3-Sulfopropyl)-3-Methylimidazolium Hydrogen Sulfate as a Catalyst for Hardwood Upgrading into Bio-Based Platform Chemicals
Reprinted from: *Catalysts* **2020**, *10*, 937, doi:10.3390/catal10080937 5

Katarzyna Glińska, Clara Lerigoleur, Jaume Giralt, Esther Torrens and Christophe Bengoa
Valorization of Cellulose Recovered from WWTP Sludge to Added Value Levulinic Acid with a Brønsted Acidic Ionic Liquid
Reprinted from: *Catalysts* **2020**, *10*, 1004, doi:10.3390/catal10091004 23

Jinyan Lang, Junliang Lu, Ping Lan, Na Wang, Hongyan Yang and Heng Zhang
Preparation of 5-HMF in a DES/Ethyl N-Butyrate Two-Phase System
Reprinted from: *Catalysts* **2020**, *10*, 636, doi:10.3390/catal10060636 39

Aleksander Grymel, Piotr Latos, Karolina Matuszek, Karol Erfurt, Natalia Barteczko, Ewa Pankalla and Anna Chrobok
Sustainable Method for the Synthesis of Alternative Bis(2-Ethylhexyl) Terephthalate Plasticizer in the Presence of Protic Ionic Liquids
Reprinted from: *Catalysts* **2020**, *10*, 457, doi:10.3390/catal10040457 51

Katia Bacha, Kawther Aguibi, Jean-Pierre Mbakidi and Sandrine Bouquillon
Beneficial Contribution of Biosourced Ionic Liquids and Microwaves in the Michael Reaction
Reprinted from: *Catalysts* **2020**, *10*, 814, doi:10.3390/catal10080814 63

Olga Bartlewicz, Izabela Dąbek, Anna Szymańska and Hieronim Maciejewski
Heterogeneous Catalysis with the Participation of Ionic Liquids
Reprinted from: *Catalysts* **2020**, *10*, 1227, doi:10.3390/catal10111227 79

Nemanja Vucetic, Pasi Virtanen, Ayat Nuri, Andrey Shchukarev, Jyri-Pekka Mikkola and Tapio Salmi
Tuned Bis-Layered Supported Ionic Liquid Catalyst (SILCA) for Competitive Activity in the Heck Reaction of Iodobenzene and Butyl Acrylate
Reprinted from: *Catalysts* **2020**, *10*, 963, doi:10.3390/catal10090963 99

Rafal Kukawka, Anna Pawlowska-Zygarowicz, Rafal Januszewski, Joanna Dzialkowska, Mariusz Pietrowski, Michal Zielinski, Hieronim Maciejewski and Marcin Smiglak
SILP Materials as Effective Catalysts in Selective Monofunctionalization of 1,1,3,3-Tetramethyldisiloxane
Reprinted from: *Catalysts* **2020**, *10*, 1414, doi:10.3390/catal10121414 117

Gabriela Dobras, Kornela Kasperczyk, Sebastian Jurczyk and Beata Orlińska
N-Hydroxyphthalimide Supported on Silica Coated with Ionic Liquids Containing $CoCl_2$ (SCILLs) as New Catalytic System for Solvent-Free Ethylbenzene Oxidation
Reprinted from: *Catalysts* **2020**, *10*, 252, doi:10.3390/catal10020252 129

Olga Bartlewicz, Magdalena Jankowska-Wajda and Hieronim Maciejewski
Highly Efficient and Reusable Alkyne Hydrosilylation Catalysts Based on Rhodium Complexes Ligated by Imidazolium-Substituted Phosphine
Reprinted from: *Catalysts* **2020**, *10*, 608, doi:10.3390/catal10060608 . . . . **143**

Magdalena Jankowska-Wajda, Olga Bartlewicz, Przemysław Pietras and Hieronim Maciejewski
Piperidinium and Pyrrolidinium Ionic Liquids as Precursors in the Synthesis of New Platinum Catalysts for Hydrosilylation
Reprinted from: *Catalysts* **2020**, *10*, 919, doi:10.3390/catal10080919 . . . . **157**

Jakub Szyling, Tomasz Sokolnicki, Adrian Franczyk and Jędrzej Walkowiak
Ru-Catalyzed Repetitive Batch Borylative Coupling of Olefins in Ionic Liquids or Ionic Liquids/scCO$_2$ Systems
Reprinted from: *Catalysts* **2020**, *10*, 762, doi:10.3390/catal10070762 . . . . **171**

About the Editor

Hieronim Maciejewski received his M.Sc. (1986) from Poznań University of Technology, Poland and his Ph.D. (1995) and D.Sc. (2004) from Adam Mickiewicz University in Poznań, Poland. Since 2014, he has been a full professor of chemistry. He is the head of the Department of Chemistry and Technology of Silicon Compounds at the Faculty of Chemistry at Adam Mickiewicz University in Poznań. Prof. Maciejewski is also the vice president of the Management Board of A. Mickiewicz University Foundation and the deputy director of Poznań Science and Technology Park. He is the author or co-author of over 180 publications, 90 patents, 11 licenses and 42 technologies as well as 12 book chapters and 1 book. Prof. Maciejewski has participated in many scientific and research projects and co-operated with various industrial partners. His research activity is focused on catalysis with organometallic and inorganometallic compounds (also with the use of ionic liquids). In addition, prof. Maciejewski is interested in properties and applications of inorganic polymers and hybrid materials (especially organosilicon compounds).

Editorial

Ionic Liquids in Catalysis

Hieronim Maciejewski [1,2]

1. Faculty of Chemistry, Adam Mickiewicz University, Uniwersytetu Poznańskiego 8, 61-614 Poznań, Poland; maciejm@amu.edu.pl
2. Adam Mickiewicz University Foundation, Poznań Science and Technology Park, Rubież 46, 61-612 Poznań, Poland

Citation: Maciejewski, H. Ionic Liquids in Catalysis. *Catalysts* 2021, 11, 367. https://doi.org/10.3390/catal11030367

Received: 8 March 2021
Accepted: 9 March 2021
Published: 11 March 2021

Publisher's Note: MDPI stays neutral with regard to jurisdictional claims in published maps and institutional affiliations.

Copyright: © 2021 by the author. Licensee MDPI, Basel, Switzerland. This article is an open access article distributed under the terms and conditions of the Creative Commons Attribution (CC BY) license (https://creativecommons.org/licenses/by/4.0/).

Ionic liquids play a larger and larger as well as more and more diversified role in catalysis [1–4], which is reflected by the fact that the number of literature reports on this subject increases annually by over 400 items. In the case of catalytic reactions, the ionic liquids enable us to obtain higher selectivities and yields and, first and foremost, they allow easy isolation of catalysts from the post-reaction mixture. Until now, the most popular application of ionic liquids is their use as a solvent and immobilizing agent. Their low vapor pressure and high thermal stability make them an alternative to classic organic solvents (the so-called green alternative). Nevertheless, at the beginning of this century, some critical voices were raised that pointed to the often complicated synthesis of the ionic liquids, which was also waste-generating and expensive. The shortcomings of ionic liquids also included their low biodegradability (or its absence) and sometimes their toxicity [5–7]. This is why a search has been made for the development of synthesis methods with a limited amount of waste (or waste-free ones), based first of all on the use of biodegradable and non-toxic starting materials, often of natural origin [8–11]. Due to the selection of new methods of synthesis, as well as the application of renewable and biodegradable raw materials, a larger and larger group of derivatives is considered safe and environmentally friendly, and the application trends include more and more areas. At present, one can notice a keen interest in two areas of the applications of ionic liquids. One of them is biotransformation (in a broad sense) that is used in the processing of biomass and lignocellulosic waste to obtain valuable chemical products [12–14]. The other widely developing trend is obtaining heterogeneous catalytic systems with the participation of ionic liquids, i.e., SILPC or SILCA (Supported Ionic Liquids Phase Catalysts or Supported Ionic Liquids Catalysts) [15–17] and SCILL (Supported Catalysts with Ionic Liquid Layer) [15,18–20] as well as catalysts in which ionic liquid makes a structural element [21–24]. In systems of this type, all problems observed in the case of catalysis in homogeneous systems are eliminated. First of all, there is no problem with the isolation of catalyst which is heterogeneous. Since in the case of SILPC and SCILL, the ionic liquid covers the supports with a thin layer, a small amount of the ionic liquid is needed which results in a considerable cost reduction (despite the high price of ionic liquids). Moreover, due to the thinness of the layer, the liquid viscosity does not influence the process course and mass transfer is not a problem. Such a form of catalyst can be applied in the continuous flow reactors what also influences the process economics.

The subject of papers published in this Special Issue reflects the two dominating trends of interest. The first one concerns biomass processing and the synthesis and application of ionic liquids from biorenewable resources. Parajo et al. presented an effective (one or two-step) method for the processing of eucalyptus wood towards producing furfural and levulinic acid with the use of acidic ionic liquid (C_3SO_3Hmim)HSO_3 as a catalyst [25]. Bengoa et al. described the processing of cellulose (obtained from municipal and industrial sewage) into valuable levulinic acid, while also using a Brønsted acidic ionic liquid as a catalyst [26]. During the conversion of cellulose to levulinic acid, 5-hydroxymethylfurfural (5-HMF) is formed as a by-product. By rehydration reaction of the latter, it can be converted into levulinic acid. The process of cellulose degradation towards 5-HMF and glucose in the presence of DES (deep eutectic solvents based on oxalic acid/choline chloride) and metal chlorides ($CrCl_3$ or $SnCl_4$) has been presented by Zhang's research group [27]. Brønsted acidic ionic liquids based on trimethylamine and sulfuric acid were employed as a solvent and catalyst in the synthesis of bis(2-ethylhexyl) terephthalate which is used as a plasticizer in the manufacture of plastics coming into contact with food. This new approach, which is characterized by the replacement of conventional acids and organometallic compounds with ionic liquids, was described by Chrobok's research group [28]. Bouquillon et al. presented the synthesis of a series of biobased ionic liquids with natural carboxylates (proline-based ionic liquids) [29]. These compounds were employed as a solvent and basic catalyst in Michael reactions.

The second group of papers concerns the synthesis and activity of heterogeneous catalytic systems. In the review article [30], the most important issues of heterogeneous catalysis with the participation of ionic liquids were presented. The methods of synthesis and exemplary trends in the application of SILP (SILCA), SCILL, and porous ionic liquids were described in the review. Subsequent papers concern the application of heterogeneous systems in particular types of reactions. Vucetic et al. described the use of $PdCl2$-containing SILCA as a catalyst for a Heck reaction [31]. The catalyst showed high stability and activity in the reaction between butylacrylate and iodobenzene and enabled its multiple use. A highly active SILP material that contained a rhodium complex was applied in the synthesis of monofunctional derivatives of disiloxanes. The above catalyst, described in the paper by Kukawka et al. [32], was characterized by high stability that enabled at least 50 catalytic runs without loss of activity. Orlińska et al. presented a new SCILL catalytic system with N-hydroxyphthalimide as an active component and applied it in the oxidation of ethylbenzene [33]. In the next two papers, heterogeneous catalysts containing ionic liquids as structural elements were described. Bartlewicz et al. [34] presented a rhodium complex ligated by imidazolium-substituted phosphine that showed high activity for hydrosilylation of alkynes. A series of anionic platinum complexes obtained by a simple reaction of piperidinium or pyrrolidinium ionic liquids with platinum compounds were presented by Jankowska-Wajda et al. [35]. The complexes prepared in such a way appeared to be very active catalysts for hydrosilylation of different olefins. It is worth mentioning that their high activity was maintained even after multiple use. Szyling et al. [36] described an effective immobilization of a ruthenium complex in ionic liquids and applied the obtained complex as a catalyst for the selective synthesis of (E)-alkenyl boronates via borylative coupling of olefins with vinyl boronic acid pinacol ester. Additional application of $scCO_2$ to the extraction of the product significantly reduced catalyst leaching, due to which it could be used many times.

All the above papers showed the great importance of ionic liquids in practice and that their catalytic application is growing wider and wider; particularly, they can play various functions in catalysis. In addition to the known advantages of the application of ionic liquids, the development of advanced methods of their synthesis from biorenewable natural resources results in their biodegradable and environmentally friendly character. For this reason, one can expect further development of their application, including in different catalytic processes.

Funding: This research received no external funding.

Data Availability Statement: Data available in a publicly accessible repository.

Conflicts of Interest: The authors declare no conflict of interest.

References

1. Hallet, J.P.; Welton, T. Room-temperature ionic liquids: Solvents for synthesis and catalysis. *Chem. Rev.* **2011**, *111*, 3508–3576. [CrossRef]
2. Dyson, P.J.; Geldbach, T.J. *Metal Catalysed Reactions in Ionic Liquids*; Springer: Dordrecht, The Netherlands, 2005.
3. Dupont, J.; Kollar, L. *Ionic Liquids (ILs) in Organometallic Catalysis*; Springer: Berlin, Germany, 2015.
4. Hardacre, C.; Parvulescu, V. *Catalysis in Ionic Liquids. From Catalysts Synthesis to Applications*; RS: Cambridge, UK, 2014.

5. Clark, J.H.; Tavener, S.J. Alternative Solvents: Shades of Green. *Org. Process. Res. Dev.* **2007**, *11*, 149–155. [CrossRef]
6. Jessop, P.G. Searching for green solvents. *Green Chem.* **2011**, *13*, 1391–1398. [CrossRef]
7. Cevasco, G.; Chiappe, C. Are ionic liquids a proper solution to current environmental challenges? *Green Chem.* **2014**, *16*, 2375–2385. [CrossRef]
8. Ozokwelu, D.; Zhang, S.; Okafor, O.C.; Cheng, W.; Litombe, N. *Novel Catalytic and Separation Processes Based on Ionic Liquids*; Elsevier: Amsterdam, The Netherlands, 2017.
9. Basudeb, S.; Fan, M.; Wang, J. (Eds.) *Sustainable Catalytic Process*; Elsevier: Amsterdam, The Netherlands, 2015.
10. Dupont, J.; Itoh, T.; Lozano, P.; Malhotra, S.V. (Eds.) *Environmentally Friendly Syntheses Using Ionic Liquids*; CRC Press: New York, NY, USA, 2015.
11. Lozano, P. *Sustainable Catalysis in Ionic Liquids*; CRC Press: New York, NY, USA, 2019.
12. Itoh, T.; Koo, Y.-M. *Application of Ionic Liquids in Biotechnology*; Springer: Cham, Switzerland, 2019.
13. Stevens, J.C.; Shi, J. Biocatalysis in ionic liquids for lignin valorization: Opportunities and recent developments. *Biotechnol. Adv.* **2019**, *37*, 107418. [CrossRef] [PubMed]
14. Quiroz, N.R.; Norton, A.M.; Nguyen, H.; Vasileiadou, E.; Vlachos, D.G. Homogeneous Metal Salt Solutions for Biomass Upgrading and Other Select Organic Reactions. *ACS Catal.* **2019**, *9*, 9923–9952. [CrossRef]
15. Fehrmann, R.; Riisager, A.; Haumann, A.M. *Supported Ionic Liquids*; Fundamentals and Applications; Wiley-VCH: Weinheim, Germany, 2014.
16. Kaur, P.; Chopra, H. Recent Advances of Supported Ionic Liquids. *Curr. Org. Chem.* **2019**, *23*, 2815–2881. [CrossRef]
17. Fehér, C.; Papp, M.; Urban, B.; Skoda-Földes, R. Catalytic Applications of Supported Ionic Liquid Phases. In *Advances in Asymmetric Autocatalysis and Related Topics*; Chapter 17; Academic Press: Cambridge, MA, USA, 2017.
18. Steinrück, H.-P.; Wasserscheid, P. Ionic Liquids in Catalysis. *Catal. Lett.* **2015**, *145*, 380–397. [CrossRef]
19. Kernchen, U.; Etzold, B.; Korth, W.; Jess, A. Solid Catalyst with Ionic Liquid Layer (SCILL)—A new concept to improve the selectivity investigated for the example of hydrogenation of cyclooctadiene. *Chem. Eng. Technol.* **2007**, *79*, 807–819.
20. Werner, S.R.L.; Szesni, N.; Kaiser, M.; Haumann, M.; Wasserscheid, P. A Scalable Preparation Method for SILP and SCILL Ionic Liquid Thin-Film Materials. *Chem. Eng. Technol.* **2012**, *35*, 1962–1967. [CrossRef]
21. Li, J.; Peng, J.; Wang, D.; Bai, Y.; Jiang, J.; Lai, G. Hydrosilylation reactions catalysed by rhodium complex with phosphine ligands functionalized with imidazolium salts. *J. Organomet. Chem.* **2011**, *696*, 263–268. [CrossRef]
22. Jankowska-Wajda, M.; Bartlewicz, O.; Szpecht, A.; Zajac, A.; Smiglak, M.; Maciejewski, H. Platinum and rhodium complexes ligated by imidazolium-substituted phosphine as efficient and recyclable catalysts for hydrosilylation. *RSC Adv.* **2019**, *9*, 29396–29404. [CrossRef]
23. Wilkes, J.S. A short history of ionic liquids—from molten salts to neoteric solvents. *Green Chem.* **2002**, *4*, 73–80. [CrossRef]
24. Jankowska-Wajda, M.; Bartlewicz, O.; Walczak, A.; Stefankiewicz, A.; Maciejewski, H. Highly efficient hydrosilylation cata-lysts based on chloroplatinate "ionic liquids". *J. Catal.* **2019**, *374*, 266–275. [CrossRef]
25. López, M.; Rivas, S.; Vila, C.; Santos, V.; Parajó, J.C. Performance of 1-(3-Sulfopropyl)-3-Methylimidazolium Hydrogen Sulfate as a Catalyst for Hardwood Upgrading into Bio-Based Platform Chemicals. *Catalysts* **2020**, *10*, 937. [CrossRef]
26. Glińska, K.; Lerigoleur, C.; Giralt, J.; Torrens, E.; Bengoa, C. Valorization of Cellulose Recovered from WWTP Sludge to Added Value Levulinic Acid with a Brønsted Acidic Ionic Liquid. *Catalysts* **2020**, *10*, 1004. [CrossRef]
27. Lang, J.; Lu, J.; Lan, P.; Wang, N.; Yang, H.; Zhang, H. Preparation of 5-HMF in a DES/Ethyl N-Butyrate Two-Phase System. *Catalysts* **2020**, *10*, 636. [CrossRef]
28. Grymel, A.; Latos, P.; Matuszek, K.; Erfurt, K.; Barteczko, N.; Pankalla, E.; Chrobok, A. Sustainable Method for the Synthesis of Alternative Bis(2-Ethylhexyl) Terephthalate Plasticizer in the Presence of Protic Ionic Liquids. *Catalysts* **2020**, *10*, 457. [CrossRef]
29. Bacha, K.; Aguibi, K.; Mbakidi, J.-P.; Bouquillon, S. Beneficial Contribution of Biosourced Ionic Liquids and Microwaves in the Michael Reaction. *Catalysts* **2020**, *10*, 814. [CrossRef]
30. Bartlewicz, O.; Dąbek, I.; Szymańska, A.; Maciejewski, H. Heterogeneous Catalysis with the Participation of Ionic Liquids. *Catalysts* **2020**, *10*, 1227. [CrossRef]
31. Vucetic, N.; Virtanen, P.; Nuri, A.; Shchukarev, A.; Mikkola, J.-P.; Salmi, T. Tuned Bis-Layered Supported Ionic Liquid Catalyst (SILCA) for Competitive Activity in the Heck Reaction of Iodobenzene and Butyl Acrylate. *Catalysts* **2020**, *10*, 963. [CrossRef]
32. Kukawka, R.; Pawlowska-Zygarowicz, A.; Januszewski, R.; Dzialkowska, J.; Pietrowski, M.; Zielinski, M.; Maciejewski, H.; Smiglak, M. SILP Materials as Effective Catalysts in Selective Monofunctionalization of 1,1,3,3-Tetramethyldisiloxane. *Catalysts* **2020**, *10*, 1414. [CrossRef]
33. Dobras, G.; Kasperczyk, K.; Jurczyk, S.; Orlińska, B. N-Hydroxyphthalimide Supported on Silica Coated with Ionic Liquids Containing CoCl2 (SCILLs) as New Catalytic System for Solvent-Free Ethylbenzene Oxidation. *Catalysts* **2020**, *10*, 252. [CrossRef]
34. Bartlewicz, O.; Jankowska-Wajda, M.; Maciejewski, H. Highly efficient and reusable alkyne hydrosilylation catalysts based on rhodium complexes ligated by imidazolium-substituted phosphine. *Catalysts* **2020**, *10*, 608. [CrossRef]
35. Jankowska-Wajda, M.; Bartlewicz, O.; Pietras, P.; Maciejewski, H. Piperidinium and Pyrrolidinium Ionic Liquids as Precursors in the Synthesis of New Platinum Catalysts for Hydrosilylation. *Catalysts* **2020**, *10*, 919. [CrossRef]
36. Szyling, J.; Sokolnicki, T.; Franczyk, A.; Walkowiak, J. Ru-Catalyzed Repetitive Batch Borylative Coupling of Olefins in Ionic Liquids or Ionic Liquids/scCO$_2$ Systems. *Catalysts* **2020**, *10*, 762. [CrossRef]

Article

Performance of 1-(3-Sulfopropyl)-3-Methylimidazolium Hydrogen Sulfate as a Catalyst for Hardwood Upgrading into Bio-Based Platform Chemicals

Mar López, Sandra Rivas, Carlos Vila, Valentín Santos and Juan Carlos Parajó *

Chemical Engineering Department, University of Vigo (Campus Ourense), Polytechnical Building, As Lagoas, 32004 Ourense, Spain; marlopezr@uvigo.es (M.L.); sandrarivas@uvigo.es (S.R.); cvila@uvigo.es (C.V.); vsantos@uvigo.es (V.S.)
* Correspondence: jcparajo@uvigo.es; Tel.: +34-988-387-033

Received: 25 July 2020; Accepted: 12 August 2020; Published: 15 August 2020

Abstract: The acidic ionic liquid 1-(3-sulfopropyl)-3-methylimidazolium hydrogen sulfate ([C3SO$_3$Hmim]HSO$_4$) was employed as a catalyst for manufacturing polysaccharide-derived products (soluble hemicellulose-derived saccharides, furans, and/or organic acids) from *Eucalyptus globulus* wood. Operation was performed in aqueous media supplemented with [C3SO$_3$Hmim]HSO$_4$ and methyl isobutyl ketone, following two different processing schemes: one-pot reaction or the solubilization of hemicelluloses by hydrothermal processing followed by the separate manufacture of the target compounds from both hemicellulose-derived saccharides and cellulose. Depending on the operational conditions, the one-pot reaction could be directed to the formation of furfural (at molar conversions up to 92.6%), levulinic acid (at molar conversions up to 45.8%), or mixtures of furfural and levulinic acid (at molar conversions up to 81.3% and 44.8%, respectively). In comparison, after hydrothermal processing, the liquid phase (containing hemicellulose-derived saccharides) yielded furfural at molar conversions near 78%, whereas levulinic acid was produced from the cellulose-enriched, solid phase at molar conversions up to 49.5%.

Keywords: acidic ionic liquid; *Eucalyptus* wood; furfural; levulinic acid

1. Introduction

Mankind is facing key challenges related to environmental issues and sustainability, caused by the massive utilization of fossil resources, which currently provide more than 90% of our energy needs and feedstocks of the chemical industry [1]. The development of new processes based on renewable vegetal biomass (the most important source of organic carbon on earth) as a feedstock is one of the strategic ways pointed out to deal with these problems. In this field, the "lignocellulose biorefinery" concept, based on the selective separation of the major components making part of lignocellulosic biomass (LB), allows the manufacture of diverse bio-based platform chemicals, including furans and levulinic acid.

The term LB is used to name the numerous types of vegetal biomass mainly made up of polysaccharides (hemicelluloses and cellulose) and an aromatic fraction (lignin). These polymeric components are known as "structural components" of LB, and they appear together with non-structural components (extractives, ash, etc.), which are not important for the objectives of this study. Hardwoods possess a high content of structural components (typically, about 90 wt %), with a hemicellulose fraction mainly made up of heteroxylan.

Eucalyptus spp. is the world's most widely planted hardwood species. Its fast, uniform growth, self-pruning, and ability to coppice make it favorable for applications as timber, pulpwood, or bioenergy feedstocks; as a result, it is one of the most productive and economically viable biomass crops in the world, with expansive commercialization on all populated continents [2].

Additionally, *Eucalyptus* woods show a number of favorable features to be used as industrial feedstocks, including compositional factors (comparative high cellulose content, hemicelluloses mainly made up by acetylated glucuronoxylan), as well as high density and growth rate [3,4]. *Eucalyptus globulus* wood is one of the most important commercial *Eucalyptus* species, and it represents a basic forest resource in the Atlantic regions of the Iberian Peninsula [5,6].

The industrial utilization of LB can be achieved on the basis of the biorefinery concept, which proposes the sustainable processing of biomass into a spectrum of marketable products and energy, including biodegradable plastics, platform chemicals and other bio-based compounds, and advanced biofuels [4,7]. In this field, furfural and levulinic acid have been included among the top platform chemicals that can be manufactured based on green chemistry [7].

Most studies dealing with lignocellulose biorefineries employ the biochemical route, based on pretreatment and further hydrolysis fermentation of cellulose and/or hemicelluloses. Alternatively, processing schemes based on the acidic processing of wood to cause the hydrolysis–dehydration or hydrolysis–dehydration–rehydration of polysaccharides, provide a suitable framework for a fully chemical lignocellulose biorefinery having furfural (FF), hydroxymethylfurfural (HMF), and/or organic acids (levulinic acid, denoted LevA; and formic acid, denoted FA) as target products. For the purposes of this study, FF, LevA, and FA are of special importance, since:

- FF is one of the most important bio-based products derived from biomass [8], with important industrial applications and huge potential as an intermediate for producing a scope of chemicals and biofuels [9–11]
- LevA is a platform chemical with a rich chemistry, which can be used for manufacturing polymers or biofuels [12–14]
- FA is a commodity chemical employed in many industries, and it may also play a role in the development of fuel cells [15].

Eucalyptus globulus can be employed as a raw material for biorefineries following diverse processing schemes, including the ones shown in Figure 1a,b.

(a) One-pot processing of wood in aqueous media containing an acidic catalyst, in order to obtain FF from hemicelluloses, and HMF or organic acids (LevA, FA) from cellulose. Ionic liquids (IL) with acidic characteristics (acidic ionic liquids, AIL) can be used as the acidic catalyst.

(b) Solubilization of hemicelluloses by hydrothermal processing with hot, compressed water (leading to the breakdown of xylan chains into soluble saccharides), with further acid-catalyzed conversion of the reaction products into FF, and the acid-catalyzed manufacture of HMF and/or organic acids from the cellulose-rich, solid phase. As before, AIL can be employed as catalysts.

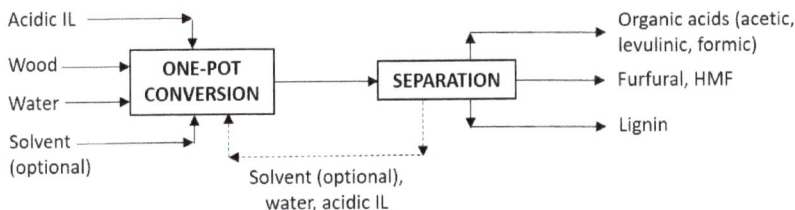

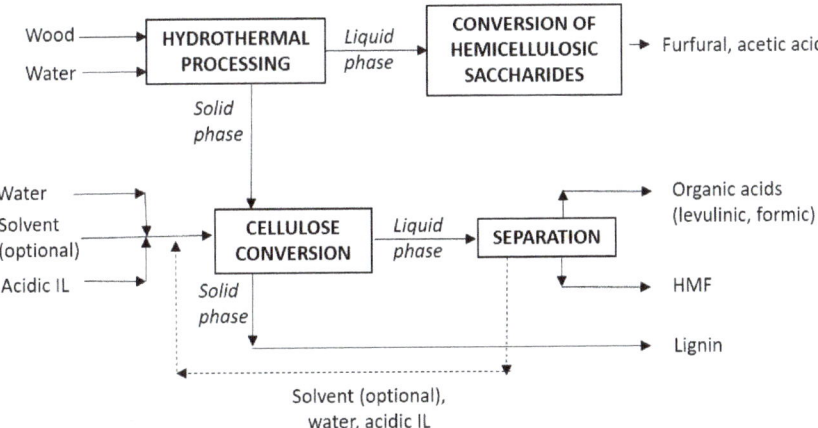

Figure 1. Processing schemes leading to the acid-catalyzed manufacture of furans and organic acids from wood. (**a**) One-pot approach. (**b**) Approach based on hemicellulose separation before furfural manufacture and cellulose conversion.

Each of these processing alternatives shows advantages and disadvantages with respect to each other: one-pot methods are easy to carry out and present a simple structure, requiring little equipment for reaction, and limited manpower [16]. However, the faster kinetics of FF generation/consumption with respect to the reactions leading to the formation/consumption of HMF and/or organic acids [17] means that some FF loss (or an incomplete formation of HMF/organic acids) has to be assumed. Alternatively, processes based on hemicellulose removal and the separate conversion of hemicellulose-derived saccharides and cellulose may increase selectivity and yields, but entail more reaction and downstream stages.

The kinetics of FF and HMF/LevA/FA formation from polysaccharides in aqueous, acidic media show complex mechanisms, which include the generation of reactive intermediates and the participation of these latter (together with the target products) into a number of series and parallel reactions, leading to the formation of humin-type products, which limit the yields of the target products [18–24].

A number of approaches have been proposed to improve the production of target products, including:

- Utilization of reaction media containing an immiscible solvent able to extract the target products, avoiding the unwanted reactions taking place in aqueous media [25–27]. This approach has been followed in this study, in which methyl isobutyl ketone (MIBK), a green solvent recommended by the CHEM21 guide [28], has been employed.

- Utilization of microwave (MW)-heated reactors enabling fast heating profiles, which are considered favorable to improve the experimental results [29], and to allow energy savings [30]. Following this idea, a stirred, MW-heated reactor has been employed in this work in experiments aiming at the manufacture of furans and/or organic acids.
- Utilization of efficient and selective catalysts. Mineral acids such as sulfuric acid have been widely used for biomass processing. Alternatively, acidic ionic liquids have been reported to have potential as catalysts with improved activity, selectivity, and stability, and to allow an easier separation and reutilization [31]. Specifically, AILs have been used as catalysts for processing LB (or fractions derived from it) in aqueous or biphasic media [9,32]. In this work, a Brønsted acidic, imidazolium-type IL has been used for manufacturing FF, LevA, and FA from *Eucalyptus* wood (or from fractions derived from it), following the general ideas shown in Figure 1. Additional literature information of the utilization of ILs in the framework of biorefineries is provided in the next paragraphs.

Owing to their "green" character and special physicochemical properties (particularly, low volatility and stability), ILs can play a number of roles in biorefineries, including physical separation, the formulation of reaction media, and utilization as catalysts [33]. In this context, the conversion of wood polysaccharides (or low molecular weight saccharides derived from hemicelluloses) in aqueous media catalyzed by Brønsted acidic ILs, eventually in the presence of an organic solvent, has been considered in the literature. SO_3H-functionalized ILs have been employed to achieve the dehydration of xylose into FF [34], whereas 1-butyl 3-methylimidazolium hydrogen sulfate ([bmim]HSO_4) has been employed to convert the hemicellulose fraction of lignocellulosic biomass into hemicellulosic sugars and FF [35], and to produce FF from xylose or soluble hemicellulose-derived saccharides obtained by the hydrothermal processing of *Eucalyptus* wood [32], eventually in the presence of a cosolvent [36]. Dealing with the conversion of cellulose, 1-(3-sulfopropyl)-3-methylimidazolium hydrogen sulfate ([$C3SO_3Hmim$]HSO_4), alone or in combination with metal chlorides, has been employed for LevA production from pure cellulose [37]; whereas [$C3SO_3Hmim$]HSO_4 and 1-(3-sulfobutyl)-3-methylimidazolium hydrogen sulfate ([$C4SO_3Hmim$]HSO_4) provided good results as catalysts for the conversion of *Pinus pinaster* wood polysaccharides [38]. Sulfonated ILs were successfully employed as catalysts to achieve the conversion of pure cellulose into LevA [26].

This work provides a quantitative assessment of the performance of [$C3SO_3Hmim$]HSO_4 as a catalyst for *Eucalyptus globulus* wood processing in biphasic media, following both the one-pot approach (Figure 1a) and a method based on hemicellulose solubilization and the further conversion of hemicellulose-derived products and cellulose into the target products (Figure 1b). Hydrolysis–dehydration or hydrolysis–dehydration–rehydration reactions were performed in a stirred, MW-heated reactor. The experimental results obtained under diverse operational conditions are discussed in terms of the volumetric concentrations and molar conversions of the target products (FF, LevA, FA, and acetic acid) and intermediates (HMF) derived from polysaccharides.

2. Results

2.1. Composition of Eucalyptus Globulus Wood

For the purposes of this work, the relevant data concerning *E. globulus* wood are the contents of the structural components (cellulose, hemicelluloses, and lignin), and the relative amounts of the diverse hemicellulose components. The wood lot employed in this study contained, in oven-dry mass basis, 22.1% ± 0.2% Klason lignin, 1.38 ± 0.07 acid soluble lignin, 44.4% ± 0.2% cellulose, 15.6% ± 0.2% xylan, 0.43% ± 0.06% arabinosyl units, 0.90% ± 0.06% mannosyl units, 1.02% ± 0.03% galactosyl units, and 2.95% ± 0.09% acetyl groups. This composition is typical for *Eucalyptus globulus*, as it can be seen in a recent review [4]. Among the above fractions, cellulose, galactosyl units, and mannosyl units are potential substrates for manufacturing HMF and/or LevA/FA; xylan and arabinosyl groups are potential substrates for FF manufacture; and acetyl groups yield acetic acid (AcH) upon hydrolysis.

2.2. One-Pot Conversion of Eucalyptus Globulus Wood

In order to assess the ability of [C3SO$_3$Hmim]HSO$_4$ as a catalyst for obtaining bio-based chemicals from *Eucalyptus globulus* wood polysaccharides, the set of experiments listed in Table 1 was performed using the MW-heated reactor. Operating at moderate catalyst loadings (0.10–0.15 g/g oven-dry wood), operation was carried out at high water-to-solid ratios (6 or 10 g/g oven-dry wood, which are markedly higher than the ones typically employed in MW reactors), keeping the relative amount of MIBK at 2 g/g aqueous phase.

The first assay (experiment 1 in Table 1) was performed under mild conditions: the reaction media, made with the highest charges of solid and catalyst considered, was heated up to 170 °C and then cooled immediately (isothermal reaction time, 0 min). The high solid yield (53.7 g/100 g wood) indicated that the conversion of polysaccharides into the target products was incomplete, which was a fact confirmed by the concentrations of reaction intermediates in aqueous phase (12.4 g glucose/L and 18.6 and g xylose/L). Interestingly, the conversion of xylan into xylose proceeded at 62.8% molar conversion, and the acetyl groups were almost quantitatively hydrolyzed into AcH, which appears distributed in both aqueous and organic phases), confirming that [C3SO$_3$Hmim]HSO$_4$ could be used for manufacturing hemicellulosic sugars. However, the concentrations of FF, HMF, FA, and LevA were low.

Based on these findings, the severity of the operational conditions was increased by raising the temperature to 180 °C and keeping the rest of experimental variables unchanged. As expected, the solid yield decreased (to 38.5%), the glucose concentration increased (28.0 g/L) owing to the higher cellulose conversion, and the xylose concentration dropped to 8.18 g/L owing to the increased generation of FF (which appeared concentrated in the organic phase, accounting for an overall molar conversion of 49%, including the contributions of the aqueous and organic phases). Although the LevA concentrations increased up to 0.98 and 0.34 g/L in the aqueous and organic phases, respectively, its overall molar conversion was still very poor (3.40%). When the reaction conditions were modified by performing an isothermal reaction stage of 16 min (experiment 3), the solid yield dropped almost by half, even when the catalyst concentration was reduced, and the xylose concentration decreased to 1 g/L, with enhanced FF concentrations in both phases (equivalent to an overall molar conversion of 78.3%). This finding confirmed that [C3SO$_3$Hmim]HSO$_4$ presents a remarkable potential for obtaining FF from xylan-containing materials. However, the concentrations of LevA were still low (equivalent to 27.0% overall molar conversion), owing to the presence in the medium of other cellulose-derived intermediates (glucose and HMF, which were obtained at 22.2% and 10.7% overall molar yields, respectively).

In order to improve the production of LevA, additional assays were carried out at a higher temperature (190 °C) for 0, 30, and 50 min of isothermal operation (experiments 4, 5, and 6 in Table 1). The results obtained in experiment 4 (26.1% solid yield, glucose concentration corresponding to 39.7% molar yield; xylose accounting for 14.9% of the stoichiometric amount) indicated that the experiment was performed under conditions too mild for practical purposes. Better results were achieved in experiment 5, with low concentrations of glucose and xylose in the aqueous phase (indicative of increased conversions into furans and/or furan-derived products), and improved concentrations of LevA and FF (which were obtained at 43.2% and 58.3% molar yields, respectively). Additionally, under these conditions, FA was obtained at 54.9% molar conversion, and acetyl groups were totally converted into AcH. Prolonging the reaction time up to 50 min (experiment 6) led to worse results (LevA and FF were obtained at 42.4% and 49.0% molar yields). In summary, the conditions of experiment 5 were the best ones for achieving the simultaneous conversion of hemicellulose and cellulose into FF and LevA/FA.

In the acidic processing of LB, increasing the relative amount of liquid with respect to the solid substrate is known to improve the experimental results, because dilution limits the participation of reactive intermediates and products in parasitic reactions. Additionally, lower catalyst charges may result in improved selectivity. In order to assess these points, experiments 7, 8, and 9 were carried out at 180 °C for 15, 30, or 45 min, respectively, in media containing 0.1 g catalyst/g oven-dry wood. The solid yields obtained in these experiments (in the range 8.1–6.9%) indicated that solubilization reactions affected both polysaccharides and lignin. The aqueous phase obtained in experiment 7 contained 16.9 g glucose/L, indicating that the severity was still not enough to achieve optimal amounts of LevA and FA (which were obtained at 13.4% and 21.2% molar conversions, respectively). In turn, these conditions were excellent for FF production, which was obtained at 89.4% molar conversion. When the reaction time was extended to 30 min (experiment 8), the FF generation decreased slightly (molar conversion, 87.4%), but the LevA production increased (molar conversion, 25.3%). Looking for an improved production of LevA/FA, the reaction was performed for 45 min (experiment 9). This modification resulted in increased FF consumption (molar yield, 80.4%), but the production of LevA and FA improved up to 43.0% and 56.3% molar conversions, respectively.

Based on the above results, experiments 10, 11, and 12 were performed to assess the possible benefits derived from operating at a higher temperature (190 °C) for 15, 30, or 45 min, keeping the rest of the operational variables unchanged. The limited solid yields (in the range 5.6–4.7%) confirmed that almost total wood liquefaction took place under the considered conditions. In experiment 10 (reaction time, 15 min), the most remarkable finding was the excellent conversion of precursors into FF, which was obtained at 92.6% molar yield. However, LevA and FA were obtained at 35.5% and 48.1% molar yields, respectively. The data suggested that the results could be improved by enhancing the conversion of intermediates (glucose and HMF, which were obtained at 2.64% and 13.5% molar yields, respectively) into the target products. This idea was assessed by increasing the reaction time to 30 min (conditions of experiment 11). As expected, the generation of LevA and FA increased (molar yields, 44.8% and 56.8%, respectively), at the cost of worsening the FF production (molar yield, 81.3%). However, the conditions of experiment 11 can be considered as a satisfactory compromise for the one-pot production of bio-based chemicals, because harsher conditions (as the ones employed in experiment 12, in which the reaction time was fixed in 45 min) resulted in a slightly improved LevA production (molar yield, 45.8%) but in an important FF loss (molar yield, 75.8%).

Table 1. Operational conditions assayed and experimental results achieved in one-pot experiments using native *E. globulus* wood as a feedstock.

EXPER	OPERATIONAL CONDITIONS					EXPERIMENTAL RESULTS												
	CC	WSR	OSR	T	t	SY	Concentrations in Aqueous Phase, g/L						Concentrations in Organic Phase, g/L					
							Glc	Xyc	FAc	AcHc	LevAc	HMFc	FFc	FAc	AcHc	LevAc	HMFc	FFc
1	0.15	6	12	170	0	53.7	12.4	18.6	2.14	3.02	0.10	0.20	0.31	0.10	1.54	0.05	0.19	2.14
2	0.15	6	12	180	0	38.5	28.0	8.18	1.06	3.11	0.98	0.68	0.61	0.28	1.54	0.34	0.54	3.52
3	0.10	6	12	180	16	19.5	17.9	1.00	3.96	3.11	6.54	1.71	0.66	1.64	1.57	3.26	1.84	5.73
4	0.15	6	12	190	0	26.1	32.2	4.38	1.44	3.07	2.99	1.17	0.74	0.59	1.55	0.88	0.85	4.12
5	0.15	6	12	190	30	18.1	0.16	0.17	5.74	3.50	10.4	0.04	0.52	2.44	1.77	5.27	0.04	4.25
6	0.15	6	12	190	50	17.9	0.17	0.20	5.40	3.41	10.1	<0.01	0.42	2.27	1.73	5.15	<0.01	3.58
7	0.10	10	20	180	15	8.10	16.9	1.16	1.27	1.95	1.95	2.11	0.43	0.58	0.99	0.95	2.30	3.89
8	0.10	10	20	180	30	8.00	9.14	0.60	2.25	2.10	3.58	1.81	0.39	0.86	1.06	1.83	2.07	3.82
9	0.10	10	20	180	45	6.90	1.00	0.29	3.55	2.17	6.08	0.62	0.40	1.46	1.11	3.13	0.69	3.49
10	0.10	10	20	190	15	5.60	1.26	0.32	2.99	2.15	4.97	1.24	0.44	1.26	1.08	2.59	1.41	4.03
								4.70										
11	0.10	10	20	190	30	5.43	0.16	0.26	3.61	2.35	6.27	0.17	0.35	1.47	1.20	3.28	0.22	3.56
12	0.10	10	20	190	45	4.69	0.14	0.29	3.58	2.31	6.44	0.03	0.35	1.53	1.17	3.34	0.03	3.30

Nomenclature: CC, catalyst concentration in g catalyst/g feedstock; WSR, water-to-solid ratio in g aqueous phase/g feedstock; OSR, organic solvent-to-solid ratio in g methyl isobutyl ketone (MIBK)/g feedstock; T, temperature in °C; t, isothermal reaction time in min; SY, solid yield in g solid after treatment/g feedstock; Glc, glucose concentration in g/L, Xyc, xylose concentration in g/L; FAc formic acid concentration in g/L; AcHc, acetic acid concentration in g/L; LevAc, levulinic acid concentration in g/L; HMFc, HMF concentration in g/L; FFc, furfural concentration in g/L.

2.3. Eucalyptus Globulus Wood Conversion Based on Hemicellulose Removal

2.3.1. Hydrothermal Processing

Figure 1b shows the general idea of a biorefinery process based on hydrothermal processing (also named autohydrolysis) and further individual utilization of the liquid phase (containing soluble hemicellulose-derived saccharides) and the solid phase (containing cellulose and lignin). The autohydrolysis of *Eucalyptus globulus* wood has been considered in the literature [39], and it is considered as a green, efficient, and selective method for hemicellulose separation from cellulose and lignin. Operating under suitable conditions, the major soluble reaction products are hemicellulose-derived saccharides (mainly of oligomeric nature), which are accompanied by minor amounts of sugars and sugar-dehydration products. The saccharides present in the aqueous phase are substrates suitable for FF manufacture [32], whereas lignin and cellulose undergo little modification and remain in solid phase, enabling their separate utilization. In this work, the treatments were performed using a water-to-solid ratio of 8 kg/kg. The media were heated up to reach 200 °C and cooled immediately, following the temperature profiles reported in Garrote and Parajó [39]. Upon hydrothermal processing, 28.9% of the wood mass was solubilized, yielding a treated solid enriched in cellulose, and a liquid phase containing xylooligomers as major products. Table 2 lists the composition of both phases.

Table 2. Composition of solid and liquid phases from hydrothermal processing.

(a) Solid Phase	
COMPONENT	CONTENT (g/100 g Treated Solid, Oven-Dry Basis)
Cellulose	58.8
Xylan	3.20
Arabinan	0.07
Acetyl groups	0.69
Klason lignin	30.6
(b) Liquid Phase	
COMPONENT	CONCENTRATION (g/L)
Glucose	0.51
Xylose	6.40
Arabinose	0.51
FA	0.66
AcH	1.41
LevA	0.02
HMF	0.09
FF	0.57
Glucosyl groups in oligomers	1.00
Xylosyl groups in oligomers	12.9
Arabinosyl groups in oligomers	0.00
Acetyl groups in oligomers	3.32

2.3.2. Conversion of the Solids from Hydrothermal Processing

Based on the above information, the cellulose-enriched solids coming from hydrothermal fractionation (see Figure 1b), mainly made up of cellulose (58.8%) and lignin (30.6%), were treated with water, [C3SO$_3$Hmim]HSO$_4$, and MIBK to yield HMF, LevA, and FA. Owing to the limited xylan and arabinan contents of the treated solids (3.20 and 0.07%, respectively), the generation of FF is not discussed in this section.

The experimental plan was established keeping in mind the results obtained in assays 7 to 12 in Table 1. Those assays were performed at 180 or 190 °C with a catalyst concentration (denoted CC) of 0.1 g catalyst/g oven-dry wood, for reaction times (t) up to 45 min. Table 3 lists the operational conditions assayed in this part of our study (similar to the above ones, except the longest reaction time, which was fixed in 75 or 60 min for operation at 180 or 190 °C, respectively). The same table includes the volumetric concentrations of the target products in the aqueous and organic phases. For clarity, Figure 2; Figure 3 show the experimental results in terms of overall molar conversions (including the amounts of the target products contained in both the aqueous and organic phases).

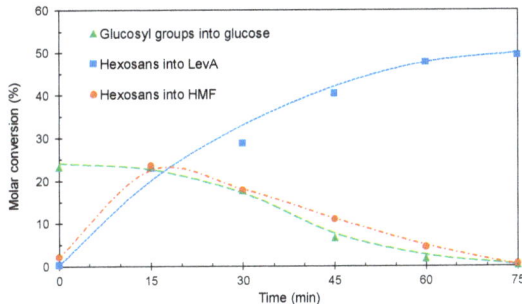

Figure 2. Overall molar conversions (including the contributions of aqueous and organic phases) achieved in experiments performed at 180 °C using the solids from hydrothermal processing as a substrate and [C3SO₃Hmim]HSO₄ as a catalyst.

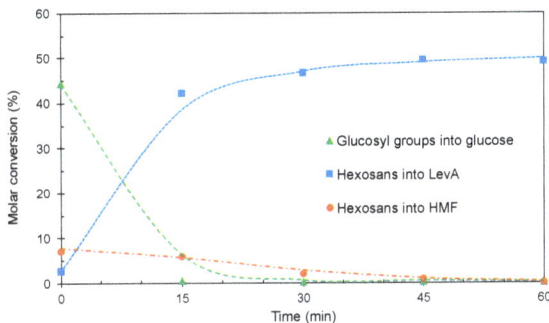

Figure 3. Overall molar conversions (including the contributions of aqueous and organic phases) achieved in experiments performed at 190 °C using the solids from hydrothermal processing as a substrate and [C3SO₃Hmim]HSO₄ as a catalyst.

The high solid yield and glucose concentration in aqueous phase obtained in experiment 1, which was performed at 180 °C with no isothermal stage (see Table 3), resulted in incomplete cellulose conversion, indicating that the conditions were too mild for the purposes of this study. Implementing an isothermal stage of 15–75 min (experiments 2–6 in Table 3) substantially decreased the solid yield to 15.3–13.4 g/100 g solid and caused a steady increase in the concentrations of LevA and FA. The effects of the reaction time on the cellulose conversion into glucose, HMF, and LevA are displayed in Figure 2, where it can be seen that glucose and HMF behaved as reaction intermediates, whereas the maximum LevA molar conversion (49.3%) was reached after 75 min.

Table 3. Operational conditions assayed and experimental results achieved in the conversion of the solids from hydrothermal processing.

EXPER	OPERATIONAL CONDITIONS					EXPERIMENTAL RESULTS								
	CC	WSR	OSR	T	t	SY	Concentrations in Aqueous Phase, g/L				Concentrations in Organic Phase, g/L			
							Glc	FAc	LevAc	HMFc	FAc	LevAc	HMFc	
1	0.1	10	20	180	0	61.7	14.9	0.17	0.03	0.25	0.03	0.02	0.27	
2	0.1	10	20	180	15	15.0	14.8	2.45	4.18	2.64	0.95	2.09	3.13	
3	0.1	10	20	180	30	15.3	11.2	2.93	5.23	2.18	1.20	2.62	2.29	
4	0.1	10	20	180	45	14.8	4.16	4.04	7.38	1.32	1.60	3.65	1.41	
5	0.1	10	20	180	60	15.2	1.10	4.51	8.69	0.52	1.86	4.34	0.58	
6	0.1	10	20	180	75	13.4	0.15	4.62	8.98	0.06	1.89	4.50	0.06	
7	0.1	10	20	190	0	42.0	28.4	0.45	0.60	0.91	0.11	0.21	0.90	
8	0.1	10	20	190	15	22.0	0.35	4.32	7.66	0.68	1.72	3.85	0.76	
9	0.1	10	20	190	30	21.9	0.19	4.54	8.41	0.23	1.82	4.31	0.27	
10	0.1	10	20	190	45	13.2	0.16	4.72	8.94	0.09	1.95	4.54	0.10	
11	0.1	10	20	190	60	12.0	0.12	4.55	8.82	<0.01	1.83	4.45	<0.01	

Nomenclature: CC, catalyst concentration in g catalyst/g feedstock; WSR, water-to-solid ratio in g aqueous phase/g feedstock; OSR, organic solvent-to-solid ratio in g MIBK/g feedstock; T, temperature in °C; t, isothermal reaction time in min; SY, solid yield in g solid after treatment/g feedstock; Glc, glucose concentration in g/L, Xyc, xylose concentration in g/L; FAc formic acid concentration in g/L; AcHc, acetic acid concentration in g/L; LevAc, levulinic acid concentration in g/L; HMFc, HMF concentration in g/L; FFc, furfural concentration in g/L..

The effects caused by increasing the temperature up to 190 °C (keeping constant the rest of operational variables) were considered in experiments 7 to 11 in Table 3. As before, high solid yields and low concentrations of intermediates and target products were obtained when no isothermal reaction stage was considered (experiment 7). Increasing the reaction time up to 60 min (experiments 8 to 11) resulted in a fast increase of the concentrations of the target products. Figure 3, which presents the results in terms of molar conversions, shows a fast consumption of intermediates and a maximum molar conversion into LevA of 49.5% (the highest value reached in this study).

2.3.3. Conversion of the Liquid Phase from Hydrothermal Processing

The liquid phase from hydrothermal processing (see Figure 1b), with the composition indicated in Table 2, was supplemented with [C3SO$_3$Hmim]HSO$_4$ and heated in the presence of MIBK to produce FF from the suitable precursors. The optimization of FF manufacture was studied using the Response Surface Methodology, based on a factorial, incomplete, and centered Box–Behnken experimental design, in which three selected independent variables (T, t, and the volumetric catalyst concentration, denoted VCC) were assayed in duplicate at three levels. The ranges considered for the operational variables were 160–180 °C for temperature and 2–30 min for t; whereas VCC varied from 1.42×10^{-2} up to 4.26×10^{-2} g/g aqueous phase. As before, the MIBK/aqueous phase mass ratio was fixed in 2 g/g. The only dependent variable considered in this part of the study was the FF molar yield (denoted FFMY), which was measured in respect to the potential substrates in the liquid phase.

The operational conditions employed in diverse experiments and the corresponding FFMY values are listed in Table 4.

Table 4. Operational conditions employed in experiments of furfural (FF) production from hemicellulose-derived saccharides, and experimental results obtained for the furfural molar yield. FFMY: FF molar yield, VCC: volumetric catalyst concentration.

Exper.	T, °C	T, min	VCC, g/g Aqueous Phase	FFMY, %
1a	170	16	2.84×10^{-2}	77.6
1b	170	16	2.84×10^{-2}	77.0
2a	160	30	2.84×10^{-2}	75.6
2b	160	30	2.84×10^{-2}	75.4
3a	180	16	1.42×10^{-2}	78.7
3b	180	16	1.42×10^{-2}	76.3
4a	170	2	4.26×10^{-2}	73.8
4b	170	2	4.26×10^{-2}	73.4
5a	160	16	1.42×10^{-2}	51.0
5b	160	16	1.42×10^{-2}	53.3
6a	170	30	1.42×10^{-2}	77.3
6b	170	30	1.42×10^{-2}	80.3
7a	160	16	4.26×10^{-2}	70.7
7b	160	16	4.26×10^{-2}	74.3
8a	180	2	2.84×10^{-2}	77.0
8b	180	2	2.84×10^{-2}	79.4
9a	180	16	4.26×10^{-2}	72.0
9b	180	16	4.26×10^{-2}	70.9
10a	170	2	1.42×10^{-2}	45.5
10b	170	2	1.42×10^{-2}	44.3
11a	160	2	2.84×10^{-2}	43.4
11b	160	2	2.84×10^{-2}	37.8
12a	180	30	2.84×10^{-2}	72.3
12b	180	30	2.84×10^{-2}	72.7
13a	170	30	4.26×10^{-2}	72.0
13b	170	30	4.26×10^{-2}	73.1

For calculation purposes, the following normalized, dimensionless variables with variation ranges (−1, 1) were defined:

$$x_1 \text{ (dimensionless temperature for experiment j)} = 2\cdot(T_j - 170)/(180 - 160) \quad (1)$$

$$x_2 \text{ (dimensionless time for experiment j)} = 2\cdot(t_j - 16)/(30 - 2) \quad (2)$$

$$x_3 \text{ (dimensionless volumetric catalyst concentration for experiment j)} = 2\cdot(VCC_j - 2.84 \times 10^{-2})/(4.26 \times 10^{-2} - 1.42 \times 10^{-2}) \quad (3)$$

The values of the FFMY data in Table 4 were correlated with the values of the correspondent normalized variables using the following second-order equation involving linear, interaction, and quadratic terms.

$$FFMY = a_0 + a_1 \cdot x_1 + a_2 \cdot x_2 + a_3 \cdot x_3 + a_{12} \cdot x_1 \cdot x_2 + a_{13} \cdot x_1 \cdot x_3 + a_{23} \cdot x_2 \cdot x_3 + a_{11} \cdot x_1^2 + a_{22} \cdot x_2^2 + a_{33} \cdot x_3^2 \quad (4)$$

Table 5 lists the values of the regression coefficients (calculated by the least squares method), their statistical significance (based on a *t*-test), the statistical significance of the model (measured using the F_{st} Fischer's test), and the correlation coefficient.

Table 5. Regression coefficients, statistical significance (based on a *t*-test), and parameters measuring the correlation and significance of the statistical model.

Coefficient	Value	Statistical Significance
a_0	77.30	>99%
a_1	7.36	>99%
a_2	7.77	>99%
a_3	4.59	>99%
a_{12}	−10.15	>99%
a_{13}	−6.59	>99%
a_{23}	−8.74	>99%
a_{11}	−4.84	>99%
a_{22}	−5.76	>99%
a_{33}	−4.08	>99%

Statistical parameters: R^2: 0.980; F_{st}: 87.3 (significance > 99%)

The model predictions were analyzed on the basis of the response surfaces shown in Figure 4; Figure 5. Figure 4 shows that at 170 °C, the FFMY was favored by operation at long reaction times with limited catalyst charges, with a clearly defined optimum zone. Figure 5 depicts a closely related situation: in assays lasting 16 min, the best results were predicted for low or intermediate catalyst loadings, with a defined optimum zone appearing near the highest temperature assayed.

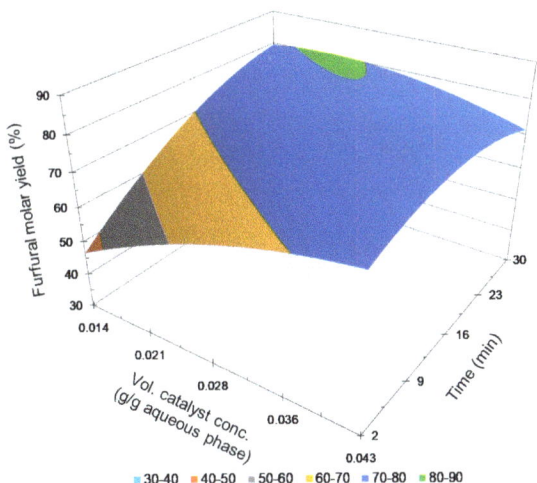

Figure 4. Dependence of the furfural molar conversion on time and volumetric catalyst concentration (data calculated for T = 170 °C).

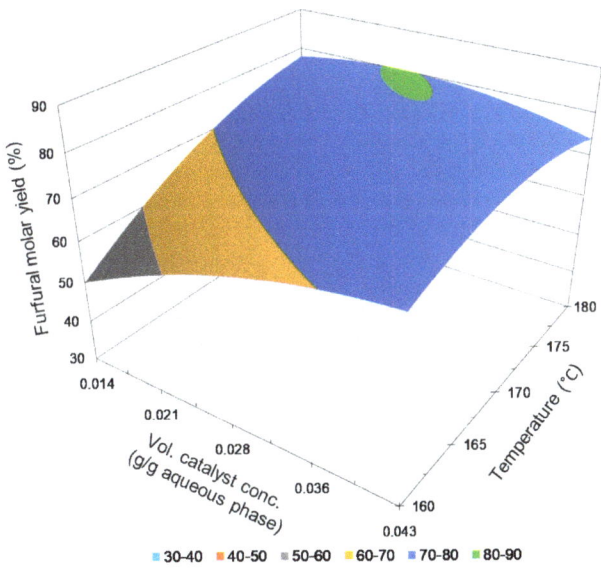

Figure 5. Dependence of the furfural molar conversion on temperature and volumetric catalyst concentration (data calculated for t = 16 min).

The mathematical optimization of the statistical model within the experimental domain (∂FFMY/∂x_i =0; $-1 \leq x_i \leq 1$; i: 1 to 3) was predicted a maximum FFMY (80.4%) for T = 171 °C, t = 30 min, and VCC = 1.97×10^{-2} g catalyst/g aqueous phase. Additional experiments performed for model validation led to an FFMY near 78%, which was close to the experimental results obtained in experiments 3, 6, and 8, and it defined a region of the experimental domain where the experimental and fitting errors overlapped the possible differences in FFMY values.

2.3.4. Comparative Analysis of Results

The literature has considered the utilization of AIL as reaction media and/or as catalysts for obtaining platform chemicals from a number of substrates, including sugars, hemicellulose-derived saccharides, purified polysacharides, LB-derived fractions, and native LB. However, comparing the results reported with the ones achieved in this study is not straightforward, owing to important differences in the types of substrates and catalysts, the diversity of operational conditions (particularly, regarding catalyst and substrate charges), and the type of target products (FF and/or HMF and/or LevA+FA).

Concerning the FF production from wood hemicellulose (with or without the simultaneous production of LevA from cellulose), Zhang and Zhao [40] treated wood in 1-butyl-3-methylimidazolium chloride catalyzed with $CrCl_3$ to obtain FF at yields up to 31%, together with HMF. The same IL was employed by Sievers et al. [41] to obtain HMF and FF (at low mass yield) from loblolly pine using trifluoroacetic acid as a catalyst. López et al. [38] employed 1-butyl-3-methylimidazolium hydrogen sulfate as a catalyst to achieve 60.5% conversion of pine wood pentosans into FF. Under the considered operational conditions, little conversion of cellulose into LevA was achieved, which is ascribed to the limited acidity of the catalyst. [C3SO$_3$Hmim]HSO$_4$ and [C4SO$_3$Hmim]HSO$_4$, two AIL of enhanced acidity (caused by the sulfonic group attached to the cation) have been successfully employed for processing purified cellulose and native LB. A near-quantitative conversion of xylosyl and arabinosyl groups in native pine wood into FF was achieved using these catalysts [38]. Concerning the production of LevA, 39.4–55% molar conversion was reached using microcrystalline cellulose as a substrate [26], [37], or 32.8% molar conversion of precursos (hexosans) in pine wood [38]. Obtaining higher LevA yields entailed the utilization of high catalyst charges (up to 4 g/g cellulose) [29], whereas the manufacture of soluble products from LB was favored when susceptible substrates of low lignin content (such as bagasse) were employed [42].

Regarding the FF production from LB-derived fractions, Peleteiro et al. [32] employed the liquid phase from *Eucalyptus globulus* wood autohydrolysis (containing xylooligosaccharides as major reaction products) as a substrate for reaction in biphasic media catalyzed with 1-butyl-3-methylimidazolium hydrogen sulfate. MIBK and toluene were employed as organic solvents, at a 4.4/1 mass ratio with respect to the aqueous phase, which was concentrated using membranes before reaction. Under the best operational conditions, the molar conversion of the suitable precursors into FF reached 68.4%. In a related study, a biorefinery solution enriched in C5 oligomers coming from the aqueous processing of agricultural LB was treated in a biphasic media (water/tetrahydrofuran) in the presence of the AIL 1-(4-sulfonylbutyl)pyridinium methanesulfonate, yielding FF at molar conversions up to 45% [34].

3. Materials and Methods

3.1. Raw Material

Eucalyptus globulus wood chips were collected in a local pulp mill, milled to pass an 8 mm screen, air-dried to a final moisture of 6.6%, and employed for reaction. The wood samples employed for analysis were further milled and screened following the NREL/TP-510-42620 standard method.

3.2. Reaction

In hydrothermal reactions, wood samples were mixed with water at the desired solid charge (8 kg liquid phase/kg oven-dry wood) and heated up to 200 °C in a 3.75 L stainless steel, stirred Parr reactor (Parr Instruments, Moline, Illinois, USA). In one-pot reactions, wood was mixed with water, [C3SO3Hmim]HSO4 (purchased from Ionic Liquids Technologies GmbH, Heilbronn, Germany) and MIBK, and reacted for the desired time at the target temperature and solid charge (see text and table headings for specific information) using a stirred MARS 6 MW reactor. Reactions involving the liquid phase from hydrothermal processing were carried out in the presence of [C3SO$_3$Hmim]HSO$_4$ and

MIBK using the same MW reactor, under the set of experimental conditions indicated in the Results section. All reaction conditions were assayed in duplicate experiments.

3.3. Analysis

Wood samples were assayed for polysaccharides (cellulose and hemicelluloses) and lignin by quantitative acid hydrolysis (NREL standard method/TP-510-42618). The target products in liquid phase (oligosaccharides, sugars, furans, and organic acids) coming from quantitative acid hydrolysis or reaction experiments were determined by HPLC-IR and HPLC-Diode Array Detector as per Rivas et al. [43].

4. Conclusions

The acidic ionic liquid [C3SO$_3$Hmim]HSO$_4$ was employed as a catalyst for *Eucalyptus globulus* wood processing in reaction media also containing water and wood (or wood-derived fractions). Operating in the presence of MIBK, two different operational modes were considered (see Figure 1): (a) one-pot reaction and (b) the separate conversion of hemicellulose-derived products and cellulose-enriched solids resulting from hydrothermal fractionation.

Following the one-pot approach, mild reaction conditions (170 °C, 0 min of isothermal reaction time, 0.15 g catalyst/g oven-dry wood) resulted in the hydrolysis of xylan into xylose at 62.8% molar conversion. Harsher conditions increased the formation of FF (up to 92.6% molar conversion) and levulinic acid (up to 45.8% molar conversion). Conditions of increased severity (190 °C, 30 min, 0.10 g catalyst/g oven-dry wood) resulted in slightly decreased FF production (81.3% molar conversion) but a better conversion of cellulose into LevA (44.8%). Concerning the separate conversion of the phases resulting from hydrothermal processing, the conversion of the cellulose-enriched solid into LevA under optimal conditions (190 °C, 45 min, 0.10 g catalyst/g oven-dry wood) proceeded at about 49.5% molar conversion, whereas the hemicellulose-derived saccharides present in the liquid phase yielded FF at near 78% molar conversion.

The experimental results confirm the suitability of [C3SO$_3$Hmim]HSO$_4$ as a versatile catalyst for the environmentally friendly production of bio-based chemicals from the polysaccharide fraction of *Eucalyptus globulus* wood.

Author Contributions: Conceptualization, M.L., S.R., C.V., V.S. and J.C.P.; methodology, M.L., S.R., C.V., V.S. and J.C.P.; validation, M.L.; investigation, M.L., S.R., C.V., V.S. and J.C.P.; resources, V.S., J.C.P.; data curation, M.L.; writing—original draft preparation, M.L., S.R., C.V., V.S. and J.C.P.; writing—review and editing, M.L., V.S. and J.C.P.; supervision, V.S. and J.C.P.; project administration, V.S. and J.C.P.; funding acquisition, V.S. and J.C.P. All authors have read and agreed to the published version of the manuscript.

Funding: This research was funded by the "Ministry of Economy and Competitiveness" of Spain (research project "Modified aqueous media for wood biorefineries", reference CTQ2017-82962-R).

Acknowledgments: M.L. thanks "Xunta de Galicia" and the European Union (European Social Fund—ESF) for her predoctoral grant (reference ED481A-2017/316). S.R. thanks "Ministerio de Ciencia, Innovación y Universidades" for her "Juan de la Cierva" contract IJC2018-037665-I.

Conflicts of Interest: The authors declare no conflict of interest.

References

1. Mika, L.T.; Cséfalvay, E.; Németh, Á. Catalytic Conversion of Carbohydrates to Initial Platform Chemicals: Chemistry and Sustainability. *Chem. Rev.* **2018**, *118*, 505–613. [CrossRef] [PubMed]
2. Oak Ridge National Laboratory. U.S.; Department of Energy. U.S. Billion-Ton Update: Biomass Supply for a Bioenergy and Bioproducts Industry. 2011. Available online: https://www1.eere.energy.gov/bioenergy/pdfs/billion_ton_update.pdf (accessed on 20 July 2020).
3. Pereira, J.S.; Linder, S.; Araujo, M.C.; Pereira, H.; Ericsson, T.; Borralho, N.; Leal, L.C. Optimization of Biomass Production in Eucalyptus Globulus Plantations—A Case Study. In *Biomass Production by Fast-Growing Trees*, 1st ed.; Pereira, J.S., Landsberg, J.J., Eds.; Springer: Dordrecht, The Netherlands, 1989; Volume 166, pp. 101–121. [CrossRef]

4. Penín, L.; López, M.; Santos, V.; Alonso, J.L.; Parajó, J.C. Technologies for Eucalyptus wood processing in the scope of biorefineries: A comprehensive review. *Bioresour. Technol.* **2020**, *311*, 123528. [CrossRef] [PubMed]
5. Rockwood, D.L.; Rudie, A.W.; Ralph, S.A.; Zhu, J.Y.; Winandy, J.E. Energy product options for Eucalyptus species grown as short rotation woody crops. *Int. J. Mol. Sci.* **2008**, *9*, 1361–1378. [CrossRef] [PubMed]
6. FAO. Planted Forests and Trees Working Paper FP38E. Global Planted Forests Thematic Study: Results and Analysis 2006. Available online: http://www.fao.org/forestry/12139-03441d093f070ea7d7c4e3ec3f306507.pdf (accessed on 20 July 2020).
7. Chandel, A.K.; Garlapati, V.K.; Singh, A.K.; Antunes, F.A.F.; da Silva, S.S. The path forward for lignocellulose biorefineries: Bottlenecks, solutions, and perspective on commercialization. *Bioresour. Technol.* **2018**, *264*, 370–381. [CrossRef]
8. Naidu, D.S.; Hlangothi, S.P.; John, M.J. Bio-based products from xylan: A review. *Carbohydr. Polym.* **2018**, *179*, 28–41. [CrossRef]
9. Peleteiro, S.; Rivas, S.; Alonso, J.L.; Santos, V.; Parajó, J.C. Furfural production using ionic liquids: A review. *Bioresour. Technol.* **2016**, *202*, 181–191. [CrossRef]
10. Cai, C.M.; Zhang, T.; Kumar, R.; Wyman, C.E. Integrated furfural production as a renewable fuel and chemical platform from lignocellulosic biomass. *J. Chem. Technol. Biotechnol.* **2014**, *89*, 2–10. [CrossRef]
11. Bohre, A.; Dutta, S.; Saha, B.; Abu-Omar, M.M. Upgrading furfurals to drop-in biofuels: An overview. *ACS Sustainable Chem. Eng.* **2015**, *3*, 1263–1277. [CrossRef]
12. Frankiewicz, A. Overview of 4-oxopentanoic (levulinic) acid production methods—An intermediate in the biorefinery process. *CHEMIK* **2016**, *70*, 203–208.
13. Kang, S.; Fu, J.; Zhang, G. From lignocellulosic biomass to levulinic acid: A review on acid-catalyzed hydrolysis. *Renew. Sustain. Energy Rev.* **2018**, *94*, 340–362. [CrossRef]
14. Pileidis, F.D.; Titirici, M.M. Levulinic acid biorefineries: New challenges for efficient utilization of biomass. *ChemSusChem* **2016**, *9*, 562–582. [CrossRef] [PubMed]
15. Aslam, N.M.; Masdar, M.S.; Kamarudin, S.K.; Daud, W.R.W. Overview on Direct Formic Acid Fuel Cells (DFAFCs) as an energy sources. *APCBEE Proc.* **2012**, *3*, 33–39. [CrossRef]
16. Badgujar, K.C.; Wilson, L.D.; Bhanage, B.M. Recent advances for sustainable production of levulinic acid in ionic liquids from biomass: Current scenario, opportunities and challenges. *Renew. Sustain. Energy Rev.* **2019**, *102*, 266–284. [CrossRef]
17. Penín, L.; Peleteiro, S.; Rivas, S.; Santos, V.; Parajó, J.C. Production of 5-hydroxymethylfurfural from pine wood via biorefinery technologies based on fractionation and reaction in ionic liquids. *BioResources* **2019**, *14*, 4733–4747.
18. Sweygers, N.; Harrer, J.; Dewil, R.; Appels, L. A microwave-assisted process for the in-situ production of 5-hydroxymethylfurfural and furfural from lignocellulosic polysaccharides in a biphasic reaction system. *J. Cleaner Prod.* **2018**, *187*, 1014–1024. [CrossRef]
19. Cai, C.M.; Zhang, T.; Kumar, R.; Wyman, C.E. THF co-solvent enhances hydrocarbon fuel precursor yields from lignocellulosic biomass. *Green Chem.* **2013**, *15*, 3140–3145. [CrossRef]
20. Climent, M.J.; Corma, A.; Iborra, S. Converting carbohydrates to bulk chemicals and fine chemicals over heterogeneous catalysts. *Green Chem.* **2011**, *13*, 520–540. [CrossRef]
21. Mok, W.S.L.; Antal, M.J.; Varhegyi, G. Productive and parasitic pathways in dilute acid-catalyzed hydrolysis of cellulose. *Ind. Eng. Chem. Res.* **1992**, *31*, 94–100. [CrossRef]
22. Patil, S.K.R.; Lund, C.R.F. Formation and growth of humins via aldol addition and condensation during acid-catalyzed conversion of 5-hydroxymethylfurfural. *Energy Fuels* **2011**, *25*, 4745–4755. [CrossRef]
23. Girisuta, B.; Janssen, L.P.B.M.; Heeres, H.J. A kinetic study on the decomposition of 5-hydroxymethylfurfural into levulinic acid. *Green Chem.* **2006**, *8*, 701–709. [CrossRef]
24. Girisuta, B.; Dussan, K.; Haverty, D.; Leahy, J.J.; Hayes, M.H.B. A kinetic study of acid catalysed hydrolysis of sugar cane bagasse to levulinic acid. *Chem. Eng. J.* **2013**, *217*, 61–70. [CrossRef]
25. Rivas, S.; Vila, C.; Alonso, J.L.; Santos, V.; Parajó, J.C.; Leahy, J.J. Biorefinery processes for the valorization of Miscanthus polysaccharides: From constituent sugars to platform chemicals. *Ind. Crops Prod.* **2019**, *134*, 309–317. [CrossRef]
26. Ren, H.; Zhou, Y.; Liu, L. Selective conversion of cellulose to levulinic acid via microwave-assisted synthesis in ionic liquids. *Bioresour. Technol.* **2013**, *129*, 616–619. [CrossRef] [PubMed]

27. Sorokina, K.N.; Taran, O.P.; Medvedeva, T.B.; Samoylova, Y.V.; Piligaev, A.V.; Parmon, V.N. Cellulose Biorefinery Based on a Combined Catalytic and Biotechnological Approach for Production of 5-HMF and Ethanol. *ChemSusChem* **2017**, *10*, 562–574. [CrossRef]
28. Prat, D.; Wells, A.; Hayler, J.; Sneddon, H.; McElroy, C.R.; Abou-Shehada, S.; Dunn, P.J. CHEM21 selection guide of classical- and less classical-solvents. *Green Chem.* **2015**, *18*, 288–296. [CrossRef]
29. Ren, H.; Girisuta, B.; Zhou, Y.; Liu, L. Selective and recyclable depolymerization of cellulose to levulinic acid catalyzed by acidic ionic liquid. *Carbohydr. Polym.* **2015**, *117*, 569–576. [CrossRef]
30. Hayes, B.L. Recent advances in microwave-assisted synthesis. *Aldrichimica Acta* **2004**, *37*, 66–77. [CrossRef]
31. Song, J.; Han, B. Green chemistry: A tool for the sustainable development of the chemical industry. *Natl. Sci. Rev.* **2015**, *2*, 255–256. [CrossRef]
32. Peleteiro, S.; Santos, V.; Parajó, J.C. Furfural production in biphasic media using an acidic ionic liquid as a catalyst. *Carbohydr. Polym.* **2016**, *153*, 421–428. [CrossRef]
33. Peleteiro, S.; Rivas, S.; Alonso, J.L.; Santos, V.; Parajó, J.C. Utilization of ionic liquids in lignocellulose biorefineries as agents for separation, derivatization, fractionation, or pretreatment. *J. Agric. Food Chem.* **2015**, *63*, 8093–8102. [CrossRef]
34. Serrano-Ruiz, J.C.; Campelo, J.M.; Francavilla, M.; Romero, A.A.; Luque, R.; Menéndez-Vázquez, C.; García, A.B.; García-Suárez, E.J. Efficient microwave-assisted production of furfural from C5 sugars in aqueous media catalyzed by Brønsted acidic ionic liquids. *Catal. Sci. Technol.* **2012**, *2*, 1828–1832. [CrossRef]
35. Carvalho, A.V.; Da Costa Lopes, A.M.; Bogel-Łukasik, R. Relevance of the acidic 1-butyl-3-methylimidazolium hydrogen sulphate ionic liquid in the selective catalysis of the biomass hemicellulose fraction. *RSC Advances* **2015**, *5*, 47153–47164. [CrossRef]
36. Peleteiro, S.; Santos, V.; Garrote, G.; Parajó, J.C. Furfural production from Eucalyptus wood using an acidic ionic liquid. *Carbohydr. Polym.* **2016**, *146*, 20–25. [CrossRef] [PubMed]
37. Shen, Y.; Sun, J.K.; Yi, Y.X.; Wang, B.; Xu, F.; Sun, R.C. One-pot synthesis of levulinic acid from cellulose in ionic liquids. *Bioresour. Technol.* **2015**, *192*, 812–816. [CrossRef] [PubMed]
38. López, M.; Vila, C.; Santos, V.; Parajó, J.C. Manufacture of Platform Chemicals from Pine Wood Polysaccharides in Media Containing Acidic Ionic Liquids. *Polymers* **2020**, *12*, 1215. [CrossRef]
39. Garrote, G.; Parajó, J.C. Non-isothermal autohydrolysis of *Eucalyptus* wood. *Wood Sci. Technol.* **2002**, *36*, 111–123. [CrossRef]
40. Zhang, Z.; Zhao, Z.K. Microwave-assisted conversion of lignocellulosic biomass into furans in ionic liquid. *Bioresour. Technol.* **2010**, *101*, 1111–1114. [CrossRef]
41. Sievers, C.; Valenzuela-Olarte, M.B.; Marzialetti, T.; Musin, I.; Agrawal, P.K.; Jones, C.W. Ionic-liquid-phase hydrolysis of pine wood. *Ind. Eng. Chem. Res.* **2009**, *48*, 1277–1286. [CrossRef]
42. Teng, J.; Ma, H.; Wang, F.; Wang, L.; Li, X. Catalytic fractionation of raw biomass to biochemicals and organosolv lignin in a methyl isobutyl ketone/H$_2$O biphasic system. *ACS Sustain. Chem. Eng.* **2016**, *4*, 2020–2026. [CrossRef]
43. Rivas, S.; González-Muñoz, M.J.; Santos, V.; Parajó, C.J. Production of furans from hemicellulosic saccharides in biphasic reaction systems. *Holzforschung* **2013**, *67*, 923–929. [CrossRef]

© 2020 by the authors. Licensee MDPI, Basel, Switzerland. This article is an open access article distributed under the terms and conditions of the Creative Commons Attribution (CC BY) license (http://creativecommons.org/licenses/by/4.0/).

Article

Valorization of Cellulose Recovered from WWTP Sludge to Added Value Levulinic Acid with a Brønsted Acidic Ionic Liquid

Katarzyna Glińska, Clara Lerigoleur, Jaume Giralt, Esther Torrens and Christophe Bengoa *

Departament d'Enginyeria Química, Universitat Rovira i Virgili, Avinguda dels Països Catalans 26, 43007 Tarragona, Spain; glinska88@gmail.com (K.G.); clara.lerigoleur@etu.univ-nantes.fr (C.L.); jaume.giralt@urv.cat (J.G.); esther.torrens@urv.cat (E.T.)
* Correspondence: christophe.bengoa@urv.cat; Tel.: +34-977-559-619

Received: 21 August 2020; Accepted: 28 August 2020; Published: 2 September 2020

Abstract: The progressive decline of using fossil sources in the industry means that alternative resources must be found to produce chemicals. Waste biomass (sewage sludge) and waste lignocellulosic resources (food, forestry, or paper industries) are ideal candidates to take over from fossil sources. Municipal sewage sludge, and especially primary sludge, has a significant proportion of cellulose in its composition. Proper treatment of this cellulose allows the production of interesting chemicals like levulinic acid that are precursors (bio-blocks or building blocks) for other organic chemical processes. Cellulose was extracted from municipal wet primary sludge and paper industry dried sludge with a commercial ionic liquid. More than 99% of the cellulose has been recovered in both cases. Extraction was followed by the bleaching of the cellulose for its purification. In the bleaching, a large part of the ash was removed (up to 70% with municipal sludge). Finally, the purified cellulose was converted in levulinic acid by catalyzed hydrothermal liquefaction. The reaction, done at 170 °C and 7 bar, catalyzed by a tailored Brønsted acidic ionic liquid produced levulinic acid and other by-products in smaller quantities. The process had a conversion of cellulose to levulinic acid of 0.25 with municipal sludge and of 0.31 with industrial sludge. These results fully justify the process but, require further study to increase the conversion of cellulose to levulinic acid.

Keywords: levulinic acid; hydrothermal liquefaction of cellulose; cellulose recovery and bleaching; paper industry sludge; municipal primary sludge; value-added chemicals; ionic liquid

1. Introduction

The fossil sources are non-renewable and, will be depleted in near future. According to International Energy Agency, coal reserves will be available up to 2112 while easy production of oil and gas will be spent by 2040 and 2042, respectively. Therefore, scientists are exploring how to produce chemicals from biomass sources in order to meet the human needs. There are 12 building blocks, or chemical platforms, essential molecules which can be converted to a wide range of chemicals or materials. These building blocks can be obtained from waste biomass without exposing supplies like food, feed, and forests and in general, biodiversity of the world. Moreover, the valorization of the waste biomass is beneficial from a point of view of sustainable waste management. Additionally, the valorization is economically attractive by eliminating waste disposal fees and giving value to wastes [1].

Building blocks are the basic components for organic chemistry. One of the building blocks is the levulinic acid (LA). LA is used as a building block for fuel additives, solvents, flavor substances, pharmaceutical agents, coating, dyes, rubber and plastic additives, and other industries [2,3]. The levulinic acid can be obtained from cellulose contained in biomass sources [4].

The production of levulinic acid from lignocellulosic material by conventional methods require the use of inorganic mineral acids (e.g., H_2SO_4, HCl, H_3PO_4, and HBr) and metal salts [5,6]. These catalytic combinations have the disadvantage of causing equipment corrosion, environmental problems, and difficulty with recyclability [7–9]. In turn, solid acid catalysts are an environmentally friendly replacement for liquid acid catalysts. They are recoverable and recyclable despite that the obtained yield is low and they need long reaction times [9,10]. Ionic liquids (ILs) are used as solvents to dissolve the cellulose. They can also be used as catalysts in the reaction to produce levulinic acid. In comparison with conventional methods, ILs are characterized by high thermal stability, low corrosive nature, easy recoverability, low vapor pressure and easy functionalization with high acidity [11]. In this study, ionic liquid with functional group (–SO_3H) was used for the catalytic conversion of cellulose to levulinic acid. According to the literature [11], the catalytic activity of the functionalized ionic liquids decreased in the following order: IL-SO_3H > IL-COOH > IL-OH. Furthermore, the type of the cation has an effect on the acidity of IL. It is shown [12], that Brønsted acidic ILs with imidazolium-based cation has better catalytic activity than ammonium or triphenylphosphonium-based cations.

Figure 1 presents the mechanism of conversion of cellulose to levulinic acid. The mechanism involving several intermediates is explained below.

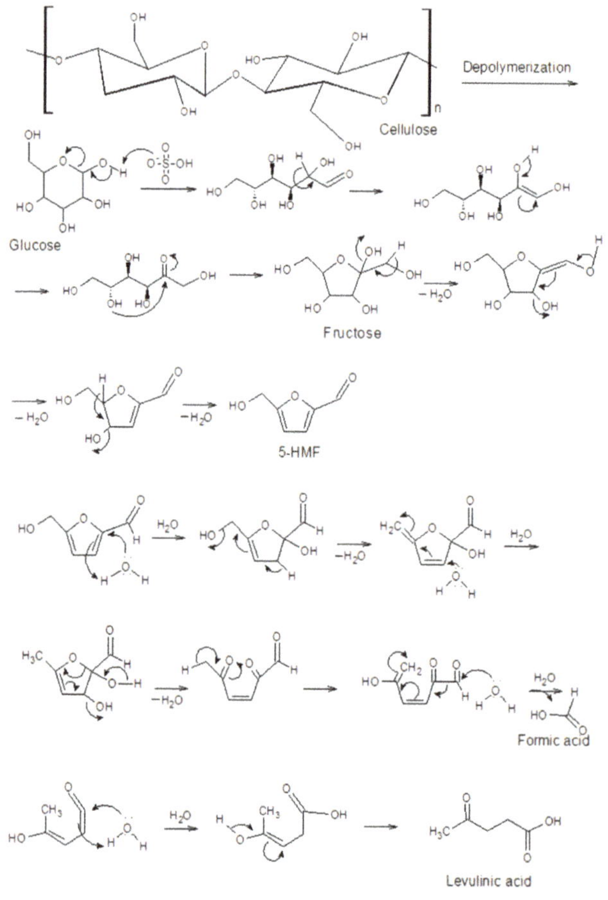

Figure 1. Mechanism of conversion of cellulose recovered from sludge to levulinic acid catalyzed by Brønsted acidic ionic liquid. Adapted from [13], copyright 2019, Elsevier.

In the first step, the cellulose is depolymerized into the monomeric gluco-pyranose sub-units. The acidic group of IL (sulfonic acid) causes the formation of a carbonyl group by the opening of pyranose ring. The carbonyl group goes through the keto-enol equilibrium and generates the 1,2-enediol. Next, the 1,2-enediol via a ring closure reaction produces fructose. Then, the keto-enol isomerization of fructose gives the 5-hydroxymethylfurfural (5-HMF) and three molecules of water are eliminated [13]. During the rehydration reaction of 5-HMF, levulinic acid is formed while a parallel reaction is forming a by-product: insoluble polymeric humins [3,14]. Humins are insoluble macromolecular components remained after carbohydrates catalysis. They can be used as composite material, building blocks, or soil amendment [15–17].

According to the literature [18], during the catalytic conversion of cellulose to levulinic acid, 5-HMF is obtained as by-product and, its concentration increases at reaction temperatures above 200 °C. Other by-products that can be obtained are furfural, formic acid and acetic acid. Due to the hydrolysis of hemicellulose to xylose followed by its dehydration allows the formation of furfural. A further dissociation of furfural let to obtain formic acid and acetic acid [19].

So far, in our view, only two papers have been published related to the production of levulinic acid from sludge or residual biomass from the paper industry [20,21]. In the first work, paper mill sludge was obtained from a packaging company and converted by hydrothermal liquefaction catalyzed by hydrochloric acid. Tests were also carried out with a microwave reactor. The conversion in levulinic acid obtained have approached 80% of the theoretical [20]. In the second work, fiber sludge from Finnish and Swedish pulp industry has been used. In this case, a microwave reactor was used, the reaction being catalyzed by Bronsted and Lewis acids. In this case, the conversion has not exceeded 35% with respect to the theoretical [21].

There are three main objectives in this work: (i) recovery of cellulose from waste biomass: municipal wastewater treatment plant (WWTP) primary wet sludge and paper industry dried mixed sludge, from primary treatment and membrane bioreactor purge. This first part was done using well-known methodology from previous studies [22–24]; (ii) the recovered cellulose from both type of sludges was purified by a bleaching procedure to decrease the amount of proteins and ashes in the solid phase after extraction; and (iii) the extracted and purified cellulose was converted to levulinic acid, value-added chemical considered as building-block by chemical industry. Hydrothermal liquefaction catalyzed by Brønsted acidic ionic liquid was used to produce the levulinic acid.

To the best of our knowledge, this is the first time that: (i) cellulose was recovered from municipal and industrial sludge by extraction with commercial ionic liquids; (ii) the recovered cellulose was bleached in two steps; and (iii) the purified cellulose was converted in levulinic acid by hydrothermal treatment catalyzed by home-made ionic liquid, all processes done consecutively.

2. Results and Discussion

Figure 2 presents the overall scheme of the process of production of levulinic acid from cellulose. In a first step, the cellulose is recovered from WWTP primary wet sludge and paper industry dried mixed sludge, from primary treatment and membrane bioreactor purge. The extraction was realized with Tetrakis ionic liquid [22–24]. In a second step, the recovered cellulose was bleached through two consecutives treatments with hydrogen peroxide and hydrochloric acid. Finally, purified cellulose was treated by catalyzed hydrothermal liquefaction to produce levulinic acid. Figure 2 also shows all the procedures of separation of products and their characterization after the overall process.

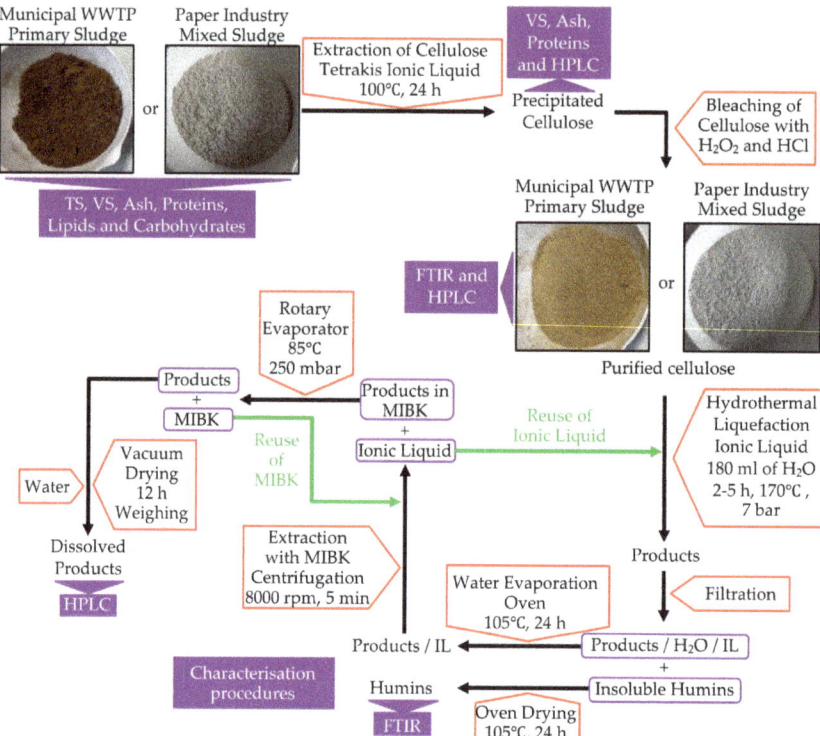

Figure 2. Scheme of full process of conversion of cellulose recovered from sludge to levulinic acid catalyzed by Brønsted acidic ionic liquid.

2.1. Characterization of Sludge

Primary sludge was collected from the municipal WWTP of Reus (Tarragona, Spain). The WWTP has a capacity to daily process 25,000 m^3 of wastewater. Primary sludge was collected after the partial gravity thickening, located later the primary decanter. The industrial sludge was provided by the international paper company Gomà-Camps, S.A., situated in La Riba (Tarragona, Spain). This sludge is a mixed sludge from physical–chemical primary treatment and purge of membrane biological reactor. Both effluents are mixed and dried with hot air. The results of the characterization of both sludges are presented in Table 1.

Table 1. Characterization of municipal primary sludge and industrial paper sludge by conventional procedures.

	Municipal Primary Sludge	Industrial Paper Sludge
TS (%, $w/w_{wet\ sludge}$)	3.8 ± 0.1	46.9 ± 1.1
Moisture (%, $w/w_{wet\ sludge}$) *	96.2 ± 0.1	53.1 ± 1.1
Ashes (%, w/w_{TS})	23.5 ± 2.3	50.5 ± 0.2
VS (%, w/w_{TS}) *	76.5 ± 1.7	49.5 ± 0.2
Lipids (%, w/w_{TS})	22.9 ± 1.8	1.2 ± 0.1
Proteins (%, w/w_{TS})	28.2 ± 1.5	5.3 ± 0.1
Carbohydrates (%, w/w_{TS})	25.8 ± 1.9	42.6 ± 0.7
Total (%, w/w_{TS})	100.4 ± 7.5	99.6 ± 0.9

Values are mean, ±SD, n = 3. * By difference.

As received, both sludges differ a lot, municipal WWTP primary sludge, henceforth municipal sludge, is essentially a liquid with solids in suspension. On the other hand, paper industry dried sludge, henceforth industrial sludge, is a solid. This difference is essentially due to the water content of each sludge.

As it can be seen in the table, the moisture of municipal sludge reaches 96.2%, whereas in industrial sludge this value is almost half-lower: 53.1%. Then, the total solids are 3.8% and 46.9% for municipal sludge and industrial sludge, respectively. After drying, both sludges also differ in their composition. Municipal sludge is constituted by more or less 25% w/w_{TS} of inorganic ashes and 25% w/w_{TS} of each organic matter: lipids, carbohydrates, and proteins. This composition has been quite constant during the last years as it can be seen comparing other works [22,25] and now. In contrast, industrial sludge is mainly constituted by ashes (50.5% w/w_{TS}) and carbohydrates (42.3% w/w_{TS}). The rest of the organic substances of the industrial sludge represent a low proportion: lipids (1.2% w/w_{TS}) and proteins (5.0% w/w_{TS}). This composition is similar than the observed in a recent work of the authors [24].

2.2. Extraction of Cellulose from Both Municipal and Industrial Sludges with IL

The process of recovery of cellulose from municipal and industrial sludges was performed using a similar method presented elsewhere [22–24]. Tetrakis ionic liquid was used to extract the cellulose. The temperature of process was fixed at 100 °C and the extraction was performed during 24 h. These values were optimized in the past [22–24]. Then methanol was added to the mixture to precipitate the cellulose. Three phases were obtained: (i) organic phase in the top of the flask; (ii) intermediate light brown aqueous phase containing ionic liquid, water, methanol, part of the proteins, and part of the ashes; and (iii) precipitated brown phase containing all cellulose and the rest of proteins and ashes. Then, hexane is added to the mixture to dissolve the organic phase and, is separated with a pipette. The operation was repeated until the organic phase is transparent. The aqueous phase and the precipitate were separated by centrifugation, 3500 rpm for 10 min. The precipitate containing cellulose and part of the proteins and ashes was separated by centrifugation at 6000 rpm during 10 min. The precipitate was cleaned with methanol and centrifuged again in the same conditions. Then the precipitate was characterized.

2.3. Characterization of the Solid Products after the Process of Recovery with IL

Table 2 presents the characterization of the solid products after the process of recovery of cellulose from both sludges. As it was expected, taking into account experimental errors, the process of recovery with Tetrakis IL was able to recover almost all carbohydrates from both sludges. However, besides of carbohydrates, the solids still contain a part of ashes and proteins.

Table 2. Characterization of the precipitates after the recovery of cellulose (100 °C, 24-h).

Composition (%, w/w_{TS})	Municipal Sludge	Industrial Sludge
Proteins	8.9 ± 1.9	4.1 ± 0.7
Ashes	16.2 ± 2.5	33.6 ± 0.5
Carbohydrates	28.7 ± 5.3	43.5 ± 1.3
Total	53.8 ± 9.7	81.2 ± 2.5

Values are mean, ±SD, $n = 3$.

On the contrary, lipids were totally removed during the separation. With municipal sludge, the process was able to decrease the quantity of ashes from 23.5 to 16.2% w/w_{TS} and the quantity of proteins from 28.4 to 8.9% w/w_{TS}. In the case of the industrial sludge, the drop-in values were from 50.5 to 33.6% w/w_{TS} and from 5.0 to 4.1% w/w_{TS} for ashes and proteins, respectively. The presence of ashes and proteins in the solid products required the need to clean them to purify the cellulose.

2.4. Bleaching of Precipitate

The procedure of the bleaching of recovered precipitate was based on the literature [26,27]. According to these works, hydrogen peroxide is commonly used as a bleaching agent responsible for lignin dissolution. In the first step of the process, a dried batch of recovered precipitate from municipal or industrial sludge was mixed with a solution of hydrogen peroxide (H_2O_2) 8% for the dissolution of lignin and protein. The mixture was stirred 24 h at room temperature. Then, the mixture was filtered. The precipitate was cleaned with distillate water. The content of protein and ashes were determined in both precipitates. The results of the bleaching process are presented in Table 3.

Table 3. Characterization of the precipitates after bleaching (8% H_2O_2 and 0.1 N HCl).

Composition (%, w/w_{TS})	Municipal Sludge	Industrial Sludge
Proteins *	6.4 ± 1.6 (−28%)	1.8 ± 0.4 (−56%)
Ashes **	4.9 ± 1.8 (−70%)	21.5 ± 0.3 (−36%)
Carbohydrates	24.4 ± 7.0 (−15%)	38.3 ± 3.7 (−12%)
Total	35.7 ± 10.4 (−34%)	61.6 ± 4.4 (−24%)

Values are mean, ±SD, n = 3. * Quantified after hydrogen peroxide treatment. ** Quantified after hydrochloric acid treatment. In brackets: rate of reduction.

Proteins were determined after H_2O_2 treatment. As it can be seen in the table, the H_2O_2 solution was able to dissolve and remove 28% and 56% of protein from municipal and industrial precipitates, respectively. Proteins dissolution in H_2O_2 can be explained by the dissociation of H_2O_2 to hydrogen (H^+) and hydroxyl (OH^-) radicals. Then, these oxidative agents readily attacked proteins and decomposed them into soluble amino acids [28]. However, H_2O_2 did not reduce the quantity of ashes from the precipitate, acid hydrolysis with 0.1 N HCl is expected to do that.

In the second step, both precipitates were hydrolyzed with hydrochloric acid (HCl) 0.1 N during 5 h in an ultrasonic bath. The initial transparent acid dissolution changed color to light brown/yellow. After separation, cleaning and drying of precipitate, ashes were quantified. The acidic treatment was able to remove 70% and 36% of the ashes from municipal and industrial precipitates, respectively. It is noticeable that during the bleaching treatment a loose of carbohydrates was also observed, 15% and 12% from municipal and industrial precipitates, respectively. According to the literature [26], the hydrolysis also makes isolation of the pure cellulose fibers by hydrolyzing traces of hemicellulose and lignin to simple sugars. In Table 4 are presented the results of high-performance liquid chromatography (HPLC) analysis of precipitates before and after bleaching. It can be seen in the table that the amount of hemicellulose and lignin reduced after bleaching. However, the amount of cellulose increased for both sludges, more in the case of the industrial sludge, from 32.8 to 42.3%. This increasing can be explained by the growth of the amount of simple sugars, such as glucose, provoked by the treatment with HCl. As the measure of cellulose is obtained from the values of glucose [29], the results show an increasing amount of cellulose.

Table 4. Characterization of the carbohydrates by high-performance liquid chromatography (HPLC) after bleaching (8% H_2O_2 and 0.1 N HCl).

Carbohydrates (%, $w/w_{precipitate}$)	Municipal Sludge		Industrial Sludge	
	Before Bleaching	After Bleaching	Before Bleaching	After Bleaching
Hemicellulose	7.7 ± 1.0	3.4 ± 0.1	7.2 ± 0.1	6.5 ± 0.0
Cellulose *	25.3 ± 0.6	26.7 ± 3.2	32.8 ± 1.3	42.3 ± 0.9
Lignin	20.0 ± 3.0	17.7 ± 0.3	10.9 ± 2.3	10.2 ± 0.3

Values are mean, ±SD, n = 3. * Amount of cellulose calculated by division of the amount of glucose by a conversion factor of 1.11 [29].

Figure 3 presents Fourier Transform Infrared (FTIR) spectra of purified cellulose recovered from municipal sludge (a) and industrial sludge (b). Both spectra present some characteristic absorbances in different frequency regions: 3300 cm^{-1} of O–H group, 2900 cm^{-1} assigned as the CH$_3$ and CH$_2$ stretching vibration of cellulose, 1160 cm^{-1} of C–O–C stretching vibration particularly associated with cellulose and the broad peak 1030 cm^{-1} of C–O stretching vibration of carbohydrates. All of those peaks demonstrate that purified solids contain cellulose. In Figure 3a there is also presented the peak at 1650 cm^{-1} associated to peptide amide groups of proteins while in Figure 3b, is absent. That confirms that after bleaching some part of proteins (6.8%) stay in the cellulose recovered from municipal sludge whereas the amount of proteins in the purified cellulose recovered from industrial sludge is too small (1.8%) to be visible in the spectra.

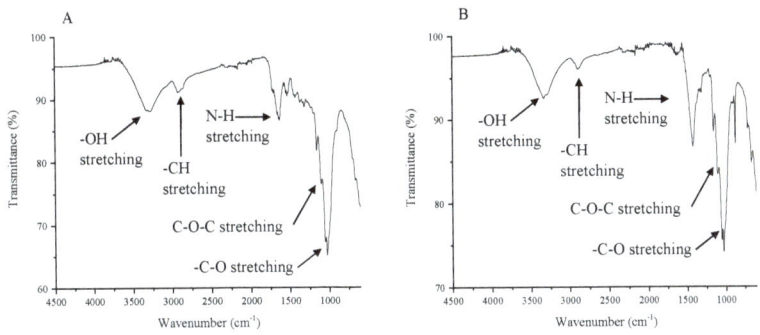

Figure 3. Fourier Transform Infrared (FTIR) spectra of purified cellulose after extraction with ionic liquid and bleaching with H$_2$O$_2$ and H$_2$SO$_4$: (**A**) municipal sludge and (**B**) industrial sludge.

2.5. Cellulose Conversion to Levulinic Acid

The procedure of conversion of cellulose to levulinic acid catalyzed by Brønsted acidic ionic liquid was directly based on a recent work [3]. In this work, authors have synthesized and tested nine different ionic liquids. The results showed that the acidity of the ionic liquid has the greater importance on the yield of the reaction. The acidity depends on the cation group and on the anion. The [mimC$_4$SO$_3$H] [HSO$_4$] ionic liquid has the higher acidity resulting in its design by the presence of the imidazolium group and by the hydrogen sulfate anion [3]. In consequence, [mimC$_4$SO$_3$H] [HSO$_4$] was selected and synthesized to convert cellulose to levulinic acid by catalyzed hydrothermal liquefaction.

After the reaction, the reactor was cleaned with deionized water and the products were separated and characterized. The first separation was realized by filtration. Residual solids were abundantly washed with deionized water, dried and weighed. The liquid phase is a mixture of water, ionic liquid and products. Water was evaporated overnight in an oven. Then, the ionic liquid was separated from the resulting liquid mixture by addition of methyl isobutyl ketone (MIBK). Two liquid phases were formed, the upper phase contain the levulinic acid dissolved in the MIBK while the lower phase the ionic liquid. Both phases were separated by centrifugation at 8000 rpm for 5 min. MIBK was evaporated from organic phase at the rotary evaporator at 85 °C and 250 mbar. Resultant products were dried under vacuum and weighed. Finally, products were characterized by HPLC. In Table 5 are presented the values of the weight of biomass, weight of cellulose contained in the biomass, weight of ionic liquid, and volume of water used in each reaction. Table 5 also presents the values of weight of soluble in MIBK products, weight and percentage of levulinic acid and finally weight and percentage of humins.

Table 5. Weight of products, levulinic acid and humins after hydrothermal treatment of cellulose with acidic ionic liquid (temperature: 170 °C, time of reaction: 5 h).

N°	Source	Biomass (g)	Cellulose (g)	IL (g)	H$_2$O (mL)	Products (g)	Levulinic Acid (g)	Levulinic Acid (%)	Humins (g)	Humins (%)
1	Sigmacell (2 h)	-	4.8	30.0	180	0.9	0.9	18.8	3.2	66.7
2	Sigmacell	-	4.5	30.0	180	2.7	2.6	57.8	0.9	20.0
3	Sigmacell	-	16.0	30.0	180	13.1	13.0	81.3	3.0	18.8
4	IS Purified	6.7	2.8	26.6	180	0.4	0.2	7.1	5.9	88.1
5	IS Purified	8.9	3.8	26.6	180	0.6	0.4	10.5	8.3	93.3
6	IS Purified	9.8	4.2	30.0	180	1.4	1.3	31.0	8.8	89.8
7	IS Purified	8.9	3.8	36.0	180	0.7	0.7	18.4	7.6	84.4
8	MS Purified	8.2	2.2	26.0	180	0.6	0.5	22.7	6.8	82.9
9	MS Purified	19.9	5.3	36.0	180	0.7	0.6	11.3	14.2	71.4
10	MS Purified	8.9	2.4	26.0	180	0.7	0.6	25.0	8.0	89.9
11	IS	12.1	4.1	26.0	180	0.01	0.01	0.2	10.2	84.3
12	MSar	200.9	1.0	6.0	0	0.03	0.02	2.0	6.8	89.5

IL: Ionic Liquid; IS: Industrial Sludge (as received); MS: Municipal Sludge; MSar: Municipal Sludge (as received, wet, TS: 3.8% $w/w_{wet\,sludge}$).

As it was expected, levulinic acid was obtained in each reaction with the three sources of cellulose: pure from provider, from municipal and from industrial sludge. About the commercial cellulose, entries 1–3, it can be observed that with a reaction time of only 2 h, the conversion to levulinic acid was only 18.8%. Increasing the reaction time to 5 h allowed a conversion of 57.8%. This is normal, increasing the reaction time produces an increase in the conversion. However, this increase in conversion reaches a maximum and then decreases, due to the appearance of condensation and recombination reactions of the products. Then, it will be necessary in a future work to confirm this fact, because it is possible that optimized time of reaction was not attained. On the other hand, the decrease of the ratio between the weight of the ionic liquid and the weight of cellulose in the sample, $w_{IL}/w_{Cellulose}$: from 6.67 to 1.88, allowed to increase the conversion to levulinic acid until 81.3%. The effect of the ratio $w_{IL}/w_{Cellulose}$ on the conversion to levulinic acid seems to be important. This will be studied in more detail in further work.

The results of conversion to levulinic acid obtained with industrial sludge are less good (entries 4–7). In the four experiments carried out, the conversion to levulinic acid was between 7.1 and 31.0%. The reduction of the conversion is normal since in the samples of industrial sludge there is presence of hemicellulose, lignin, ash and other materials that disturb the reaction of conversion of the cellulose. As it was the case with commercial cellulose (Sigmacell), the decrease in the ratio $w_{IL}/w_{Cellulose}$ causes an increase in the conversion to levulinic acid. Indeed, in entries 4 and 7 the ratio is 9.5 and the conversions are 7.1 and 18.4%, respectively. In entries 5 and 6 the ratio is 7.0 and conversions are 10.5 and 31.0%, respectively. However, there is a disparity between the results obtained with the same ratio. There is another variable that seems to be important: the water to cellulose ratio, $w_{Water}/w_{Cellulose}$. Indeed, in entries 4 and 7 the $w_{Water}/w_{Cellulose}$ ratios were 64.3 and 47.4, respectively. On the other hand, in entries 5 and 6 the $w_{Water}/w_{Cellulose}$ ratios were 47.4 and 42.9, respectively. The decrease in the $w_{Water}/w_{Cellulose}$ ratio causes a consequent increase in the conversion to levulinic acid. These trends will have to be confirmed in subsequent work.

In the case of municipal sludge, entries 8–10, the results of conversion to levulinic acid were quite similar: between 11.3 and 25.0%. Again, impurities in the samples were responsible of the decrease of the conversion to levulinic acid. However, with municipal sludge, the decrease in the ratio $w_{IL}/w_{Cellulose}$ does not seem to present a clear trend. Indeed, in entries 8, 10, and 9, the $w_{IL}/w_{Cellulose}$ ratios are 11.8, 10.8, and 6.8, respectively, while the conversions to levulinic acid are 22.7, 25.0, and 11.3% respectively. It can be seen that between entries 8 and 10 there is an increase in the conversion but, then, it decreases sharply. The effect of the $w_{Water}/w_{Cellulose}$ ratio on the conversion was the same.

In entries 8, 10, and 9, the $w_{Water}/w_{Cellulose}$ ratios are 81.8, 75.0, and 34.0, respectively. The explanation for this negative result must be sought in the design of the experiment. Indeed, the experiment in entry 9 was carried out with a mass of sludge of almost 20 g, while in experiments 8 and 10 this mass was between 8 and 9 g. While the amount of cellulose available is higher, the amount of impurities even more. This must have disturbed the cellulose conversion reaction and maybe the mixing into the reactor. On the other hand, observing the mass balances of these experiments it can be seen that the balance for entry 9 was only 75%, while the mass balances for entries 8 and 10 were 90 and 98%, respectively. This fact could explain the bad result obtained with the experiment in entry 9.

The experiments realized with industrial and municipal sludge as received, entries 11 and 12 respectively, gave even smaller conversions to levulinic acid: 0.2 and 2.0%, respectively. The cause of this great decrease can be found in the fact that both sludges contain large amounts of ashes, 23.5 and 50.5%, respectively.

As it can be seen in Table 5, levulinic acid represents more of the 90% of the total products. The other products are lactic acid, formic acid, HMF, and furfural. These other products were also found elsewhere [3]. In Figure 4, two chromatograms from HPLC analysis of products obtained from hydrothermal processing of cellulose are presented from industrial sludge (a) and municipal sludge (b).

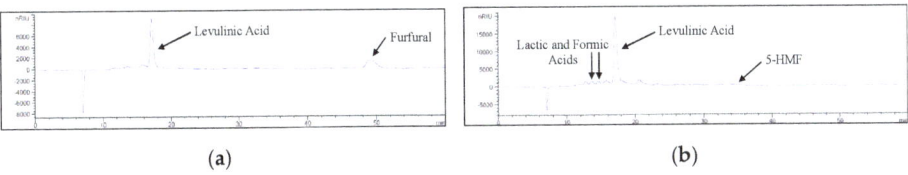

Figure 4. Chromatograms of products obtained from catalyzed hydrothermal liquefaction of cellulose with acidic ionic liquid: (**a**) industrial sludge and (**b**) municipal sludge.

In both chromatograms it is clearly showed that the higher peak (retention time 17.1 min) corresponds to levulinic acid. The other small peaks are the co-products identified in laboratory: furfural, lactic acid, formic acid, and 5-hydroxymethylfurfural (5-HMF). Further analysis of the products by LC–MS allowed to identify more substances in the products. Table 6 presents the list of substances detected by LC–MS, the retention time of each compound and molecular mass of positive and negative ion electrospray analysis.

Table 6. Products identified by LC–MS after hydrothermal treatment.

Entry	Retention Time (min)	m/z +	m/z −	Compound
1	8.093	117.0188	115.0031	Fumaric Acid
2	10.709	135.0293	133.0137	Malic Acid
3	11.625	181.0712	N.D.	Mannose
4	11.653	N.D.	179.0556	Glucose
5	13.558	91.0395	89.0239	Lactic Acid
6	13.791	119.0344	117.0188	Succinic Acid
7	13.924	93.0552	91.0395	Glycerol
8	14.491	97.0290	95.0133	Furfural
9	14.602	175.0243	173.0086	cis-Aconitic Acid
10	14.849 *	N.D.	59.0133	Acetic Acid
11	16.124	111.0421	87.0446	Butyric Acid
12	19.473	117.0552	115.0395	Levulinic Acid
13	23.863	151.0243	149.0086	Tartaric Acid
14	43.633	127.0395	125.0239	5-HMF

N.D. not detected. * Only detected in products obtained from municipal sludge.

All compounds, except acetic acid, were detected in the products obtained from both types of sludge. Acetic acid was only presented in products obtained from municipal sludge. Almost all compounds are short-chain organic acid. These are the expected products of hydrolysis of the cellulose [30]. Apart from organic acids, products of conversion of both sludges contain also glucose, furfural, and 5-HMF.

The peak of furfural has appeared in the chromatogram of conversion products of industrial sludge, Figure 4a, but not in that of municipal sludge, Figure 4b. LC–MS analysis confirmed the peak of furfural in both cases. Furfural comes from the dehydration of xylose. On the other hand, xylose makes up most of the hemicellulose. In Table 4, it could be observed that the hemicellulose content in industrial sludge is double that in municipal sludge. For this reason, the furfural peak is observed in the chromatogram of industrial sludge and not in that of municipal sludge. In the case of 5-HMF, the conversion reactions were carried out at temperatures below 200 °C and therefore large amounts of 5-HMF were not produced. A 5-HMF peak cannot be seen in the chromatogram of conversion products of industrial sludge, Figure 4a. However, in the chromatogram of conversion products of the municipal sludge, Figure 4b, a very small one appears, as a result of which the 5-HMF was not completely dehydrated to levulinic acid.

On the other hand, Table 5 also presents the weights and the yields of residual solids. Calculation was made based on the weight of biomass used in each reaction. The residual solids were in water insoluble dark-brown solids. The solids after filtration were abundantly washed with water to eliminate the remaining ionic liquid. Then, they were dried and weighted. As it can be seen in the table, depending on the source of cellulose used for the reaction of conversion to levulinic acid, the obtained yield of humins differ. Commercial cellulose (Sigmacell) obtained conversions to humins of ~20% except in the case of the experiment carried out for two hours, where the conversion reached was 66%. This increment was caused by insufficient time to convert cellulose to products. In the case of all the other experiments, purified celluloses from municipal and industrial sludges, or from both sludges used as received, the conversions to humins were higher and quite similar, between 80 and 90%. The yields of residual wastes formed during the reactions with cellulose from sludges was due to the content of ash and proteins in municipal and industrial sludge. After cellulose extraction and purification, a part of ashes and proteins still remained in the cellulose. Subsequently, during the reaction, those organic and inorganic matters pass to the solid residue, at the same time, increasing the yield of humins. Figure 5 shows the FTIR spectra of humins obtained after reaction of purified cellulose from both sludges.

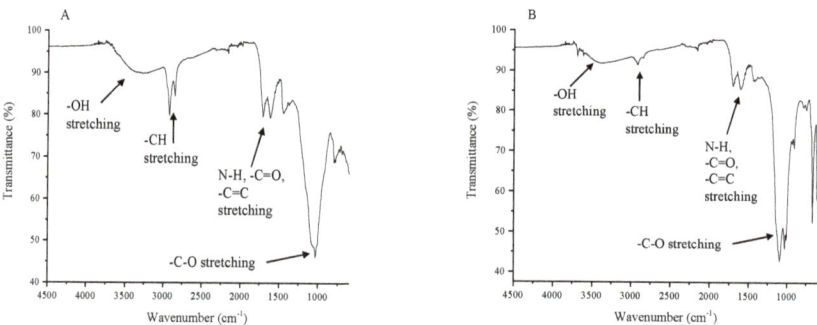

Figure 5. FTIR spectrum of humins obtained from hydrothermal processing of cellulose with acidic ionic liquid: (**A**) municipal primary sludge and (**B**) industrial sludge.

In both spectra, it can be observed some characteristic absorbances in different frequency regions. The peak at 3400 cm^{-1} is associated with hydroxyl group stretching vibrational bands. The peaks at 2950 cm^{-1} and 1000–1250 cm^{-1} are attributed to –CH and –CO stretching vibrations, respectively. It is also possible to observe strong peaks at 1700 and 1625 cm^{-1}, corresponding to carbonyl group conjugated to an alkene group. This is because carbohydrates are converted into humins according to the reaction pathway: cellulose → 5-HMF → 2,5-dioxo-6-hydroxyhexanal → humins [14,31]. Although, the peak at 1625 cm^{-1} is also associated with peptide amide groups of proteins. The FTIR spectra confirm that except humins, ashes, proteins and part of unreacted cellulose goes to insoluble solids.

To summarize, the full process of production (extraction, bleaching, and catalyzed hydrothermal liquefaction) of levulinic acid from municipal or paper industry has been a success. It is true that the conversions obtained are not high, but it must be taken into account that the entire study is in fact a technological feasibility study. A scale-up of the process can only be carried out when some aspects of the process units will have been solved: (i) optimizing the extraction of cellulose from both sludges through an improved design of the ionic liquid; (ii) improve the cleaning of the cellulose after its extraction, although if more severe methods are used, a greater part of it can be hydrolyzed; and (iii) optimize the reaction operations in the conversion of cellulose to levulinic acid, reaction time and temperature, cellulose/ionic liquid/water ratios, or improvement in the design of the ionic liquid.

3. Materials and Methods

3.1. Reagents

All chemical reagents used in this work were supplied by Sigma-Aldrich (Barcelona, Spain). Tetrakis (hydroxymethyl) phosphonium chloride (P(CH$_2$OH)$_4$]Cl, hydrated ionic liquid, 80% in water, ref: 404861) was used to recover the cellulose from municipal primary sludge and from industrial sludge. Hydrogen peroxide solution (≥30% ref: 95321) and 0.1 N hydrochloric acid solution (ref: 2104) were applied to purify recovered cellulose. 1-methylimidazole (purity ≥ 99%, ref: 336092), 1,4-butanesultone (purity ≥ 99%, ref: B85501) and sulfuric acid (95.0–97.0%, ref: 30743) were used to synthesize the Brønsted acidic IL used in the hydrothermal conversion of recovered cellulose to levulinic acid. Methyl isobutyl ketone (MIBK) (purity ≥ 99.5%, ref: 293261) was used to extract levulinic acid from Brønsted acidic ionic liquid after the process of conversion. Pure cellulose Sigmacell (ref: S3504) was used to realize comparative experiments of hydrothermal liquefaction. Glucose (purity ≥ 99.5%, ref: 49139), levulinic acid (98 wt.%, ref: L2009), 5-hydroxymethylfurfural (purity ≥ 98.0%, ref: 53407), formic acid (purity ≥ 98%, ref: 33015-M), DL-lactic acid (purity 90%, ref: 69785) and furfural (purity 99%, ref: 185914) were used to prepare calibration curves for high-performance liquid chromatography (HPLC).

3.2. Sludge Collection and Managing

Cellulose was extracted from two types of sludge: municipal primary sludge and paper industry sludge. The primary sludge (96% water content) was collected after partial gravity thickening of the Wastewater Treatment Plant (WWTP), located in Reus, Spain. The industrial dehydrated sludge (53% water content) was provided by the paper international company Gomà-Camps, S.A., La Riba, Tarragona, Spain. After collection, both types of sludge were stored in a fridge at 4 °C and were used for characterization as received. Prior recovering of cellulose, both sludge were dried in an oven at 105 °C during 48 h. Then, they were directly used for the rest of the procedures.

3.3. Sludge Characterization

A full characterization of each sludge was realized in triplicate. Determination of total solids (TS), volatile solids (VS), and content of ash, proteins, lipids and carbohydrates were proceeded by conventional methods mentioned in previous studies [22,25].

Total solids (TS) of each sludge were determined by the standard method 2540 B [32]. Wet primary sludge and dried paper industry sludge with known quantity (between 0.5 to 2.0 g) were placed in a previously weighted ceramic capsule and then they were put in the oven at 105 °C for 24 h. Total solids were calculated with the formula:

$$\%TS = \frac{(\text{Weight dry capsule plus dry sample g} - \text{Weight dry capsule g})}{\text{Weight sample as received g}} \times 100$$

Volatile solids (VS) and ash content were determined by the standard method 2540 E [32]. The weighted dry capsule with dry sample was placed in the muffle at 550 °C for 1 h. Ash and volatile solids were calculated with the formulas:

$$\%Ash = \frac{(\text{Weight of capsule and sample after 550 °C g} - \text{Weight dry capsule g})}{(\text{Weight dry capsule plus dry sample g} - \text{Weight dry capsule g})} \times 100$$

$$\%VS = 100 - \%Ash$$

Lipids content of each sludge was determined by the standard method 5520 E [32]. There was 20 g of sample of each sludge acidified with 0.3 mL of HCl. Then, 25 g of $MgSO_4 \cdot H_2O$ was added in order to dewater acidified samples of sludge. The samples were mixed and then, they were crushed until they became a fine powder. Then, the samples were transferred into the cellulose Soxhlet extraction thimble and covered with a glass wool. The extraction was done in a Soxhlet extractor using hexane as a solvent at a velocity of 20 cycles/hour for 4 h. After all, the hexane was evaporated through a rotary evaporation. The lipids were kept in a desiccator overnight to remove remained traces of hexane and weighed the next day. The flasks were measured before and after the experiment. Mass of lipids was calculated by weighing and with the formula:

$$\%Lipids = \frac{(\text{Weight Lipids} + \text{flask g} - \text{Weight of flask g})}{\text{Weight of TS g}} \times 100$$

Proteins content of each sludge was determined by the Lowry method using Folin phenol reagent [33]. Firstly, the calibration curve of 0.5 g Bovine serum albumin/L patron was prepared. Then, diluted samples of each sludge were heated with 2 M solution of NaOH in order to dissolve proteins. The absorbance of patron as well as sludge samples were measured at 750 nm. The amount of proteins was calculated by the colorimetric method.

Carbohydrates determination was carried out based on phenol sulfuric method of Dubois [34]. Calibration curve of 0.1 g glucose/L was prepared as a patron. Diluted samples of each sludge were mixed with phenol solution and then with concentrated sulfuric acid. The absorbance of patron as well as sludge samples were measured at 480 nm. The amount of carbohydrates was calculated by the colorimetric method.

3.4. Recovery of Cellulose from Sludge

The recovery of cellulose from both types of sludge was carried out using the method described in previous studies [22–24], with small modifications. Briefly, instead of wet sludge, dry sludge (DS) was used. This was necessary to prepare a big batch of recovered cellulose. The use of wet sludge (96% or 53% water content) would require working with too big volumes of sludge and ionic liquid. Dried sludge was placed in a round bottom flask and Tetrakis IL was added in ratio 1 g_{DS}: 10 mL_{IL}. The mixture was stirred and heated to 100 °C with an oil bath maintained 24 h. After this time, the oil bath was removed and the flask cooled down to ambient temperature. At that point, 10 mL of water for each gram of dry sludge were added to the solubilized cellulose. The rest of the procedure was the same as in the previous studies. After reaction, IL was recovered by water evaporation in a rotary evaporator at 80 °C and 300 mbar. Then, IL was ready to be reused for other successive recovering.

3.5. Characterization of Recovered Cellulose

The cellulose was analyzed to determine the volatile solid and ash content according to standard method 2540 E [32]. The amount of proteins was calculated as a difference between proteins in sludge as received and proteins determined in ionic liquid phase by Lowry method [33]. Cellulose, hemicellulose and lignin were analyzed by high performance liquid chromatography (HPLC) after strong hydrolysis following method presented elsewhere [24].

3.6. Bleaching of Recovered Cellulose

Bleaching method was carried out according to literature [26]. Dried recovered cellulose from both sludge was placed in a round bottom flask. Then, 250 mL of freshly prepared 8% H_2O_2 (*v/v*) was added. The mixture was kept under agitation at room temperature during 24 h. After that, H_2O_2 and solids were separated by centrifugation (8000 rpm, 5 min). H_2O_2 was kept in order to determine the content of proteins by Lowry method [33] and, the ash content by the conventional method 2540 E [32]. On the other side, the solid was washed with distilled water and centrifuged (8000 rpm, 5 min). Then, the solid was hydrolyzed with 250 mL of 0.1 N HCl in an ultrasonic bath (Bandelin Electronic DT 514H, Berlin, Germany) at a temperature of 60 ± 1 °C for 5 h. Afterward, the liquid was removed by centrifugation (8000 rpm, 5 min) and the solid was washed several times with distilled water until neutralization of the washing liquid of the solid. At the end, the solid was dried overnight in the oven at 105 °C and analyzed for ash, proteins, cellulose, hemicellulose and lignin content as it was described before. Additionally, purified cellulose was characterized by Fourier Transform Infrared (FTIR) spectroscopy. The samples, without any further preparation, were scanned using a Fourier Jasco FT/IR-600 Plus spectrometer (Barcelona, Spain) with a diamond golden gate ATR reflectance cell.

3.7. Synthesis of the Brønsted Acidic Ionic Liquid

The Brønsted acidic ionic liquid 4-(3-methylimidazolium) butanesulfonic acid hydrogenosulfate [mimC$_4$SO$_3$H] [HSO$_4$] was prepared according to literature [25]. Briefly, equimolar quantities of 1-methylimidazole and 1,4-butanesultone were added to a round bottom flask fitted with a condenser and stirred at 80 °C for 10 h under nitrogen purge. Then, the obtained solid was washed with toluene and diethyl ether and, dried under vacuum overnight. A stoichiometric amount of acid H_2SO_4 was added dropwise to the white solid and stirred overnight at 60 °C under reflux condenser and nitrogen purge. The obtained viscous IL was dried under vacuum and characterized by ^{1}H NMR spectroscopy (Varian NMR System 400, Palo Alto, CA, USA).

The IL was characterized by ^{1}H NMR spectroscopy, to confirm its structure. ^{1}H NMR spectrum (400 MHz, d^4-methanol), δ (ppm): 1.79 (p, 2H, CH$_2$, J = 7.2), 2.05 (p, 2H, CH$_2$, J = 7.6), 2.9 (t, 2H, CH$_2$–SO3H, J = 7.6), 3.94 (s, 3H, N–CH$_3$), 4.27 (t, 2H, N–CH$_2$, J = 7.2), 7.59 (t, 1H, CH, J = 1.2), 7.66 (t, 1H, CH, J = 1.6), 8.96 (s, 1H, N–CH–N).

3.8. Conversion of Purified Cellulose to Levulinic Acid

3.8.1. Reaction Experimental Procedure

The procedure of conversion of cellulose to levulinic acid, presented in Figure 1, is based on a recent work [3]. Different amounts of purified cellulose (2.2 to 16.0 g), Brønsted acidic ionic liquid (26.0–36.0 g) and 180 mL of distilled water were utilized. The substances were introduced in a 1 L stainless steel autoclave (Autoclave Engineers model EZE Seal) with heating shell and MagneDrive® stirrer (magnetically coupled, packless rotary impeller system). The amounts of purified cellulose and IL were chosen based on the best conditions described in the work of Ren et al., 2015 [3]. Two last experiments were carried out with wet primary sludge as received (96% water content) and paper industry sludge as received (53% water content). These two experiments, considered as blank assays, allow the comparison of performances with the other experiments realized with bleached sludge. After closing the reactor, the temperature was settled up to 170 °C. The reactor took ~30 min to reach the desired temperature. The pressure in the reactor at that temperature was over 7 bars and was constant during all the reaction time. Two different times of reaction were used after reaching the process temperature: 2 and 5 h. After the reaction, the heating shell was removed and the system was cooled in cold water. The cooling time was ~35 min. After cooling, the reactor went back to atmospheric pressure.

3.8.2. Cleaning of the Reactor and Separation of Products

After opening of the reactor, the content was poured into a beaker and the reactor interior itself was cleaned with deionized water to recover all products: a black mixture of insoluble solids and liquid. Then, the dark brown insoluble solids, also called humins [3,13], were separated from the liquid phase by filtration with a Büchner funnel. The solids were washed with water several times. Then, they were oven dried at 105 °C for 24 h and weighed. The yield of insoluble solids (%) was calculated with the initial weight of biomass used in the process. Furthermore, the solids were characterized by FTIR spectroscopy.

The liquid phase, mixture of water, IL and products, was left overnight in the oven at 105 °C to evaporate the water. Then, the organic products were solubilized with methyl isobutyl ketone (MIBK) and separated from the ionic liquid by centrifugation at 8000 rpm for 5 min. The formation of two phases, MIBK and products (upper layer) and ionic liquid (down layer), allowed an easily separation of products from the ionic liquid. The upper layer was gathered and then, MIBK was evaporated at the rotary evaporator at 85 °C and 250 mbar. At that point, products were dried under vacuum and weighted. Products were characterized by HPLC (Agilent Technologies 1100 Series HPLC System, Barcelona, Spain). For this, they were dissolved in water and analyzed by HPLC equipped with a Refractive Index Detector (RID) and a Hi-Plex H column (300 × 7.7 mm), filled by a robust sulfonated cross-linked styrene-divinylbenzene gel in hydrogen form. Chromatograph conditions were as follow: injection volume 20.0 µL, mobile phase 0.005 M H_2SO_4, flow rate 0.7 mL/min, column temperature 60 °C, RID temperature 55 °C. The concentration of each compound was determined using calibration curves obtained with standard solutions with known concentrations. The yield of levulinic acid (%) was calculated from the initial weight of cellulose used in the reaction. Additionally, products were analyzed by LC–MS with an Agilent 1200 liquid chromatograph, coupled to a 6210 Time of Flight (TOF) mass spectrometer (Agilent Technologies, Waldbronn, Germany) and with an electrospray ionization (ESI) interface. Experiments were performed using a Hi-Plex H column (300 × 7.7 mm) provided by Agilent Technologies under isocratic conditions (H_2O + 0.1% HCOOH) and a constant flow of 0.7 mL/min. The HPLC eluate was directly pumped into ESI interface without flow splitting. The ESI conditions were as follow: gas temperature 300 °C, drying gas 12 L/min, nebulizer 40 psi, fragmentor 120 V, capillary voltage (3000 V), and mass range 50–1200.

4. Conclusions

The process of production of levulinic acid from municipal or paper industry sludges has allowed to obtain very interesting results: (i) the Tetrakis (hydroxymethyl) phosphonium chloride ionic liquid allows the extraction of all the cellulose in both sludges, however, the precipitated cellulose still contains proteins and ashes; (ii) the bleaching process by hydrogen peroxide and hydrochloric acid does not work as well as expected, after the process the cellulose continues to have impurities that will be annoying in the reaction; (iii) catalyzed hydrothermal liquefaction of cellulose is a process that makes it possible to produce levulinic acid, but the operating conditions have to be optimized; and (iv) it is possible to carry out an integrated process for the production of levulinic acid from residual biomass, in this case of sludge from municipal or industrial treatment plants. Finally, it should be noted that the obtained conversions to levulinic acid are not high, up to 31% with industrial sludge and up to 25% with municipal sludge. However, it must be emphasized that the entire project is a technological feasibility study. The use of waste from wastewater treatment plants can lead to an increase in their income when it is well known that the cost of disposal their wastes is the biggest part of their operational costs.

Author Contributions: Conceptualization, C.B.; Data curation, J.G. and C.B.; Funding acquisition, J.G. and C.B.; Investigation, K.G. and C.L.; Methodology, K.G., E.T., and C.B.; Project administration, J.G.; Resources, J.G. and C.B.; Supervision, C.B.; Validation, E.T. and C.B.; Visualization, K.G. and C.B.; Writing—original draft, K.G. and C.L.; Writing—review & editing, K.G., E.T., and C.B. All authors have read and agreed to the published version of the manuscript.

Funding: This research was funded by Universitat Rovira i Virgili (2017PFR-URV-B2-33 and 2016PMF-PIPF-29), Spanish Ministerio de Economía y Competitividad and the FEDER grant (CTM2015-67970).

Acknowledgments: The authors wish to acknowledge the public company Gestió Ambientali Abstament S.A. (WWTP of Reus, Spain) and the company Gomà-Camps, S.A. (La Riba, Tarragona, Spain) for their kind collaboration during this project. Katarzyna Glińska thanks for the Martí Franquès pre-doctoral scholarship provided by the Universitat Rovira i Virgili. The authors are recognized by the Comissionat per a Universitats i Recerca del DIUE de la Generalitat de Catalunya (2017-SGR-396).

Conflicts of Interest: The authors declare no conflict of interest.

References

1. Xiong, X.; Yu, I.K.M.; Tsang, D.C.W.; Bolan, N.S.; Ok, Y.S.; Igalavithanan, A.D.; Kirkham, M.B.; Kim, K.H.; Vikrant, K. Value-added chemicals from food supply chain wastes: A critical review. *Chem. Eng. J.* **2019**, *375*, 121983. [CrossRef]
2. Liu, S.; Wang, K.; Yu, H.; Li, B.; Yu, S. Catalytic preparation of levulinic acid from cellobiose via Brønsted-Lewis acidic ionic liquids functional catalysts. *Sci. Rep.* **2019**, *9*, 1810. [CrossRef]
3. Ren, H.; Girisuta, B.; Zhou, Y.; Liu, L. Selective and recyclable depolymerization of cellulose to levulinic acid catalysed by acidic ionic liquid. *Carbohyd. Polym.* **2015**, *117*, 569–576. [CrossRef] [PubMed]
4. Elumalai, S.; Agarwal, B.; Sangwan, R.S. Thermo-chemical pretreatment of rice straw for further processing for levulinic acid production. *Biores. Technol.* **2016**, *98*, 1448–1453. [CrossRef] [PubMed]
5. Yu, I.K.M.; Tsang, D.C.W. Conversion of biomass to hydroxymethylfurfural: A review of catalytic systems and underlying mechanisms. *Biores. Technol.* **2017**, *238*, 716–732. [CrossRef] [PubMed]
6. Cao, L.; Yu, I.K.M.; Liu, Y.; Ruan, X.; Tsang, D.C.W.; Hunt, A.J.; Ok, Y.S.; Song, H.; Zhang, S. Lignin valorization for the production of renewable chemicals: State-of-the-art review and future prospects. *Biores. Technol.* **2018**, *269*, 465–475. [CrossRef] [PubMed]
7. Yang, F.; Fu, J.; Mo, J.; Lu, X. Synergy of Lewis and Brønsted acids on catalytic hydrothermal decomposition of hexose to levulinic acid. *Energy Fuel* **2013**, *27*, 6973–6978. [CrossRef]
8. Zhi, Z.; Li, N.; Qiao, Y.; Zheng, X.; Wang, H.; Lu, X. Kinetic study of levulinic acid production from corn stalk at relatively high temperature using $FeCl_3$ as catalyst: A simplified model evaluated. *Ind. Crop. Prod.* **2015**, *76*, 672–680. [CrossRef]
9. Sun, Z.; Xue, L.; Wang, S.; Wang, X.; Shi, J. Single step conversion of cellulose to levulinic acid using temperature-responsive dodeca-aluminotungstic acid catalysts. *Green Chem.* **2016**, *18*, 742–752. [CrossRef]
10. Kumar, K.; Pathak, S.; Upadhyayula, S. 2nd generation biomass derived glucose conversion to 5-hydroxymethulfurfural and levulinic acid catalyzed by ionic liquid and transition metal sulfate: Elucidation of kinetics and mechanism. *J. Clean. Prod.* **2020**, *256*, 120292. [CrossRef]
11. Kumar, K.; Parveen, F.; Patra, T.; Upadhyayula, S. Hydrothermal conversion of glucose to levulinic acid using multifunctional ionic liquids: Effect of metal ion cocatalysts on the product yield. *New J. Chem.* **2018**, *42*, 228–236. [CrossRef]
12. Matsagar, B.M.; Dhepe, P.L. Effect of cations, anions and H^+ concentration of acidic ionic liquids on valorization of polysaccharides into furfural. *New J. Chem.* **2017**, *41*, 6137–6144. [CrossRef]
13. Badgujar, K.C.; Wilson, L.D.; Bhanage, B.M. Recent advances for sustainable production of levulinic acid in ionic liquids from biomass: Current scenario, opportunities and challenges. *Renew. Sust. Energ. Rev.* **2019**, *102*, 266–284. [CrossRef]
14. Velaga, B.; Parde, R.P.; Soni, J.; Peela, N.R. Synthesized hierarchical mordenite zeolites for the biomass conversion to levulinic acid and the mechanistic insights into humins formation. *Micropor. Mesopor. Mat.* **2019**, *287*, 18–28. [CrossRef]

15. Rasmussen, H.; Sørensen, H.R.; Meyer, A.S. Formation of degradation compounds from lignocellulosic biomass in the biorefinery: Sugar reaction mechanisms. *Carbohydr. Res.* **2014**, *385*, 45–57. [CrossRef] [PubMed]
16. Mija, A.; van der Waal, J.C.; Pin, J.M.; Guigo, N.; de Jong, E. Humins as promising material for producing sustainable carbohydrate-derived building materials. *Constr. Build. Mater.* **2017**, *139*, 594–601. [CrossRef]
17. Kang, S.; Fu, J.; Zhang, G. From lignocellulosic biomass to levulinic acid: A review on acid catalysed hydrolysis. *Renew. Sustain. Energ. Rev.* **2018**, *94*, 340–362. [CrossRef]
18. Girisuta, B.; Janssen, L.P.B.M.; Heeres, H.J. Kinetic study on the acid-catalyzed hydrolysis of cellulose to levulinic acid. *Ind. Eng. Chem. Res.* **2007**, *46*, 1696–1708. [CrossRef]
19. Gürbüz, E.I.; Gallo, J.M.R.; Alonso, D.M.; Wettstein, S.G.; Lim, W.Y.; Dumesic, J.A. Conversion of hemicellulose into furfural using solid acid catalyst in γ-valerolactone. *Angew. Chem. Int. Ed.* **2013**, *52*, 1270–1274. [CrossRef]
20. Galletti, A.M.R.; Antonetti, C.; De Luise, V.; Licursi, D.; Di Nasso, N.N.O. Levulinic acid production from waste biomass. *BioResources* **2012**, *7*, 1824–1834.
21. Lappalainen, K.; Kuorikoski, E.; Vanvyve, E.; Dong, Y.; Kärkkäinen, J.; Niemelä, M.; Lassi, U. Brønsted and Lewis acid catalyzed conversion of pulp industry waste biomass to levulinic acid. *BioResources* **2019**, *14*, 7025–7040.
22. Olkiewicz, M.; Plechkova, N.V.; Fabregat, A.; Stüber, F.; Fortuny, A.; Font, J.; Bengoa, C. Efficient extraction of lipids from primary sewage sludge using ionic liquids for biodiesel production. *Sep. Purif. Technol.* **2015**, *153*, 118–125. [CrossRef]
23. Olkiewicz, M.; Caporgno, M.P.; Font, J.; Legrand, J.; Lepine, O.; Plechkova, N.V.; Pruvost, J.; Seddon, K.R.; Bengoa, C. A novel recovery process for lipids from microalgae for biodiesel production using a hydratedphosphonium ionic liquid. *Green Chem.* **2015**, *17*, 2813–2824. [CrossRef]
24. Glińska, K.; Ismail, M.; Goma-Camps, J.; Valencia, P.; Stüber, F.; Giralt, J.; Fabregat, A.; Torrens, E.; Olkiewicz, M.; Bengoa, C. Recovery and characterization of cellulose from industrial paper mill sludge using tetrakis and imidazolium based ionic liquids. *Ind. Crop. Prod.* **2019**, *139*, 111556. [CrossRef]
25. Olkiewicz, M.; Plechkova, N.V.; Earle, M.; Fabregat, A.; Stüber, F.; Fortuny, A.; Font, J.; Bengoa, C. Biodiesel production from sewage sludge lipids catalysed by Brønsted acidic ionic liquids. *Appl. Catal. B Environ.* **2016**, *18*, 738–746. [CrossRef]
26. Ahuja, D.; Kaushik, A.; Singh, M. Simultaneous extraction of lignin and cellulose nanofibrils from waste jute bags using one post-treatment. *Int. J. Biol. Macromol.* **2018**, *107*, 1294–1301. [CrossRef]
27. Wu, Y.; Wu, J.; Yang, F.; Tang, C.; Huang, Q. Effect of H2O2 bleaching treatment on the properties of finished transparent wood. *Polymers* **2019**, *11*, 776. [CrossRef]
28. Lin, N.; Zhu, W.; Fan, X.; Wang, C.; Chen, C.; Zhang, H.; Chen, L.; Wu, S.; Cui, Y. Key factor on improving secondary advanced dewatering performance of municipal dewatered sludge: Selective oxidative decomposition of polysaccharides. *Chemosphere* **2020**, *249*, 126108. [CrossRef]
29. Sambusti, C.; Monlau, F.; Barakat, A. Bioethanol fermentation as alternative valorisation route of agricultural digestate according to a biorefinery approach. *Biores. Technol.* **2016**, *212*, 289–295. [CrossRef]
30. Li, S.; Deng, W.; Wang, S.; Wang, P.; An, D.; Li, Y.; Zhang, Q.; Wang, Y. Catalytic transformation of cellulose and its derivatives into functionalized organic acids. *ChemSusChem* **2018**, *11*, 1995–2028. [CrossRef]
31. Kumar, S.; Ahluwalia, V.; Kundu, P.; Sangwan, R.S.; Kansal, S.K.; Runge, T.M.; Elumalai, S. Improved levulinic acid production from agri-residue biomass in biphasic solvent system through synergistic catalytic effect of acid and products. *Biores. Technol.* **2018**, *251*, 143–150. [CrossRef] [PubMed]
32. Rice, E.W.; Baird, R.B.; Eaton, A.D.; Clesceri, L.S. *Standard Methods for the Examination of Water and Wastewater*, 22nd ed.; APHA AWWA WEF: Washington, DC, USA, 2012.
33. Lowry, O.H.; Rosebrough, N.J.; Farr, A.L.; Randall, R.J. Protein measurement with the Folin phenol reagent. *J. Biol. Chem.* **1951**, *193*, 265–275. [PubMed]
34. Dubois, M.; Gilles, K.A.; Hamilton, J.K.; Rebers, P.A.; Smith, F. Colorimetric method for determination of sugars and related substances. *Anal. Chem.* **1956**, *28*, 350–356. [CrossRef]

© 2020 by the authors. Licensee MDPI, Basel, Switzerland. This article is an open access article distributed under the terms and conditions of the Creative Commons Attribution (CC BY) license (http://creativecommons.org/licenses/by/4.0/).

Article

Preparation of 5-HMF in a DES/Ethyl N-Butyrate Two-Phase System

Jinyan Lang [1], Junliang Lu [1], Ping Lan [2], Na Wang [1], Hongyan Yang [1] and Heng Zhang [1,2,3,*]

- [1] College of Marine Science and Biological Engineering, Qingdao University of Science & Technology, Qingdao 266042, China; ljy17806248212@sina.com (J.L.); juling_lu@sina.com (J.L.); wlalala21@163.com (N.W.); kdjh401@163.com (H.Y.)
- [2] Guangxi Key Laboratory of Polysaccharide Materials and Modification, School of Chemistry and Chemical Engineering, Guangxi University for Nationalities, Nanning 530008, China; gxLanping@163.com
- [3] Key Laboratory of Biomass Chemical Engineering of Ministry of Education, Zhejiang University, Hangzhou 310027, China
- * Correspondence: hgzhang@qust.edu.cn

Received: 21 May 2020; Accepted: 4 June 2020; Published: 7 June 2020

Abstract: In this paper, a two-phase system, formed by oxalic acid/choline chloride-based deep eutectic solvent (DES) and chosen extractants, was used as a dissolution–reaction–separation system, and metal chloride was used as a catalyst to study the degradation of cellulose to produce 5-hydroxymethylfurfural (5-HMF) and glucose. The effects of the amount of organic solvent and the reaction temperature on product yield, the repeated recycling of DES, the comparison between a two-phase system and a homogeneous system, and the mechanism of cellulose degradation to 5-HMF were investigated. The results show that ethyl n-butyrate has the best extraction effect on 5-HMF. Compared with the homogeneous system, the yield of 5-HMF and glucose in the two-phase system is significantly improved. At a temperature of 140 °C and a reaction time of 120 min, the yields of glucose and 5-HMF reached the maximum, which were 23.5% and 29.8%, respectively. After DES was reused three times, the yields of glucose and 5-HMF decreased greatly, indicating that the recycling rate of DES was low.

Keywords: cellulose; deep eutectic solvents; 5-HMF; biphasic system

1. Introduction

5-HMF is a very important new type of platform compound [1]. It not only contains a variety of functional groups, but also has good reactivity. Therefore, it is widely used in the manufacture of polymer materials [2], fine chemicals [3], biofuel processing [4–6], and other fields. Chemical products synthesized using 5-HMF as a raw material, such as 2,5-furandicarboxylic acid [7], 2,5-dimethylfuran, etc., have very high application value in the pharmaceutical, optoelectronic materials, and biofuel industries.

At present, obtaining 5-HMF through the direct degradation of biomass resources has become one of the research hotspots for the high value utilization of biomass. Cellulose is an abundant and renewable biomass resource in nature. It is a chain-like polymer compound linked by D-glucosyl groups with β-1,4 glycoside bonds. There have been many reports that 5-HMF is prepared from cellulose through certain transformation pathways [8,9]. First, cellulose is hydrolyzed to glucose under an acidic catalyst and glucose is converted into fructose through isomerization. Finally, fructose is formed into a specific intermediate under the catalyst. The intermediate produces 5-HMF by removing three molecules of water.

As a green dissolving solvent, ionic liquids are all organic salts composed of ions. Most of them are liquids at room temperature. They have good solubility for cellulose and other polymers [10].

Therefore, in the process of the catalytic degradation of cellulose to 5-HMF, ionic liquids are widely used [11]. The ionic liquid has some advantages, in that it has a good effect in dissolving cellulose and a high yield in the preparation of 5-HMF, but the ionic liquid has some disadvantages, in that its preparation is relatively complicated, its reuse problem remains to be solved, and its price is high, so there are few examples of industrialized production at present. As a result, researchers have replaced some or all of the ionic liquids with low-cost solvents. Mixed solvents are solvents which retain the dissolving properties of ionic liquids and reduce costs after replacing some or all of the ionic liquids in the mixed system [12]. The result is that, although the reaction cost is reduced, the yield of 5-HMF is correspondingly reduced, and 5-HMF is difficult to extract and efficiently separate, which reduces the yield.

Abbott [13] found that choline chloride and urea can form a transparent solution by stirring at room temperature, and then named this solvent deep eutectic solvent (DES). It is a eutectic mixture composed of a hydrogen bond acceptor and a hydrogen bond donor in the form of hydrogen bonding, and its melting point is lower than that of any single component. This solvent has many similarities with ionic liquids in melting point, viscosity, and many other physical and chemical-related properties, so it is also called an ionic liquid. Compared with ionic liquids, DES has a low price, simple preparation, 100% atomic utilization rate, biodegradability, safety and non-toxicity, and truly realizes the advantages of green chemistry.

Studies at home and abroad show that DES can be used as a solvent for carbohydrate degradation. In 2014, Xing et al. [14] reported that, since DES was able to dissolve cellulose, preliminary studies have begun in the study of the degradation of cellulose-based macromolecular compounds into 5-HMF. Liu et al. [15] studied the dissolution of cellulose by DES synthesized from several different solid organic matters and opened up new ideas for the preparation of 5-HMF from biomass resources. Liu et al. [16] used $FeCl_3/AlCl_3$ as catalyst to dissolve cellulose in a low crystalline region to obtain a 5-HMF yield of 49% in a H_2O-ChCl/MIBK two-phase system, and to dissolve cellulose in a high crystalline region to obtain a 5-HMF yield of 27%. However, there are many limitations to obtaining 5-HMF by simply degrading cellulose through a single-phase reaction system. The products after the reaction are difficult to separate, so recovery is difficult and the yield is likely to decrease. Therefore, selecting a suitable extractant to constitute a two-phase or multi-phase system can effectively improve the yield of 5-HMF.

Therefore, the cellulose dissolution, catalytic conversion process, and separation mechanism in a new type of green and environmentally friendly solvent (deep eutectic solvent) system are studied in this paper. Although its current research is less, it provides new approaches for the development of new catalytic separation systems. Therefore, deep eutectic solvents may become a new type of high-efficiency catalytic system instead of ionic liquid transition metals. It is necessary to carry out in-depth research in terms of its structural design, reaction mechanism, and reaction method, and to fundamentally understand the deeply eutectic solvent-catalyzed reaction mechanism and the process and mechanism to improve reaction yield.

2. Results and Discussion

2.1. Selection of Extractant

At room temperature, a small amount of 5-HMF was added to an equal amount of DES, and then a sufficient amount of tetrahydrofuran (THF), ethyl butyrate, methyl isobutyl ketone (MIBK), ethyl acetate, and toluene were added, shocked it until fully dissolved, then left to stand and the clock was started. At the same time, the color of the reaction solution was observed. The stratification status was noted and timing was stopped. The extraction ability of 5-HMF was investigated by observing the stratification. The results are shown in Table 1 below.

Table 1. The solubility of DES and organic solvents.

Organic Solvents	V1/mL	V2/mL	DES	Time of Stratification	Upper Color	Primary Extraction Rate/%
THF	5	4.86	-	30 min The stratification was not obvious	Dark brown	58.7
Ethyl butyrate	5	4.96	-	6 min Distinct stratification	Light yellow	82.1
MIBK	5	4.94	-	10 min Distinct stratification	Brown	81.5
AcOEt	5	4.82	-	3 min Distinct stratification	Light yellow	83
Toluene	5	4.90	-	5 min Distinct stratification	Light yellow	84.4

(Note: V_1: Initial volume of organic phase; V_2: Volume of organic phase after extraction; -: indicates that there is no DES residue in the organic phase after the rotary evaporation).

It can be seen from Table 1 that there is no residual DES after the evaporation of the five extractants selected in this experiment. The remaining volume of ethyl acetate and tetrahydrofuran is relatively small, because the boiling points of ethyl acetate and tetrahydrofuran are relatively low, and parts of them were evaporated at high temperature. The volume of the other four extractants is almost equal to the original volume.

It can be seen from Table 1 that the five extraction agents selected in this experiment all have a certain extraction effect on 5-HMF. However, when tetrahydrofuran was used as an extractant, the stratification was not obvious. The extraction rate was low, while the extraction effect was not ideal after standing for 30 min. Although methyl isobutyl ketone has better effect, its upper layer color is brown, which is different from the light yellow 5-HMF, indicating that the upper layer liquid contains other dark-colored impurities. The extraction rates of ethyl acetate, ethyl butyrate, and toluene are relatively high and nearly the same rate. Comparing these three extractants, although toluene has the highest extraction rate and it is flammable and easily forms explosive mixtures with air at high temperatures. Toluene is low toxicity, but inhaling toluene can cause obvious eye and upper respiratory tract irritation symptoms, which is harmful to humans. In addition, the boiling point of ethyl acetate (about 77 °C) is far lower than the temperature in this experiment, which makes it is easy to volatilize and it cannot form a binary liquid phase system. Ethyl butyrate has a boiling point of about 120 °C and is harmless to the environment. It is an environmentally friendly solvent. Moreover, the extraction rate of 5-HMF by ethyl n-butyrate as an extractant is not significantly different from that with 5-HMF by ethyl acetate as an extractant.

In summary, ethyl n-butyrate, with the higher boiling point, has a good extraction effect and is not harmful to the human body or the environment. It is an environmentally friendly solvent and meets the requirements of this experiment. Therefore, ethyl n-butyrate is selected as the extractant in this experiment.

2.2. Effect of the Amount of Organic Solvent on Product Yield

The addition of an organic solvent has an important effect on the product. The organic solvent not only has an extraction effect on 5-HMF, but also can reduce the viscosity of the reaction system and has a swelling effect on cellulose. It is important to discuss the amount of organic solvent added to the experiment.

One gram of cellulose was added to 25 g DES solvent, then the catalyst $SnCl_4$ was added and at 140 °C, ethyl n-butyrate and DES were successively added to produce mass ratios of 0.5, 1, 1.5, 2, 2.5, 3, and 3.5. During the reaction process, the product was taken out every half an hour to measure its yield. The experimental results are shown in Figure 1.

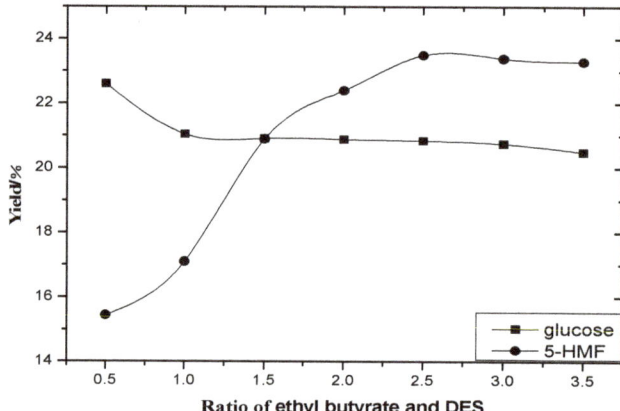

Figure 1. Effect of the ratio of ethyl butyrate and DES on products yields.

It can be concluded from Figure 1 that the amount of ethyl butyrate added has almost no effect on the yield of glucose, because the extraction agent is added to extract 5-HMF, which has little relation with glucose. The yield of 5-HMF increased with the dosage ratio of ethyl butyrate to DES. When the dosage ratio of ethyl butyrate to DES was 0.5, the yield of 5-HMF was 15.44% and that of glucose was 22.6%. The yield of 5-HMF continued to increase until the dosage ratio of ethyl butyrate to DES was 2.5. The yield of 5-HMF was 23.5% and that of glucose was 20.89%. On this basis, the dosage ratio was increased again, and the yield of the two is basically unchanged. When the addition of ethyl butyrate is too small, the amount of 5-HMF is large, the extractant reaches saturation, and 5-HMF is not completely extracted. As the reaction proceeds, 5-HMF further generates other substances, resulting in a reduced yield. When the dosage is increased to 2.5:1, the yield of 5-HMF reaches the maximum. On the one hand, while the amount of organic solvent is increased and the viscosity of the system is reduced, the cellulose is moistened and expanded [8], and the intermolecular and intramolecular hydrogen bonds are easy to break, which is conducive to the reaction. On the other hand, almost all the generated 5-HMF was extracted into the organic phase, indicating that ethyl n-butyrate played an extraction role in the reaction and promoted the reaction in a positive direction. The ratio of 5-HMF yield has little effect on the continuous increase of the dosage. Therefore, the ratio of ethyl butyrate to DES was 2.5.

In summary, the ratio of ethyl butyrate to DES was 2.5. Under this condition, the yield of glucose was 22.6% and the yield of 5-HMF was 23.5%.

2.3. Effect of Reaction Temperature on Product Yield

Fifty grams of ethyl n-butyrate was added in 25 g DES solvent and 1 g cellulose was added. Then the catalyst SnCl$_4$ was added. With the system setting off a chemical reaction under three different temperatures (120 °C, 130 °C, and 140 °C), the sample was taken out every half an hour and the yield was measured. The results are shown in Figure 2.

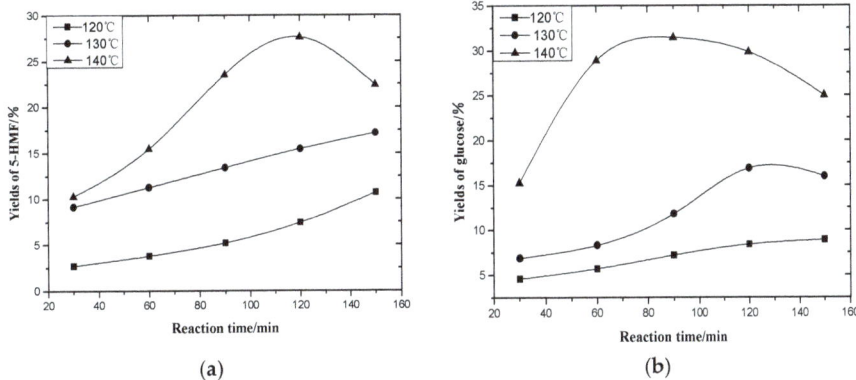

Figure 2. Effect of reaction time and temperature on (**a**) 5-HMF yield and (**b**) glucose yield.

It can be seen from Figure 2 that the yields of glucose and 5-HMF increase with time at 120 °C, but the yields of glucose and 5-HMF are not too high due to the low temperature. The yields of glucose and 5-HMF are only 2.7% and 4.6% at 30 min. Additionally, they are also only 10.7% and 8.9% at 150 min, not reaching the highest values at this time.

It can be seen from Figure 2 that the yields of glucose and 5-HMF also increase with time before 120 min at 130 °C, but their growth rate is faster. Therefore, the high temperature can accelerate the reaction within limits. The glucose yield reaches a maximum (29.8%) at 120 min, rises gradually with temperature before 120 min, then decreases. However, the 5-HMF yield rises between 30 to 150 min. Compared with 120 °C, the yields of glucose and 5-HMF greatly increase, but the yield of 5-HMF still does not reach the maximum shown in Figure 2 because the temperature at which DES completely catalyzes cellulose degradation cannot be reached.

As can be seen from Figure 2, when the temperature is 140 °C, the yields of glucose and 5-HMF are increased significantly, but the time taken to reach the maximum is different. Compared to 120 °C and 130 °C, glucose reaches its highest yield earlier. The highest yield is 31.5% at 90 min, then glucose quickly converts to 5-HMF. The highest yield of 5-HMF is 23.6% at 140 °C, then the value begins to go down after 120 min. When the reaction temperature is 140 °C, the organic solvent carrying 5-HMF begins to evaporate, so some products are not returned to the reactor. It results in a small loss of the product. In this experiment, it is further found that ethyl n-butyrate at 150 °C can be completely evaporated in less than 10 min, and cannot perform the extraction function.

In conclusion, the optimal reaction time and temperature to prepare glucose and 5-HMF in the two-phase system (DES-ethyl n-butyrate) are 2 h and 140 °C, respectively. The yield of glucose is 29.8% and that of 5-HMF is 23.5%.

2.4. Deep Eutectic Solvents Recovery

The reaction solution was extracted by an extractant and the subnatant was removed. Then the subnatant was filtered with distilled water to get the filtrate. The filtrate was evaporated in a rotary evaporator to get the recovery solution, which was dried at 70 °C for 24 h in a vacuum oven. After that, the sample, which was prepared by the smear method for infrared analysis, was obtained.

The first recycled solution dissolved cellulose as a reaction solution, and $SnCl_4$ was selected as the catalyst at a dosage of 1.42 wt%, then the reaction was carried out for 2 h at 140 °C.

The second recycled solution dissolved cellulose as a reaction solution, and $SnCl_4$ was selected as the catalyst at a dosage of 1.42 wt%, then the reaction was carried out for 2 h at 140 °C.

Figure 3 is an infrared spectrum of DES which contrasts the first recycled cellulose and multiply recycled cellulose. The association of hydroxyl groups to form hydrogen bonds at 3400–3200 cm^{-1}, which is the O–H vibration absorption speak of DES's characteristic peak. We can conclude that the

O–H vibration absorption, which was recycled many times, moved from 3400 cm^{-1} to 3390 cm^{-1}. The infrared absorption peak of the C–N stretching vibration of choline chloride weakened. That is because Cl– is very electronegative and it can induce a reaction. There is a reaction between N- and cellulose and the reaction destroyed the hydrogen bonds of the cellulose. This reaction weakened the peak. The C=O vibration absorption peak of oxalic acid moved to 1744.9 cm^{-1}, the peak band became narrower and weaker, and its content reduced. There is almost nothing changed about the construction, but the moisture content increased after being recycled many times. Because DES can absorbed water easily, the moisture content increases after being recycled many times, which easily reduces the yield of the product. Therefore, it is not suitable for recycling many times.

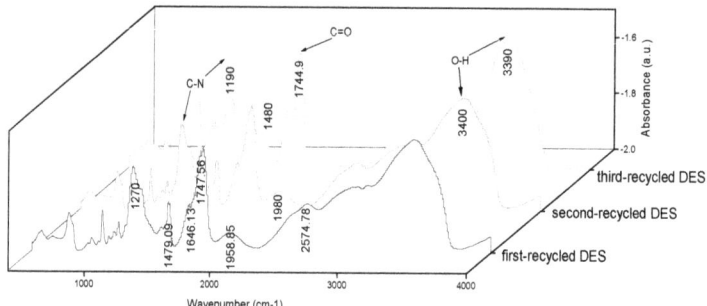

Figure 3. IR spectrum of multiply recycled DES.

Here is the test result:

Cellulose was dissolved by the once recycled solution as a reaction solvent, SnCl$_4$ was selected as the catalyst at a dosage of 1.42 wt%, then the reaction was carried out for 2 h at 140 °C. After the reaction, ethyl n-butyrate was separated in the reaction and the content of 5-HMF in the glucose and ethyl n-butyrate was measured after adding a certain amount of ethyl n-butyrate to extract the reaction solution several times.

Cellulose was dissolved by the second recycled solution as a reaction solvent, SnCl$_4$ was selected as the catalyst at a dosage of 1.42 wt%, then the reaction was carried out for 3 h at 140 °C. After the reaction, ethyl n-butyrate was separated in the reaction and the content of 5-HMF in the glucose and ethyl n-butyrate was measured after adding a certain amount of ethyl n-butyrate to extract the reaction solution several times.

Cellulose was dissolved by the third recycled solution as a reaction solvent, SnCl$_4$ was selected as the catalyst at a dosage of 1.42 wt%, then the reaction was carried out for 5 h at 140 °C. After the reaction, ethyl n-butyrate was separated in the reaction and the content of 5-HMF in the glucose and ethyl n-butyrate was measured after adding a certain amount of ethyl n-butyrate to extract the reaction solution several times.

From Figure 4 we can conclude that, even though the reaction time is prolonged, the yield of glucose and ethyl butyrate both reduce with the increasing duration of recycling. After being recycled three times, the yield of 5-HMF declined to 5.4% from 10.79% and the yield of glucose declined to 15.2% from 22.8%. On the one hand, the recovery rate of DES reduced and the cellulose that was dissolved also declined with the increasing duration of the extractions. One the other hand, the amount of humin and the viscosity of the reaction system were increased, and the yield of the reaction product declined with the increase in the by-product.

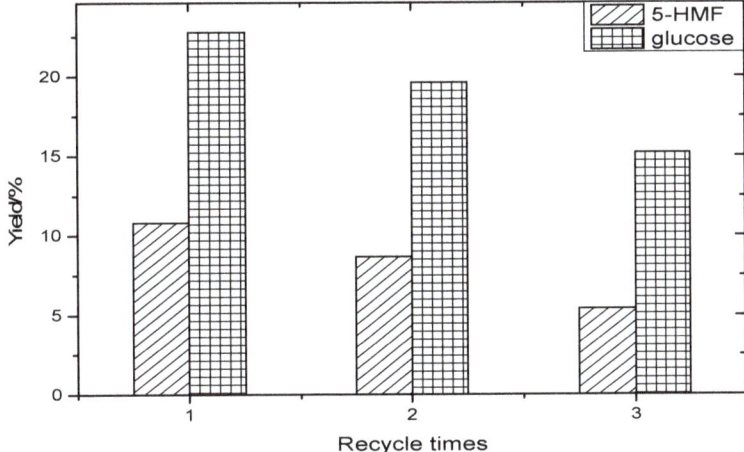

Figure 4. Recycling durations of DES.

In conclusion, DES can be recycled and reused, but the yield of 5-HMF is only 5.4%, and the yield of glucose is 15.2% after being recycled three times.

2.5. Comparisons between Two-Phase and Homogeneous Systems

Compared with the homogeneous system composed of a reaction system only using DES without an extraction agent, the results are shown in Table 2.

Table 2. Comparisons between one-phase and biphase systems.

	Homogeneous System	**Two-Phase System**
Optimum reaction temperature/°C	160	140
Optimum reaction time/min	90	120
DES repeatable number	Reuse zero times	Reuse three times
Optimal yield of product/%	Glucose: 22%; 5-HMF: 11%	Glucose: 29.8%; 5-HMF: 23.5%

As can be seen from Table 2:

(1) Optimum reaction temperature

Compared with the homogeneous system, in the two-phase system, the reaction temperature that is required for the target product to achieve the maximum yield is lower. In the homogeneous system, the reaction temperature needs to rise to 160 °C to reach the maximum value for 5-HMF. In the two-phase system, the maximum value can be reached at 140 °C.

(2) Optimum reaction time

The reaction time for achieving the maximum volume of 5-HMF in the two-phase system is longer than that in the homogeneous system. Due to the addition of a large number of organic solvents, and considering that the organic solvents will evaporate to carry part of 5-HMF after reaching the boiling point, which affects the experimental results, the temperature is decreased, and if we reduce the temperature, cellulose dissolution is slow and the product yield is delayed.

(3) DES repeatable number

In the two-phase system, DES can be reused three times, while the homogeneous system cannot be reused. The reason for this is that if the temperature is too high and the homogeneous system's viscosity is too high, the reaction solution sticks to the wall, the by-product increases, and the recovery rate of DES is very low. In the two-phase system, due to the addition of a large number of organic solvents, DES is diluted and DES can be successfully recovered after the completion of the reaction, and there is no sticky wall phenomenon.

(4) Optimal yield of product

Compared with the homogeneous system, the yield of glucose in the two-phase system was not significantly different from that of the homogeneous system, while the yield of 5-HMF increased by 18.8%. Glucose is just an intermediate product which exists in the DES system. However, 5-HMF was not extracted in time due to the high temperature and the long time in the homogeneous system, resulting in a too low yield of 5-HMF. In the two-phase system, due to the addition of an extraction agent, 5-HMF was extracted in time with the progress of the reaction, so the yieldincreased.

It can be seen from Table 3 that the reaction temperature of the homogeneous system is high and the yield of 5-HMF is low. This is because the viscosity of the reaction system is relatively large, which is not conducive to the contact between the reaction solvent and cellulose, resulting in less cellulose glycosidic bond and hydrogen bond breakage, and less yield of 5-HMF generated. In this case, only by increasing the temperature to break the glycosidic bonds and hydrogen bonds of cellulose, the side reaction increases while 5-HMF is produced. 5-HMF is further hydrolyzed into other rot substances, increasing the viscosity of the reaction system, resulting in a decrease in product yield. In the two-phase system, because of the addition of the organic solvent, on the one hand, the organic solvent itself can swell the cellulose and increase the dissolution rate of cellulose. With the addition of the organic solvent, the viscosity of the reaction system decreases and the molecular motion is intense. It is conducive to the destruction of cellulose glycosidic bonds and hydrogen bonds by the solvent, so excessive temperature is not required to dissolve the cellulose. The lower the temperature, the lower the yield of by-products, and increase the yield of 5-HMF. On the other hand, the organic solvent has a certain extraction effect on 5-HMF. During the reaction, 5-HMF can be extracted by the organic solvent in time, reducing the further hydrolysis of 5-HMF to other substances and increasing the yield of 5-HMF.

Table 3. The conversion of cellulose to 5-HMF in different systems.

Number	Solvent	Catalyst	Temperature/°C	Time	Conversion Rate/%	Yield/%
1	MIBK/water	TiO_2	270	2 min	-	35
2	water	H_2CO_3	250	30 min	-	16.2
3	[BMIM]Cl	$CrCl_3$	160	1 h	95	31
4	[BMIM]Cl/Toluene	$CrCl_3$	130	3 h	57	55
5	ChCl/oxalic acid	$SnCl_4$	160	90 min	60	11
6	ChCl/oxalic acid-ethyl n-butyrate	$SnCl_4$	140	2 h	-	23.5 (This study)

In summary, the two-phase system is more suitable for the preparation of glucose and 5-HMF by cellulose degradation than the homogeneous system. Compared with other studies, the advantage of this experiment is that the choline chloride/oxalic acid solvent used is inexpensive and non-polluting. The organic solvent ethyl n-butyrate used is an environmentally friendly solvent. 5-HMF can also be removed from the reaction as it is extracted from it, which provides a new idea for the future large-scale direct preparation of platform compounds from renewable resources, such as cellulose.

2.6. Mechanism of the Catalytic Degradation of Cellulose in DES to 5-HMF

Cellulose is a kind of chain polymer with stable physical properties which is dehydrated with glucose and connected by $β-1,4$ glycoside bonds. $SnCl_4$ was used as the catalyst and choline chloride/oxalic acid was used as the solvent to catalyze the degradation of cellulose to produce 5-HMF, the mechanism for which is shown in Figure 5.

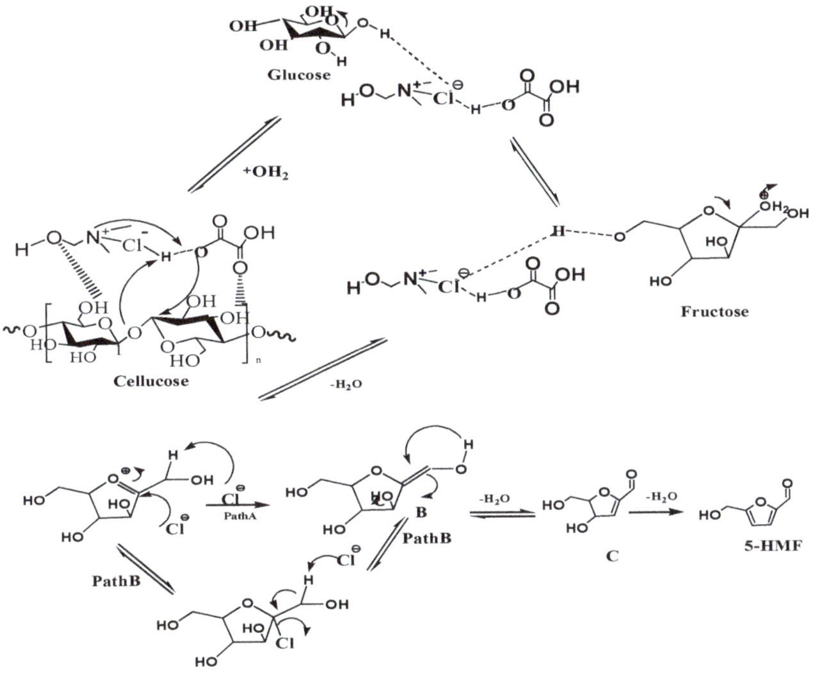

Figure 5. Mechanism of the conversion of cellulose to 5-HMF in a DES system.

For the dissolution of cellulose by choline chloride/oxalic acid, firstly, the O atoms of choline chloride/oxalic acid, oxalic acid, and cellulose −OH can form new hydrogen bond between the original cellulose hydrogen bonds, and secondly, the N atoms of choline chloride haves a high electropositive activity, which reduces the strength of the hydrogen and oxygen bonds on the carboxyl group and increases the acidity of the system and breaks the glycoside bonds [17].

For the process of glucose isomerization into fructose, firstly, the O atoms on the glycoside bond attack the hydrogen in the oxalate carboxyl group to form the hydroxyl group, thus breaking the glycoside bond. Under the action of the metal chloride $SnCl_4$, the Cl atoms in $SnCl_4$ can interact with the H atoms in the glucose hydroxyl group, and then transfer H atoms. Then the Sn atoms in $SnCl_4$ can interact with the O atoms in the glucose to promote the formation of enol-type intermediates and the isomerization into fructose [18]. Secondly, the chloride ion attacks the oxygen atom of the glucose hydroxyl bond and captures the hydrogen atom on the hydroxyl group, thus causing the oxygen of the hydroxyl group to show an electronegative property, and the C_1 position forms the aldehyde group. Then the oxygen on C_5 attacks C_2, causing the hydroxyl group on C_2 to fall off and form a five-membered ring. Later, due to the oxidizability of the carbonyl group, it combines with the hydrogen in oxalic acid to form an enol intermediate, resulting in the five-membered ring of fructose. Fructose takes off three molecules of water to get 5-HMF.

3. Materials and Methods

3.1. Materials

Ethyl acetate was purchased from Henan Lianchuang Chemical Co., Ltd (Jiyuan, China). Ethyl butyrate, choline chloride, and HPLC grade 5-Hydroxymethylfurfural (5-HMF) were purchased from Shandong West Asia Chemical Industry Co., Ltd (Linyi, China), Oxalic acid was purchased from Tianjin Damao chemical reagent factory (Tianjin, China). Medicinal grade microcrystalline cellulose

(MCC) was obtained from Chengdu Cologne Chemical Reagent Factory (Chengdu, China). Tin (IV) chloride ($SnCl_4$) was obtained from Tianjin Beichen Fangzheng Reagent Factory (Tianjin, China). Tetrahydrofuran (THF) was obtained from Suzhou Yake Chemical Reagent Co., Ltd (Suzhou, China). Methyl isobutyl ketone (MIBK) was purchased from Shanghai Runjie Chemical Reagent Co., Ltd (Shanghai, China). and methylbenzene from Sinopharm Chemical Reagent Co., Ltd (Huaian, China). All chemical reagents were used with no further purification.

3.2. Methods

The studied reagents were dried for 48 h at 50 °C in a vacuum drying oven to remove moisture, then they were added into a four-necked flask with a hydrogen bond donor and hydrogen bond acceptor in a fixed mole ratio. The mixture was heated in a thermostatic oil bath and protected by nitrogen to fully mix it to a transparent liquid. The synthesized DES was dried in a vacuum drying oven for 48 h at 70 °C.

Five milliliters of DES reagent and 5 mL extracting solvent (recorded as V_1) were successively added into a separatory funnel. Then the top stopper was plugged and the separatory funnel was gently shaken up and down to make full contact between the two phases. The product was kept still for thirty minutes and the delamination of the solvent in the test tube was observed. After separation, the two-phase solvent was separated and the volume of the separated organic phase was measured (recorded as V_2). The two recorded volumes were compared to determine the mutual solubility of organic solvents and DES.

Five milliliters of DES reagent and 5 mL extracting solvent (recorded as V_1) were successively added into a separatory funnel. Then the top stopper was plugged and the separatory funnel was gently shaken up and down to make full contact between the two phases. The product was kept still while timing was started and the delamination phenomenon and the color of the upper liquid was observed. The concentration of the reaction solution before and after extraction was measured with a UV spectrophotometer (JY10001 German Bruker, Karlsruhe, German) at 284 nm to calculate the extraction rate. The extraction rate is the percentage content of the original concentration of 5-HMF minus the concentration of 5-HMF in the sample, which is then compared with the original concentration of 5-HMF.

A syringe was used to draw a certain amount of solution from the upper layer every half an hour. The reaction was stopped and a small amount of solution was drawn from the lower layer. The upper solution was diluted with ethyl n-butyrate and the supernatant was analyzed using an ultraviolet spectrophotometer at 284 nm to measure the concentration of 5-HMF. The lower solution was diluted with deionized water to a constant volume and a small amount of the solution and was placed under an ultraviolet spectrophotometer at 540 nm to measure the glucose concentration. The lower layer of the solution was centrifuged, and the lower layer of the solution was analyzed using an ultraviolet spectrophotometer to calculate the corresponding yield.

3.3. Calculation of Product Yield

The prepared DES made by the processes above was weighed into a four-necked flask and it was then filled with N_2 and dissolved at 100 °C and adjusted to the temperature required for the reaction. Then the specified mass of microcrystalline cellulose was added, the catalyst was added after stirring well, and finally the specified volume of ethyl n-butyrate was added into a four-necked flask.

Some of the upper layer solution was sucked up every half an hour. Then the reaction was stopped and a small volume of the lower layer solution was sucked up. The supernatant was analyzed using a UV spectrophotometer after diluting the upper layer solution and to a constant volume with ethyl n-butyrate. The lower layer solution was diluted to a constant volume with deionized water, then the solvent of the constant volume was sucked up and placed in a UV spectrophotometer to measure the glucose concentration. Finally, the lower layer solution was centrifuged and was analyzed using a UV spectrophotometer to calculate the corresponding yield.

The formula of glucose yield was calculated as shown in the following equation:

$$Y_G = \frac{C_G \times V}{m} \times 100\% \qquad (1)$$

where C_G is the concentration (mg/mL) of glucose, V is the volume (mL) of the reaction solution after a constant volume, and m is the original dosage (mg) of cellulose.

The formula of 5-HMF yield was calculated as shown in the following equation:

$$Y_{5-HMF} = \frac{C_{5-HMF} \times 180 V}{126 m} \times 100\% \qquad (2)$$

where the C_{5-HMF} above is the mass concentration (mg/mL) of 5-HMF, which is measured using Wang's method [19].

After the reaction was over, the reaction solution was evaporated on a rotary evaporator, and the organic solution was recovered to obtain DES for reuse.

4. Conclusions

Glucose and 5-HMF were prepared by using microcrystalline cellulose as a raw material and metal chloride stannous tetrachloride as a catalyst. Ethyl n-butyrate was selected as the extraction agent. When the temperature was 140 °C and the reaction time was 120 min, the yield of the product reached the maximum value. At that time, the yield of glucose was 23.5% and the yield of 5-HMF was 29.8%. The recovery rate of DES was studied, and the two-phase system was more suitable for the preparation of glucose and 5-HMF by cellulose degradation than the homogeneous system.

Author Contributions: J.L. (Jinyan Lang): write manuscripts and modify manuscripts, data collection, analysis and interpretation; J.L. (Junliang Lu): data analysis and interpretation; P.L.: literature search; N.W.: making charts; H.Y.: research design, data collection; H.Z.: the concept of the article content to be proposed, the manuscript to be revised. All authors have read and agreed to the published version of the manuscript.

Funding: This work was supported by the Shandong Provincial Natural Science Foundation of China (Grant No. ZR2017MC032), the Open Fund of Guangxi Key Laboratory of Polysaccharide Materials and Modification (Grant No. GXPSMM18YB-03), the Shandong Provincial Key Research and Development Program (SPKR&DP) (Grant No. 2019GGX102029) and the Foundation of Key Laboratory of Biomass Chemical Engineering of the Ministry of Education, Zhejiang University (Grant No. 2018BCE005).

Conflicts of Interest: The authors declare no conflict of interest.

References

1. Roman-Leshkov, Y.; Barrett, C.J.; Liu, Z.Y.; DumeSic, J.A. Production of dimetbylfuran for biomass-derived cabobydrates. *Nature* **2007**, *447*, 982–985. [CrossRef] [PubMed]
2. Yang, L. Conversion of Carbohydrates into 5-Hydroxymethylfurfural Catalyzed by Solid Acid in Deep Eutectic Solvents. Master's Thesis, Zhejiang University of Technology, Hangzhou, China, 2016.
3. Zhao, B.Y. Study on the Preparation, Properties of Deep Eutectic Solvents and Its Application to the Extraction of Rutin. Master's Thesis, South China University of Technology, Guangzhou, China, 2016.
4. Feng, R.; Zhao, D.; Guo, Y. Revisiting characteristics of ionic liquids: A review for further application development. *J. Environ. Prot.* **2010**, *1*, 95–104. [CrossRef]
5. Avelino, C.; Sara, I.; Alexandra, V. Chemieal routes for the transformation of biomass into chemieals. *Cheminform* **2007**, *38*, 2411–2502.
6. Zhou, L.L.; He, Y.M.; Ma, Z.W.; Liang, R.; Wu, T.; Wu, Y. One-step degradation of cellulose to 5-hydroxymethylfurfural in ionic liquid under mild conditions. *Carbohydr. Polym.* **2015**, *117*, 694–700. [CrossRef] [PubMed]
7. Wang, H.; Zhu, C.; Li, D.; Liu, Q.; Tan, J.; Wang, C.; Cai, C.; Ma, L. Recent advances in catalytic conversion of biomass to 5-hydroxymethylfurfural and 2,5-dimethylfuran. *Renew. Sustain. Energy Rev.* **2019**, *103*, 227–247. [CrossRef]

8. Zhang, H.; Li, S.; Song, X.; Li, P.; Li, J. Preparation of 5-HMF by the catalytic degradation of cellulose in an ionic liquid/organic biphasic system. *BioResources* **2016**, *11*, 5190–5203. [CrossRef]
9. Zhang, H.; Li, S.; Xu, L.; Sun, J.; Li, J. Kinetic Study of the Decomposition of Cellulose to 5-Hydroxymethylfurfural in Ionic Liquid. *BioResources* **2016**, *11*, 4268–4280. [CrossRef]
10. Wu, Q.; Zhang, H.Q.; Wang, X.Z. Relationship between the structure of ionic liquids and their physical properties. *J. Hebei Natl. Teach. Coll.* **2015**, *35*, 65–69.
11. Yang, Y.P.; Sheng, M.G.; Shang, S.B.; Song, Z.Q. Research progress of cellulose catalytic conversion for preparation of 5-HMF in different solvents. *Biomass Chem. Eng.* **2016**, *50*, 47–52.
12. Hu, L.; Wu, Z.; Lin, L.; Zhou, S.; Liu, S. Recent advances in catalytic transformation of biomass-derived 5-Hydroxymethylfurfural into the innovative fuels and chemicals. *Renew. Sustain. Energy Rev.* **2017**, *74*, 230–257. [CrossRef]
13. Abbott, A.P.; Capper, G.; Davies, D.L.; Rasheed, R.K.; Tambyrajah, V. Novel solvent properties of choline chloride/urea mixtures. *Chem. Commun.* **2003**, *1*, 70–71. [CrossRef] [PubMed]
14. Xing, Y.J.; Liu, Y.T.; Zhang, J.B.; Zhang, H.H. Method for Pyrolyzing Cellulose Raw Material by Eutectic Solvent. Patent CN104178527A, 9 July 2014.
15. Liu, H.R.; Zhou, E.P.; Zhang, X.H.; Zhang, X.C. Study of deep eutectic solvents prepared from solid organic compounds and their application on dissolution for cellulose. *Mater. Guide* **2013**, *27*, 95–98.
16. Liu, F.; Audemar, M.; Vigier, K.; Cartigny, D.; Clacens, J.-M.; Gomes, M.F.C.; Pádua, A.A.H.; Campo, F.D.; Jérôme, F. Selectivity enhancement in the aqueous acid-catalyzed conversion of glucose to 5-hydroxymethylfurfural induced by choline chloride. *Green Chem.* **2013**, *15*, 3205–3213. [CrossRef]
17. Zhang, H.; Lang, J.Y.; Lan, P.; Yang, H.; Lu, J.; Lu, Z. Study on the dissolution mechanism of cellulose by ChCl-based deep eutectic solvents. *Materials* **2020**, *13*, 278. [CrossRef] [PubMed]
18. Zuo, M.; Le, K.; Li, Z.; Jiang, Y.; Zeng, X.; Tang, X.; Sun, Y.; Lin, L. Green process for production of 5-hydroxymethylfurfural from carbohydrates with high purity in deep eutectic solvents. *Ind. Crop. Prod.* **2017**, *99*, 1–6. [CrossRef]
19. Wang, Z. Preparation of Furfural Platform Compounds from Biomass Resources in Deep Eutectic Solvents. Master's Thesis, Qingdao University of Science and Technology, Qingdao, China, 2019.

© 2020 by the authors. Licensee MDPI, Basel, Switzerland. This article is an open access article distributed under the terms and conditions of the Creative Commons Attribution (CC BY) license (http://creativecommons.org/licenses/by/4.0/).

Article

Sustainable Method for the Synthesis of Alternative Bis(2-Ethylhexyl) Terephthalate Plasticizer in the Presence of Protic Ionic Liquids

Aleksander Grymel [1], Piotr Latos [2], Karolina Matuszek [3], Karol Erfurt [2], Natalia Barteczko [2], Ewa Pankalla [1] and Anna Chrobok [2,*]

1. Grupa Azoty Zakłady Azotowe Kędzierzyn, S.A., Mostowa 30A, 47-220 Kędzierzyn-Koźle, Poland; Aleksander.Grymel@grupaazoty.pl (A.G.); Ewa.Pankalla@grupaazoty.pl (E.P.)
2. Department of Chemical Organic Technology and Petrochemistry, Silesian University of Technology, Krzywoustego 4, 44-100 Gliwice, Poland; Piotr.Latos@polsl.pl (P.L.); Karol.Erfurt@polsl.pl (K.E.); Natalia.Barteczko@polsl.pl (N.B.)
3. School of Chemistry, Monash University, Clayton, VIC 3800, Australia; Karolina.Matuszek@monash.edu
* Correspondence: Anna.Chrobok@polsl.pl; Tel.: +48-32-237-20-14

Received: 2 April 2020; Accepted: 21 April 2020; Published: 23 April 2020

Abstract: Inexpensive Brønsted acidic ionic liquids based on trimethylamine and sulfuric acid are proposed as both solvents and catalysts in the synthesis of alternative plasticizer bis(2-ethylhexyl) terephthalate, which has a broad spectrum of applications in plasticization processes. The utilization of 50 mol % of Brønsted ionic liquid led to the full conversion of terephthalic acid after 8 h of reaction at 120 °C. Additionally, a 100% selectivity of bis(2-ethylhexyl) terephthalate was obtained. The advantage of the presented reaction system is based on the formation of a biphasic system during the reaction. The bottom phase consists of an ionic liquid and water, and the upper phase is created by the ester and unreacted alcohol. This phenomenon helps overcome the equilibrium of the reaction and drives it towards a high yield of product. The presented new approach is proposed as a safe, cost-effective, and alternative method to conventional processes with organometallic compounds that, in turn, leads to greener and a more economically viable technology.

Keywords: plasticizers; acidic catalysis; terephthalate esters; ortho-phthalate esters; esterification; solvents; ionic liquids

1. Introduction

In the recent years, strict regulations on environmental protection has intensified pending restrictions on plasticizers production, limiting the use of some ortho-phthalates esters, due to the concerns about negative effects on human health [1]. Plasticizers are important additives to improve the flexibility, plasticity, and processability of polymers [2]. These additives, which make polyvinylchloride (PVC) flexible, are massively used for the production of medical tubing and bags, toys for children, wires and cables, and many others. Non-phthalate alternative plasticizers include terephthalates, citrates, phosphates, polyesters, halogenated alkanes, and epoxy compounds [3]. Terephthalate esters have been proven to have different toxicological profiles [4] compared to ortho-phthalates, and they may be used as direct replacements, thus creating an emerging alternative market. Today, the most commercially used terephthalate ester is bis(2-ethylhexyl) terephthalate, which is used as a plasticizer in materials applied as containers for food and drinking water.

The main group of catalysts for the production of terephthalates esters are compounds that are soluble in a reaction mixture containing a sulfone group (–SO_3H). In this group, methanesulfonic and para-toluenesulfonic acids, as well as ion exchange resins in hydrogen form, are crucial [5,6].

However, organometallic compounds are currently the most dominant catalysts used at the industrial scale [7–12]. Among them, titanium, zirconium, tin, antimony, and zinc compounds have been claimed in patents [5–12]. Most of the patents belong to Mitsubishi Chemical and Eastman Chemical representing the Asian and American markets, who are the largest consumers of soft PVC products. The subjects of the patent claims are process conditions, the compositions of catalysts, or specific design of the reactor. The advantage of titanium compounds is their commercial availability in large quantities and low price. Titanium compounds are catalysts for not only esterification reactions but also isomerization, transesterification, and disproportionation reactions. The selectivity of the catalyst is crucial for pure and efficient synthesis of terephthalates esters. Water formed during the process is removed from the reaction mixture in the form of an azeotrope distillation with alcohol, which is separated after being condensed and recycled in part or in whole to the reaction. Elevated temperatures can increase the reaction rate, but, simultaneously, side reactions can occur, and these cause the formation of colored by-products. Reaction temperatures using titanium- or zirconium-based catalysts are in the range of 160–270 °C. The reaction temperature can be controlled by increasing or decreasing the pressure in the reaction system, which is preferable for low-boiling alcohols.

It should be noted that some companies receive bis(2-ethylhexyl) terephthalate via two methods [13]. The first method is based on the esterification of terephthalic acid with an excess of octanol, while the second one assumes the transesterification of dimethyl terephthalate (DMT) with octanol to dioctyl terephthalate, both in the presence of organometallic compounds as catalysts. In the transesterification reaction, methanol is a byproduct instead of water. The disadvantages of the transesterification reaction are a higher temperature compared to the esterification reaction and a significant dilution of the reaction mixture with octanol. Additionally, dimethyl terephthalate is more expensive than terephthalic acid, and this can influence the cost of the esterification process.

The low solubility of terephthalic acid in common solvents makes its industrial production very difficult and not environmentally benign. In our previous work, the solubility of terephthalic acid in ionic liquids was studied, and the results showed that terephthalic acid is partially soluble in ionic liquids, such as 1-ethyl-3-methylimidazolium diethylphosphate, 1-butyl-3-methylimidazolium acetate, and dialkylimidazolium chlorides up to four times higher than in DMSO (20 g of terephtalic acid per 100 g DMSO at 25 °C) [14]. Even though the solubility of terephthalic acid increased in the ionic liquid, the problem was not entirely solved. The potential application of ionic liquids for esterification process can still cause acid crystallization and clogging in the production plant.

The state-of-the-art process assumes the use of ionic liquids for the synthesis of terephthalates as solvents together with the addition of acidic catalysts (sulfuric, phosphoric, methanesulfonic acid, ion exchange resins, and zeolites) or organometallic compounds [15–18]. The reaction can be carried out using a classical method with the azeotropic distillation of water in the presence of an excess of alcohol to shift the equilibrium towards the product formation. The most often used ionic liquids are quaternary ammonium or imidazolium salts, where the production via quaternization reaction is expensive, environmentally burdensome, and, therefore, economically unfavorable.

Using a new family of Brønsted acidic protic ionic liquids, based on sulfuric acid for esterification processes of acetic [19] and lactic acid [20], we have already achieved competitive advantages over the classical methods. Highly acidic protic ionic liquids were obtained in the simple reaction between amine (e.g., triethylamine or 1-methylimidazole) and sulfuric acid. Using an excess of sulfuric acid to amine (molar ratio of 1:2–1:3; mole fraction $\chi_{H_2SO_4}$ = 0.67 and 0.75) led to the formation of a hydrogen-bonded network of sulfuric acid molecules and hydrogen sulfate anions in an anion structure $[(HSO_4)(H_2SO_4)_x]^-$ (x = 1 or 2) with a protonated base:

$$H_2SO_4 + Et_3N \rightarrow [Et_3NH][HSO_4] \qquad (\chi_{H_2SO_4} = 0.50)$$

$$H_2SO_4 + Et_3N \rightarrow [Et_3NH][HSO_4(H_2SO_4)] \qquad (\chi_{H_2SO_4} = 0.67)$$

$$H_2SO_4 + Et_3N \rightarrow [Et_3NH][HSO_4(H_2SO_4)_2] \qquad (\chi_{H_2SO_4} = 0.75)$$

Ionic liquids synthesized from equimolar amounts of amine and sulfuric acid, based on [HSO$_4$]$^-$ anions are poorly active in the esterification process [21]. When protic ionic liquids ($\chi_{H_2SO_4}$ = 0.67 and 0.75) are used for the esterification reaction, the formation of biphasic systems was the driving force of this process and caused a shift in the equilibrium to the product formation. The insolubility of ester was a key parameter for the success of this method.

Herein, we propose inexpensive Brønsted acidic ionic liquids as solvents and catalysts for the sustainable synthesis of bis(2-ethylhexyl) terephthalate from terephthalic acid and 2-etyl-1-hexanol. This new approach is safe, cost-effective, and can be attractive as a replacement of conventional acids and organometallic compounds for industrially relevant processes.

2. Results and Discussion

In order to lower the cost of ionic liquids, inexpensive feedstocks such as sulfuric acid and simple amines were combined into a protic ionic liquids containing hydrogen sulfate anions [19]. Proper amine selection is crucial, as the production cost of ionic liquids mainly depends on its price [22].

As presented on Figure 1, the acidity of protic ionic liquids based on sulfuric acid, expressed by Gutmann acceptor number (AN) [23], does not depend on the pK$_a$ of amine (triethylamine, 1-methylimodazol, 2-methylpyridine, and 1-methylpyrrolidine) [19]. Therefore, for this study, we chose trimethylamine, which is the cheapest raw material.

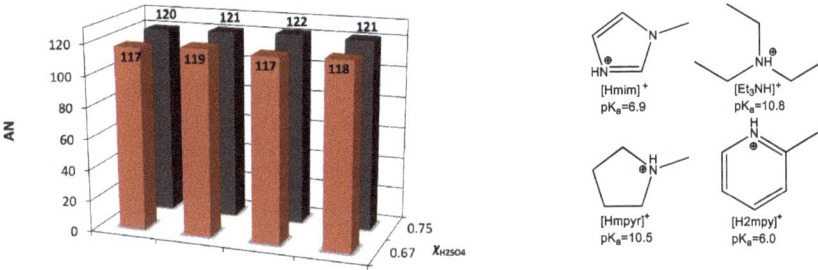

Figure 1. Gutmann acceptor number (AN) values measured for protic ionic liquids (left), based on four different amines (right) and excess of H$_2$SO$_4$ (mole fraction $\chi_{H_2SO_4}$) [19].

Ionic liquids based on the trimethylammonium cation [Et$_3$HN]$^+$ and, for comparison, the more expensive 1-methylimidazolium cation [Hmim]$^+$, were synthesized by mixing the amine with an excess of sulfuric acid (molar ratios of 1:1.5; 1:2, 1:2.7, and 1:3). All ionic liquids were slightly yellow liquids at room temperature and characterized by NMR analysis (NMR spectra available in supplementary materials).

The key to a successful esterification, utilizing protic ionic liquids, is based on the formation of a biphasic system during the reaction. The bottom phase should consist of an ionic liquid and water, and the upper phase should be created by the ester and unreacted alcohol. We assumed, based on previous studies [14–18], that this system would overcome the equilibrium and obtain a product with a high yield.

Unfortunately, the solubility of terephthalic acid in the protic ionic liquids based on sulfuric acid is very low. To resolve this, in the beginning of the process, terephthalic acid was suspended in the ionic liquid. In the next step, 2-ethyl-1-hexanol was added (Scheme 1). The reaction was carried out in a 250 mL rector with a thermostatic cooling/heating jacket, a mechanical stirrer, and a condenser. The samples were taken from the reaction mixture and were analyzed using gas chromatography (GC). We started the optimization of process parameters by setting out the excess of alcohol per terephthalic acid (molar ratio of 8:1). The reaction was carried at 120 °C. In preliminary experiments, the influence of the structure of the amine and the composition of ionic liquids were studied. As presented in Figure 2, ionic liquids based on [Et$_3$HN]$^+$ and [Hmim]$^+$ cations (the molar ratios of amine to sulfuric acid being 1:3 and 1:2.7) were equally active in the esterification process, yielding quantities of ester after 4 h of reaction. However, when lowering the acidity of the ionic liquid by reducing the contribution of sulfuric acid in the ionic liquid structure (with molar excesses of 1:2 and 1:1.5, respectively, for [Et$_3$HN]$^+$ and [Hmim]$^+$ cations), a decrease in the activity of catalyst was observed. In both cases, after a longer reaction time, the whole conversion of terephthalic acid was achieved (after 5 and 14 h, respectively, for the [Et$_3$HN]$^+$ and [Hmim]$^+$ cations).

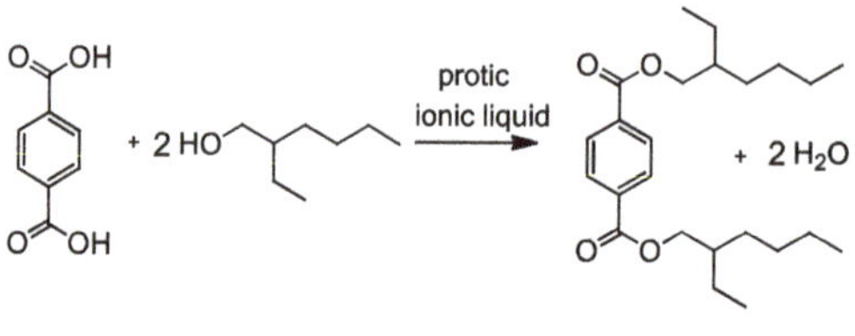

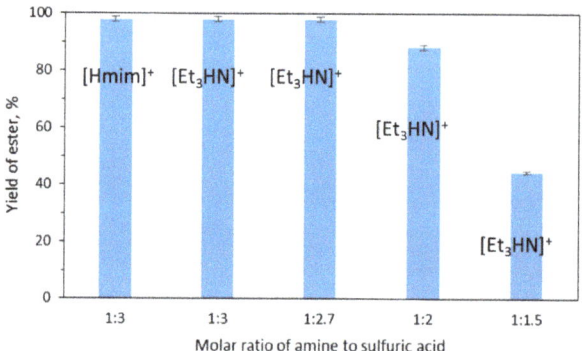

Scheme 1. Esterification of terephthalic acid with 2-ethyl-1-hexanol.

Figure 2. The influence of ionic liquids composition on the yield of bis(2-ethylhexyl) terephthalate obtained in the reaction of terephthalic acid (0.12 mol) with 2-ethyl-1-hexanol (0.96 mol) in the presence of 50 mol % of the protic ionic liquid relative to terephthalic acid at 120 °C after 4 h.

The application of the composition of the ionic liquid below molar ratio of 1:3 was more favorable for two reasons. First, the color of the post reaction mixtures (upper phase) was almost colorless (Scheme 2), and using the ionic liquid below the ratio of 1:3 neared and then achieved a completely colorless mixture from the 1:2 ratio.

Scheme 2. The color of the upper phase of post reaction mixture obtained in the reaction of terephthalic acid (0.12 mol) with 2-ethyl-1-hexanol (0.96 mol) in the presence of 50 mol % of the ionic liquid based on Et$_3$N relative to terephthalic acid at 120 °C after 4 h.

The second reason was the low content of the bis(2-ethylhexyl) ether created as a by-product. The high temperature and the acidity of the ionic liquid were also favorable factors for the etherification of alcohol. Tables 1 and 2 summarize the data on the influence of temperature and ionic liquid composition on the formation of the ether. The composition of the ionic liquid (1:2.7) could be the best solution to overcome these issues. At 120 °C, the reaction time required to obtain the product with a high yield (ionic liquid (1:2.7); yield 99.1%) was the same as for the ionic liquid (1:3), but the level of ether was much lower than for the higher composition (1:3). In the case of ionic liquids (1:3), higher selectivity could be obtained by decreasing the temperature to 110 °C. However, the yield of the ether was still higher than for the reaction with the ionic liquid (1:2.7) at 120 °C. Additionally, the color of the post reaction mixture (1:2.7) was acceptable. Therefore, an ionic liquid based on Et_3N (composition 1:2.7) at a temperature of 120 °C was chosen for the next few experiments.

Table 1. Influence of ionic liquid composition on the formation of bis(2-ethylhexyl) ether during the reaction at 120 °C.

Molar Ratio $Et_3N:H_2SO_4$	1:3	1:2.7	1:2	1:1.5
Yield of ester, %	99.5	99.1	99.2	99.7
(time, h)	(4)	(4)	(5)	(14)
Yield of ether, %	23.1	2.5	2.2	1.7
(time, h)	(4)	(4)	(4)	(4)

Table 2. Influence of temperature on the formation of bis(2-ethylhexyl) ether after 4 h.

Temp., °C	Yield of Bis(2-Ethylhexyl) Ether, % Ionic Liquid 1:3	Yield of Bis(2-Ethylhexyl) Ether, % Ionic Liquid 1:2.7
90	0.7	0.6
100	5.3	2.1
110	5.2	2.2
120	23.1	2.5

The influence of the ionic liquid loading (30–100 mol % per terephthalic acid) was tested at 120 °C with a fixed molar ratio of alcohol (1:8) (Figure 3). The optimum conditions were found below equimolar amounts of the ionic liquid (40–50 mol %). The reaction conditions and catalyst loadings were selected to achieve the full conversion of terephthalic acid in less than 2 h. Increasing the ionic liquid loading up to the equimolar amounts caused the decrease of the reaction rate. This can be explained as a result of the dilution of the reaction system. Ionic liquids also play the roles of water extractants, so the amount of the ionic liquid used is important to consider.

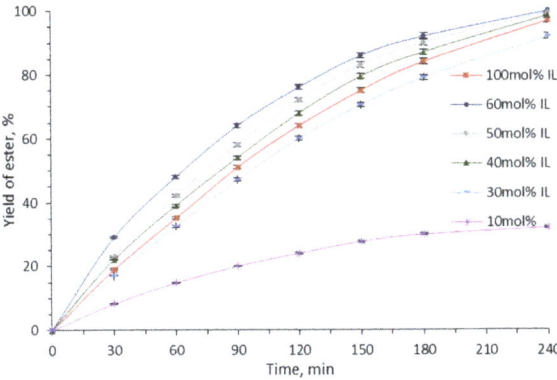

Figure 3. The influence of the ionic liquid based on Et_3N (1:2.7) loading on the yield of bis(2-ethylhexyl) terephthalate obtained in the reaction of terephthalic acid (0.12 mol) with 2-ethyl-1-hexanol (0.96 mol) at 120 °C.

The influence of reaction temperature on the reaction time required for full acid conversion was also tested (Figure 4). Reactions were carried out in a temperature range from 90 to 120 °C. The reaction rate increased with temperature, as expected, and reached full conversion in 8 h at 110 °C and only 4 h at 120 °C. Below 110 °C, the reaction times were much longer. Additionally, in order to check if this biphasic reaction was influenced by the mixing, the experiments were conducted at various mixing speeds (200–1000 rpm) at 120 °C. In doing so, it was observed that varying mixing speeds did not influence the reaction. All experiments were therefore carried out at 800 rpm.

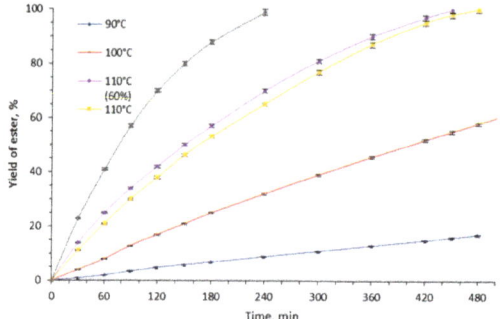

Figure 4. The influence of temperature on the yield of bis(2-ethylhexyl) terephthalate obtained in the reaction of terephthalic acid (0.12 mol) with 2-ethyl-1-hexanol (0.96 mol) in the presence of 50 mol % of the ionic liquid based on Et$_3$N (1:2.7) (24.66 g) relative to terephthalic acid.

In the next step, the influence of the molar ratio of alcohol to terephthalic acid was determined (Figure 5). Lowering the amount of alcohol is beneficial from an economical and sustainability standpoint (smaller plant size and less energy to recover and recycle). As presented in the Figure 5, the optimum amounts of alcohol was 8:1. It was possible to carry out the reaction with a molar ratio of 6:1, but the full conversion of acid was reached after 3 h and 40 min, which was a longer reaction time.

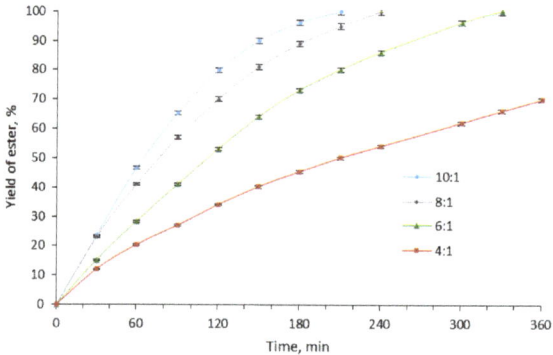

Figure 5. The influence of an excess of 2-ethyl-1-hexanol on the yield of bis(2-ethylhexyl) terephthalate obtained in the presence of 50 mol % of the ionic liquid based on Et$_3$N (1:2.7) (24.66 g) relative to terephthalic acid at 120 °C.

The isolation of the product after the reaction was based on the simple separation of the biphasic system (Scheme 3). The upper phase was first washed with water to remove traces of the ionic liquid, and then the mixture was dried and concentrated on a rotary evaporator. Next the ester, ether, and unreacted alcohol were isolated via distillation. The water was evaporated from the bottom layer, and the ionic liquid was reused five times without becoming lost in activity (Figure 6). The recovery percentage of the ionic liquid after each cycle was approximately 97%.

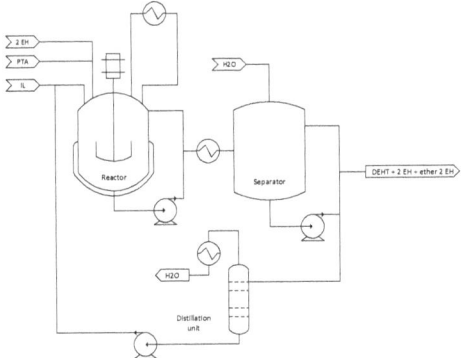

Scheme 3. The proposed block scheme for the production of bis(2-ethylhexyl) terephthalate.

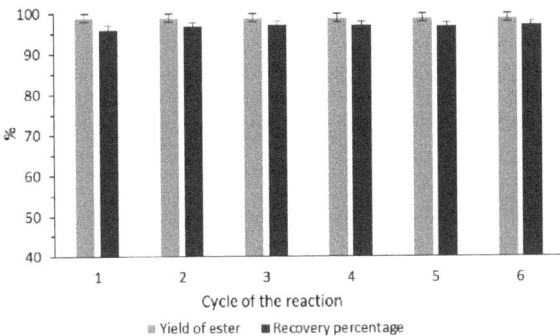

Figure 6. Recycling of ionic liquid study.

The presented esterification method was very useful for the production of bis(2-ethylhexyl) terephthalate. However, the modification of the alcohol structure to 1-butanol disrupted the reaction system. The created ester dibutyl terephthalate was partially soluble in the ionic liquid, which resulted in an only 50% conversion of terephthalic acid after 4 h of esterification at 120 °C (50 mol % of the ionic liquid relative to terephthalic acid; a 1:8 molar ratio of terephthalic acid to 1-butanol). The prolongation of reaction time caused a very slow increase in the amount of ester (up to 80% after 12 h). Additionally, 1-butanol was very prone to etherification in the reaction conditions, and, as an outcome, 25% of alcohol was converted to dibutylether as a by-product. This phenomenon confirmed that the success of the proposed reaction system was based on the creation of two phases during the reaction.

It is worth noting that the synthesis of the 'forbidden' ortho-phthalate esters from phthalic anhydride would be very simple using the presented method. Phthalic anhydride is soluble in ionic liquids, and the formed bis(2-ethylhexyl) phthalate is insoluble in the ionic liquid phase (Figure 7). As a result, after the optimization of the reaction parameters using 20 mol % of ionic liquid relative to phthalic anhydride and a 1:3 molar ratio of phthalic anhydride to 2-ethyl-1-hexanol at 80 °C, it was possible to obtain full conversion of anhydride to bis(2-ethylhexyl) phthalate after 90 min. A lower temperature and a lower amount of the ionic liquid was necessary to obtain a full conversion of phthalic anhydride with the lack of ether as a by-product.

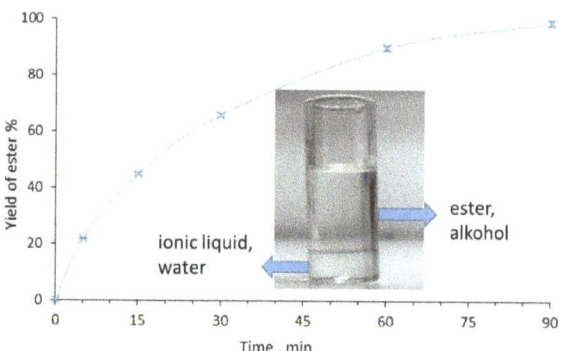

Figure 7. Yield of bis(2-ethylhexyl) phthalate obtained in the reaction of phthalic anhydride (0.12 mol) with 2-ethyl-1-hexanol (0.96 mol) in the presence of 20 mol % of the ionic liquid based on Et$_3$N (1:2.7) (9.86 g) relative to phthalic anhydride at 80 °C.

3. Materials and Methods

3.1. Materials

Decane, triethylamine, 1-methylimidazole, sulfuric acid (95%), phthalic anhydride (99%), and activated carbon were purchased from Sigma-Aldrich (Merck, Darmstadt, Germany). Terephthalic acid (99%), n-butanol, and 1-ethyl-1-hexanol were produced at the plants of Grupa Azoty Zakłady Azotowe Kędzierzyn S.A. NaHCO$_3$ and MgSO$_4$ were obtained from Chempur (Krupski Młyn, Poland) and were used without additional purification.

3.2. Methods

GC analyses were performed on a Shimadzu GC-2010 Plus (Kyoto, Japan) (carrier gas helium), with an oven temperature program as follows: start 80 °C for 1 min, next reach 300 °C with rate 40 °C/min), and then 5 min at 300 °C. The Shimadzu GC-2010 Plus was equipped with a flame ionization detector (FID) and a Zebron ZB-5MSi column (30 m × 0.25 mm 0.25 µm film). Details presented in supplementary materials.

The NMR spectra of the products were recorded using a Varian 500 spectrometer (Palo Alto, CA, USA.) at the following operating frequencies: ^{1}H 400 MHz and ^{13}C 150 MHz. Chemical shifts are reported as parts per million (ppm) in reference to tetramethylsilane (TMS) for 0.020 g of the sample.

3.3. Synthesis

A protic ionic liquid (molar ratio of amine to H$_2$SO$_4$ 1:3, 1:2.7, 1:1.5): triethylamine (25.30 g, 0.25 mol, 1 mol eq.) was introduced into the batch jacket reactor (250 mL) with a thermometer, a condenser, and a mechanical stirrer. The water solution of sulfuric acid (38.72–77.43 g, 0.38–0.75 mol, 1.5–3 mol eq.; water 6.40–10.27 g, and 0.36–0.57 mol) was added while keeping the temperature at 40 °C (400 rpm, 20 min). Next the rector content was mixed for 1 hour at room temperature. In the end, the ionic liquid was dried (80 °C, 1–2 mbar, 12 h). The ionic liquid was obtained with a 98% yield as a colorless liquid. ^{1}H NMR (400 MHz, DMSO) δ 8.95 (s, 1H), 7.51 (bs, 5H), 3.07 (kw, 6H, J = 6.4 Hz), 1.15 (t, 9H, J = 7.9 Hz). ^{13}C NMR (150 MHz, DMSO) δ 48.02, 12.20.

The protic ionic liquid (molar ratio of amine to H$_2$SO$_4$ 1:3): 1-methylimidazole (20.53 g, 0.25 mol, 1 mol eq.) was introduced into the batch jacket reactor (250 mL) with a thermometer, a condenser, and a mechanical stirrer. The water solution of sulfuric acid (77.43 g, 0.75 mol, 3 mol eq.; water (9.80 g, 0.54 mol) was added while keeping the temperature at 40 °C (400 rpm, 20 min. Next, the rector content was mixed for 1 hour at room temperature. In the end, the ionic liquid was dried (80 °C, 1–2 mbar, overnight). The ionic liquid was obtained with a 99% yield as a colorless liquid. ^{1}H NMR (400 MHz,

DMSO) δ 9.01 (s, 1H), 8.68 (bs, 6H), 7.66 (s, 1H), 7.61 (s, 1H), 3.84 (s, 3H). ^{13}C NMR (150 MHz, DMSO) δ 136.23, 123.60, 120.14, 35.82.

Bisalkyl terephthalate: protic ionic liquid (molar ratio of amine to H_2SO_4 1:3, 1:2.7, 1:1.5) (mol (23.47–24.66 g, 0.06 mol, 50 mol %), alcohol (71.16–125.02 g, 0.96 mol), terephthalic acid (19.94 g, 0.12 mol, molar ratio of acid to alcohol 1:8), and decane (internal standard for GC, 3 g) were introduced into the batch jacket reactor (250 mL) with a thermometer, a condenser, and a mechanical stirrer. Next, the reaction was mixed at 120 °C for 4–12 h. To control the reaction progress using GC analysis, the samples were taken from the upper phase. In the end, the upper phase, after separation, was washed with water (10 mL) to get rid of the ionic liquid. The organic layer was concentrated. During the distillation of the organic layer, the alcohol was removed at 63–65 °C (10 mmHg) for 2-ethyl-1-hexanol and 117–119 °C for 1-butanol. After the evaporating of the alcohol, the ether was removed at 130–131 °C (10 mmHg) for bis(2-ethylhexyl) ether and 76–77 °C (100 mmHg) for dibutyl ether. Finally, the product was decolorized using activated carbon. Bis(2-ethylhexyl) terephthalate was obtained with a 95% yield (46.40 g) as a colorless liquid: ^{1}H NMR (400 MHz, DMSO) δ: 8.08 (s, 4H), 4.24 (td, J = 7.6, 3.7 Hz, 4H), 1.74–1.65 (m, 2H), 1.27–1.45 (m, 16H), 0.91 (t, J = 7.5 Hz 6H), 0.88 (t, J = 7.4 Hz, 6H). ^{13}C NMR (150 MHz, DMSO) δ: 165.01, 133.74, 129.44, 67.12, 38.25, 29.91, 28.37, 23.40, 22.37, 13.87, 10.88.

Dibutyl terephthalate was obtained as a colorless liquid with a 75% yield (26.72 g). ^{1}H NMR (400 MHz, CDCl$_3$) δ 8.07 (s, 4H), 4.30 (t, J = 6.5 Hz, 4H), 1.70 (quin, J = 7.5 Hz, 4H), 1.42 (sext, J = 7.4 Hz, 4H), 0.93 (t, J = 7.3 Hz, 6H). ^{13}C NMR (150 MHz, DMSO) δ: 165.03, 133.73, 129.43, 64.86, 30.16, 18.71, 13.57.

Bis(2-ethylhexyl) phthalate: protic ionic liquid (molar ratio of amine to H_2SO_4 1:2.7) (9.86 g, 0.02 mol, 20 mol %), 2-ethyl-1-hexanol (125.02 g, 0.96 mol), phthalic anhydride (17.77 g, 0.12 mol, molar ratio of acid to 2-ethyl-1-hexanol 1:8), and decane (internal standard for GC, 3 g) were introduced into the batch jacket reactor (250 mL) with a thermometer, a condenser, and a mechanical stirrer. The reaction mixture was stirred at 80 °C for 90 min. To control the reaction progress using GC analysis, the samples were taken from the upper phase. In the end, the upper phase, after separation, was washed with water (10 mL) to get rid of the ionic liquid. The organic layer was concentrated. During the distillation of the organic layer, the alcohol was removed at 63–65 °C (10 mmHg). Bis(2-ethylhexyl) phthalate was obtained with a 96% yield (46.40 g) of as a colorless liquid. ^{1}H NMR (400 MHz, DMSO) δ 7.73–7.65 (m, 4H), 4.13 (td, J = 5.7, 3.5 Hz, 4H), 1.68–1.58 (m, 2H), 1.24–1.43 (m, 16H), 0.88 (t, J = 6.7 Hz, 6H), 0.85 (t, J = 6.7 Hz, 6H).

3.4. Recycling Studies

In order to recycle the catalyst, after the reaction was carried out in the presence of 50 mol % of the ionic liquid, the upper phase was decanted, and the lower, ionic liquid phase was washed with water and dried (80 °C, 1–2 mbar, 12 h).

4. Conclusions

In conclusion, in this work, an efficient method for the synthesis of an alternative bis(2-ethylhexyl) terephthalate plasticizer was presented. The activity of the proposed protic ionic liquids in 120 °C was far above conventional organometallic compounds working at 160–270 °C, which, in turn, led to a both greener and mote economically viable process. Protic ionic liquids can compete with the cheapest pretreatment chemicals, such as sulfuric acid and triethylamine, in terms of effectiveness and process cost, removing ionic liquid cost as a barrier to the economic viability of ionic liquids-based esterification. In order to translate these findings to an industrially viable proposition, scaled-up experiments are necessary. This stage of research is currently under investigation in our group.

Supplementary Materials: The following are available online at http://www.mdpi.com/2073-4344/10/4/457/s1, NMR analysis of esters and conditions of GC analysis of reaction mixtures.

Author Contributions: Conceptualization, A.C.; data curation, A.G., P.L., K.M., K.E. and N.B.; formal analysis, E.P. and A.C.; funding acquisition, E.P. and A.C.; investigation, A.G. and P.L; methodology, A.G., P.L., K.M. and

K.E; software, P.L.; supervision, E.P.; visualization, A.G.; writing—original draft, A.C.; writing—review and editing, A.G., P.L. and K.M. All authors have read and agreed to the published version of the manuscript.

Funding: This research was funded by the company Grupa Azoty Zakłady Azotowe Kędzierzyn S.A. P.L. also acknowledges the National Agency for Academic Exchange of Poland (under the Academic International Partnerships program, grant agreement PPI/APM/2018/1/00004).

Conflicts of Interest: The authors declare no conflict of interest.

References

1. Jamarani, R.; Erythropel, H.C.; Nicell, J.A.; Leask, R.L.; Maric, M. How green is your plasticizer? *Polymers* **2018**, *10*, 834. [CrossRef] [PubMed]
2. Jia, P.; Xia, H.; Tang, K.; Zhou, Y. Plasticizers derived from biomass resources: A short review. *Polymers* **2018**, *10*, 1303. [CrossRef] [PubMed]
3. Kumar, S. Recent developments of biobased plasticizers and their effect on mechanical and thermal properties of Poly(vinyl chloride): A review. *Ind. Eng. Chem. Res.* **2019**, *58*, 11659–11672. [CrossRef]
4. Ball, G.L.; McLellan, C.J.; Bhat, V.S. Toxicological review and oral risk assessment of terephthalic acid (TPA) and its esters: A category approach. *Crit Rev. Toxicol.* **2012**, *42*, 28–67. [CrossRef] [PubMed]
5. Nakajima, I.; Takai, M. Method for Producing Terephthalic Acid Diester. JP 2004300078, 28 October 2004. Available online: https://patents.google.com/patent/JP2004300078A/en (accessed on 1 April 2020).
6. Nakajima, I.; Takai, M. Method for Producing Terephthalic Acid Diester. JP 2004339075, 2 December 2004. Available online: https://patents.google.com/patent/JP2004339075A/en?oq=JP+2004339075 (accessed on 1 April 2020).
7. Matsumoto, S. Catalyst for Esterification and Transesterification and Process for Producing Ester. U.S. Patent 2005/0176986, 18 April 2005.
8. Ikeda, H.; Inoue, T.; Takahashi, K. Method for Producing Diester of Terephthalic Acid. JP 2005306759, 4 November 2005. Available online: https://patents.google.com/patent/JP2005306759A/en?oq=JP+2005306759 (accessed on 1 April 2020).
9. Matsuo, T.; Nakajima, I.; Takai, M. Method for Producing Terephthalic Acid Diester. JP 2006273799, 12 October 2006. Available online: https://patents.google.com/patent/JP2006273799A/en?oq=JP+2006273799 (accessed on 1 April 2020).
10. Osborne, V.H.; Turner, P.W.; Cook, S.L. Production of Terephthalic Acid Di-Esters. WO 2008/094396, 7 August 2008. Available online: https://patents.google.com/patent/WO2008094396A1/en?oq=WO+2008%2f094396 (accessed on 1 April 2020).
11. Nakajima, I.; Takai, M. Method for Producing Terephthalic Acid Diester. JP 2009185040, 20 August 2009. Available online: https://patents.google.com/patent/JP2009185040A/en?oq=JP+2009185040 (accessed on 1 April 2020).
12. Sutor, E.; Grzybek, R.; Grymel, A.; Janik, L.; Trybuła, J.; Tkacz, B.; Fiszer, R.; Krueger, A.; Jasienkiewicz, J.; Filipiak, B.; et al. Process for the Preparation of Dioctyl Terephthalate. PL 216179, 31 March 2014. Available online: https://patents.google.com/patent/PL216179B1/en?oq=PL+216179 (accessed on 1 April 2020).
13. Grass, M. Dialkyl Terephthalates and Their Use. U.S. Patent 20070179229, 2 August 2007.
14. Matuszek, K.; Pankalla, E.; Grymel, A.; Latos, P.; Chrobok, A. Studies on the Solubility of Terephthalic Acid in Ionic Liquids. *Molecules* **2020**, *25*, 80. [CrossRef] [PubMed]
15. Kaller, A.S.M.; Bronneberg, R.; Stammer, J.; Das, M.; Harnischmacher, G. Method for Producing Diesters of Terephthalic Acid with A Dehydration of Recirculated Alcohol. WO 2016046118A1, 31 March 2016. Available online: https://patents.google.com/patent/WO2016046118A1/en?oq=WO+2016046118A1 (accessed on 1 April 2020).
16. Lin, J.Q.; Fang, G.; Jin, C.; Sun, Y.; Zuo, S.; Qian, C.; Lin, W.; Zang, X.; Liu, L.; Chen, Y. Method for Synthesizing Dioctyl Terephthalate Through Esterification. CN 102001948A, 6 April 2011. Available online: https://patents.google.com/patent/CN102001948A/en?oq=CN102001948A (accessed on 1 April 2020).
17. Xuesong, P. Production Method of Dioctyl Terephthalate. CN 102701984A, 3 October 2012. Available online: https://patents.google.com/patent/CN102701984A/en?oq=CN102701984A (accessed on 1 April 2020).

18. Hongyun, G.; Maolin, T.; Lingling, L.; Chenxi, L.; Maohua, D. Applications of Acidic Functionalized Ion Liquid in Esterification Reaction. CN 103752340A, 30 April 2014. Available online: https://patents.google.com/patent/CN103752340A/en?oq=CN103752340A (accessed on 1 April 2020).
19. Matuszek, K.; Chrobok, A.; Coleman, F.; Seddon, K.R.; Swadźba-Kwaśny, M. Tailoring ionic liquid catalysts: Structure, acidity and catalytic activity of protonic ionic liquids based on anionic clusters, $[(H_SO_4)(H_2SO_4)x]^-$ (x = 0, 1, or 2). *Green Chem.* **2014**, *16*, 3463–3471. [CrossRef]
20. Dorosz, U.; Barteczko, N.; Latos, P.; Erfurt, K.; Pankalla, E.; Chrobok, A. Highly efficient biphasic system for the synthesis of alkyl lactates in the presence of acidic ionic liquids. *Catalysts* **2020**, *10*, 37. [CrossRef]
21. Chiappe, C.; Rajamani, S.; D'Andrea, F. A dramatic effect of the ionic liquid structure in esterification reactions in protic ionic media. *Green Chem.* **2013**, *15*, 137–143. [CrossRef]
22. George, A.; Brandt, A.; Tran, K.; Zahari, S.M.S.N.S.; Klein-Marcuschamer, D.; Sun, N.; Sathitsuksanoh, N.; Shi, J.; Stavila, V.; Parthasarathi, R.; et al. Design of low-cost ionic liquids for lignocellulosic biomass pretreatment. *Green Chem.* **2015**, *17*, 1728–1734. [CrossRef]
23. Gutmann, V. *The Donor-Acceptor Approach to Molecular Interactions*; Plenum Press: New York, NY, USA, 1978.

© 2020 by the authors. Licensee MDPI, Basel, Switzerland. This article is an open access article distributed under the terms and conditions of the Creative Commons Attribution (CC BY) license (http://creativecommons.org/licenses/by/4.0/).

Article

Beneficial Contribution of Biosourced Ionic Liquids and Microwaves in the Michael Reaction

Katia Bacha, Kawther Aguibi, Jean-Pierre Mbakidi and Sandrine Bouquillon *

Institut de Chimie Moléculaire de Reims UMR CNRS 7312, Université de Reims Champagne-Ardenne, Boîte n° 44, B.P. 1039, F-51687 Reims, France; katia.bacha@etudiant.univ-reims.fr (K.B.); kawther.aguibi@etudiant.univ-reims.fr (K.A.); jean-pierre.mbakidi@univ-reims.fr (J.-P.M.)
* Correspondence: sandrine.bouquillon@univ-reims.fr; Tel.: +33-(0)3-26-91-89-73; Fax: +33-(0)3-26-91-31-66

Received: 24 June 2020; Accepted: 20 July 2020; Published: 22 July 2020

Abstract: We developed a synthesis of chiral ionic liquids from proline and one of its derivatives. Nine chiral ionic liquids were synthesized with yields from 78% to 95%. These synthesized ionic liquids played two roles in Michael reactions, as solvents, and as basic catalysts, where the ionic phase could also be reused at least five times without loss of activity. The yields up to 99% were improved by increasing the amount of dimethylmalonate from 1.2 equivalents to 3 or 4 equivalents. Furthermore, the reaction time could be reduced from 24 h to 45 min through microwaves activation.

Keywords: ionic liquids; biomass; microwaves; Michael reaction; chalcone; dimethylmalonate

1. Introduction

The impact of chemical solvents on the environment is an increasingly debated subject. Indeed, most of these solvents are volatile organic compounds and can, therefore, be easily dispersed in the environment. Risks often accompany this because they are flammable. In addition, they are generally harmful from an ecological and health point of view [1]. Over the past fifteen years, green chemistry, which has become a priority area for academic and industrial research, has undergone considerable development. In this context, much research is focused on alternative, environmentally friendly solvents that can replace some highly volatile and environmentally harmful solvents [2].

These include water [3–9], supercritical CO_2 [10–14], and ionic liquids (ILs) [15–21]. Ionic liquids are of particular interest because they are a new class of solvents that offer interesting opportunities as a reaction medium for cleaner chemistry due to their multiple properties.

Ionic liquids are salts consisting solely of ions, whose melting point is below an arbitrary temperature, often 100 °C. They consist of generally bulky organic cations and organic or inorganic anions, which gives them an asymmetry of shape and charge, as well as great flexibility [22,23]. In general, cations have a voluminous and asymmetric nature [15]. The most commonly represented are composed of ammoniums, phosphoniums, or heteroaromatic systems such as imidazolium, pyridinium, pyrrolidinium, oxazolium, thiazolium, pyridazinium, triazolium, and tetraalkylammonium. Different ionic liquids can be formed either by the appropriate combination of cations and anions or by the chemical modification of the cation or anion. This offers a choice of considerable combinations, and the literature now reports the possible synthesis of at least 10^6 different ILs [19].

Their synthesis is generally carried out in two steps: the formation of the cation followed by an anion exchange step called anionic metathesis [24].

The properties of the ILs are also very large [19,25,26]: low vapor tension, high decomposition temperature, high solubilizing power, and a wide variety of the structures that allow different polarities [27,28].

The ILs are commonly used in the domain of (bio)catalysis [18,29], organic synthesis [30,31], electrochemistry [32], or extraction [33] as recently described in the appropriate reviews [34–36]. Catalyses (hydrogenation, oxidation, Pd-catalyzed couplings, Friedel-Crafts, Diels-Alder, etc.) in ILs are also largely developed [37] and the use of these new solvents generally led to higher kinetics and better selectivity's.

However, the low biodegradability and the toxicity of the ionic liquids, in general, led the scientific community to reduce their use or to find other greener alternatives by using renewable resources as starting materials such as acids, amino acids, amino alcohols, and sugars which could improve the green character of ILs [38–42]. Catalyses (hydrogenation, heck reactions) have also been performed in such ILs [43–45].

The Michael reaction is a conjugated nucleophilic addition between a nucleophile and an α,β-unsaturated electrophile. The nucleophiles are thiols, anions, or amines, and the electrophiles are alkenes, alkynes in α and β position of carbonyls, amides, or nitriles. Recently, Yadav et al. reviewed innovative catalysis in Michael addition reactions, including the use of organocatalysts and heterogeneous processes [46]. Gu et al. in 2019 [47] also described the role of functionalized quaternary ammonium salt ionic liquids (FQAILs) as an economical and efficient catalyst for the synthesis of glycerol carbonate from glycerol and dimethyl carbonate.

In 2003, Salunkhe and coll. studied the use of hydrophobic or hydrophilic ionic liquids ([bmim]PF$_6$, [bmim]BF$_4$, [BuPy]BF$_4$) as solvents in the conjugated addition reaction of dimethyl malonate to chalcone in the presence of a quaternary ammonium salt derived from quinine (Scheme 1) [48]. This last compound used as a chiral phase transfer agent led to moderate ee's.

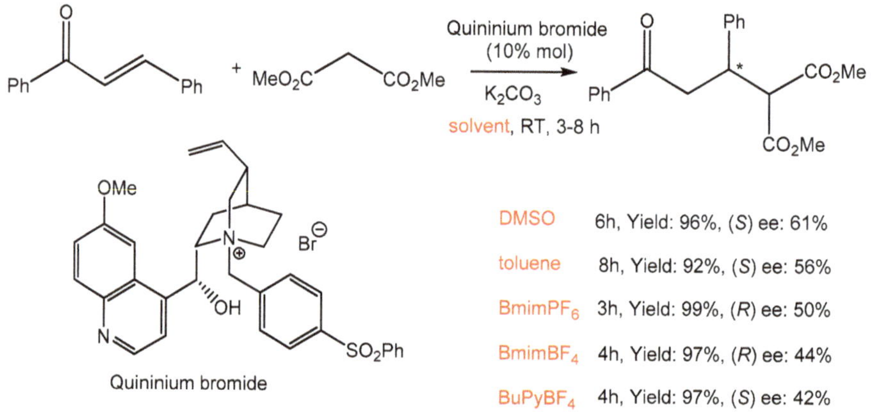

Scheme 1. Enantioselective reactions in chiral ionic liquids [48].

The yields were quantitative and the ee up to 56%. Next, in 2005, Ranu and coll. performed the Michael addition of activated methylene compounds on ketones, esters, and conjugated nitriles in the presence of the ionic liquid [Bmim]OH (Scheme 2). This one was used as a solvent and catalyst for the reaction. The reaction products were obtained with good yields; furthermore, the compound issued from the double addition could also be easily obtained in one step, which was surprising when compared to conventional methods [49].

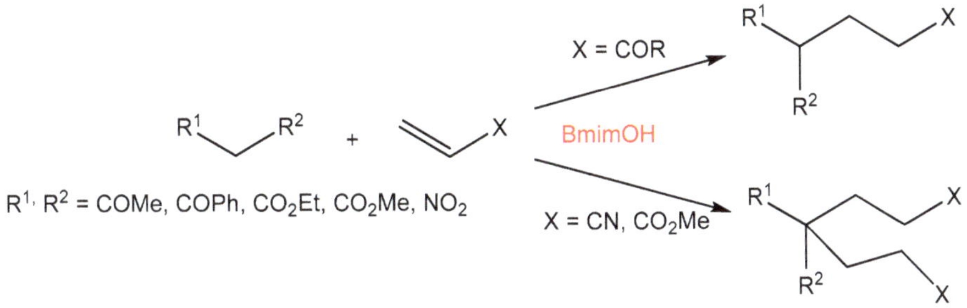

Scheme 2. Michael addition of activated methylene compounds in BmimOH [49].

In 2005, Rao and Jothilingam described how microwaves could drastically decrease the reaction time for the Michael addition of active methylene compounds to α,β-unsaturated carbonyl compounds in the presence of a large excess of K_2CO_3 as a base [50]. The reaction took place on the surface of potassium carbonate under microwave irradiation (450 W) and led to good to high yields (50% to 90%) in short times (5–10 min).

As previously mentioned, we prepared some biobased ionic liquids with natural carboxylates and used these in hydrogenation and Heck couplings [38–45]. Among these ILs, we prepared tetrabutylphosphonium and tetrabutylammonium prolinate or hydroxyprolinate. This paper aims to show how these proline-based ionic liquids could be used as solvents and basic catalysts in a Michael model reaction. Furthermore, this study will be extended to cholinium cation, and we will demonstrate that microwaves could generate greener activation conditions for this reaction in ionic liquids.

2. Results and Discussion

Biosourced ionic liquids were prepared using an acid-base reaction between the tetrabutyl-ammonium or -phosphonium hydroxides (TBAOH and TBPOH, respectively) or cholinium hydroxide ChOH with natural chiral amino acids (S-proline, R-proline, and trans-4-hydroxy-S-proline) (Scheme 3). The synthetic methodology was inspired by Ohno's work [51] in 2005 and previously reported works for some phosphonium or ammonium derivatives [38–45].

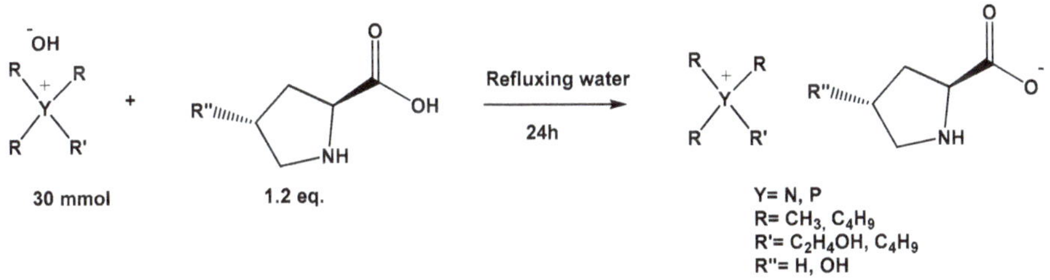

Scheme 3. Synthesis of chiral ionic liquids.

Nine ionic liquids were prepared easily with high yields and purity (Table 1), and the cholinium-based ionic liquids were synthesized following similar procedures previously used with ammonium or phosphonium-based ionic liquids.

The decomposition temperature values obtained by TGA confirmed the good thermal stability of all synthesized ionic liquids (Tdec ≥ 198 °C). We noticed that the thermal stability of ionic liquids was more dependent on the nature of the cation, particularly with phosphonium cations, as described in the literature [41]. Indeed, the ionic liquid TBP^+ prolinate was the most stable. However, we could remark that the cholinium based ionic liquids were more stable than the ammonium-based ones when prolinate or hydroxyprolinate were present as a counter anion (Figure 1); this observation is probably due to the formation of hydrogen bonds which could explain this slightly higher thermal stability [52].

Table 1. Synthetized ionic liquids.

Acid	Cation	Ionic Liquid	Yield (%)	Dec. Temperature Tdec (°C)	Viscosity (cp) at 60 °C	Ref.
S-proline	TBP+	1	78	340	75.33	[41]
	Ch+	2	95	230	98.36	This work
	TBA+	3	85	200	68.04	[40,42]
R-proline	TBP+	4	89	//	76.75	[41]
	Ch+	5	77	//	98.65	This work
	TBA+	6	78	//	68.18	[40,42]
Trans-4-hydroxy-S-proline	TBP+	7	93	326	245.6	[41]
	Ch+	8	82	222	387.2	This work
	TBA+	9	84	198	188.2	[40,42]

TBA: Tetrabutylammonium TBP: Tetrabutylphosphonium Ch: Cholinium

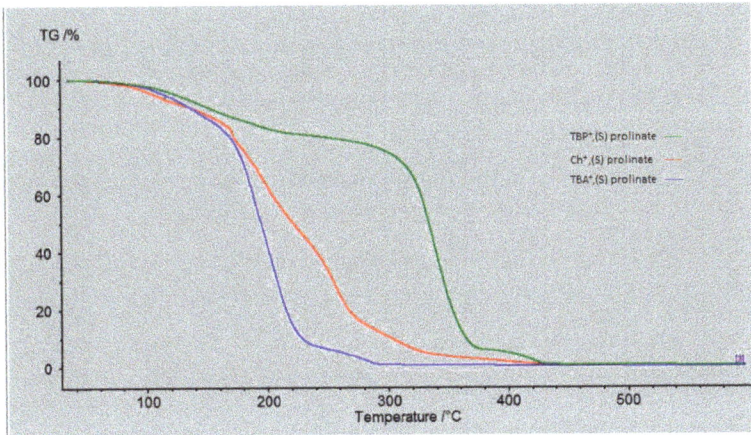

Figure 1. Thermal decomposition of ionic liquids 1–3.

The viscosities also show the importance of the anion (Table 1). In fact, with (R)-prolinate or (S)-prolinate as counter-anions, the viscosities at 60 °C are relatively low between 70 and 100 cp. With the same cation, we observed that the viscosity of choline-based ionic liquids was higher, probably due to the OH group present in this cation involved in hydrogen bonds. The same conclusions could be deduced when *trans*-4-hydroxy-S-prolinate was used as anion following much higher viscosity values due, in this case, to the presence of two OH groups (Table 1).

These ionic liquids have been used as a catalyst and solvent in Michael reaction models involving dimethyl malonate and chalcone as starting materials. Preliminary results were obtained with a slight excess of dimethyl malonate towards the chalcone (1.2 eq.) at 50 °C for 24 h (Table 2). DMF was used as a unique solvent or was associated with ionic liquid in a lower amount to reduce the viscosity of the reacting medium.

Table 2. Michael reactions under thermic conditions.

Entry	Solvent	Conversion of 10 (%)	Isolated Yield in 12 (%)
1 [a]	DMF	13	8
2 [b]	DMF + (S) prolinate TBP	69	46
3 [b]	DMF + (S) prolinate Ch	49	28
4 [b]	DMF + (S) prolinate TBA	46	33
5 [b]	DMF + (R) prolinate TBP	54	41
6 [b]	DMF + (R) prolinate Ch	58	49
7 [b]	DMF + (R) prolinate TBA	72	64
8 [b]	DMF + *trans*-4-hydroxy-S-prolinate TBP	24	13
9 [b]	DMF + *trans*-4-hydroxy-S-prolinate Ch	76	68
10 [b]	DMF + *trans*-4-hydroxy-S-prolinate TBA	47	29

[a] conditions: 1 eq. chalcone, 1.2 eq dimethylmalonate, DMF (10 mL), K_2CO_3 (1.5 eq.), 50 °C, 24 h; [b] conditions: 1 eq. chalcone, 1.2 eq dimethylmalonate, DMF (2 mL), 1.2 eq. ionic liquid, 50 °C, 24 h.

The results obtained proved the positive impact of the use of ionic liquids for this Michael reaction while higher conversions of chalcone were observed (Table 2, entries 2–7, 9 and 10 vs. entry 1). Nevertheless, the conversions were not complete, and further experiments were conducted using an excess of dimethylmalonate. Indeed, with 3 or 4 equivalents of dimethylmalonate, the conversions obtained are much higher; furthermore, the reaction time could also be reduced to 4 h (Table 3, entry 4).

Table 3. Michael addition with an excess of dimethylmalonate (conditions: 1 eq. chalcone, 1.2 eq. (S) prolinate TBP$^+$, 2 mL DMF, 50 °C).

Entry	Equivalent of 11	Time (h)	Conversion of 10 (%)
1	1.2	24	69
2	3	24	89
3	4	24	84
4	4	4	97

Based on these results, Michael's reaction was carried out using the various synthesized ILs as solvents (without the addition of DMF) under microwaves activation.

The conditions were adjusted after several experiments; at the beginning, the reactions were carried out with a power of 220 W leading to significant temperature variations depending on the ionic liquid. This aspect was unfavorable to the stability of the reactants or products. In addition, we observed that the presence of water in the ILs could also influence the conversion of the chalcone. Indeed, according to the results obtained in GC, the conversion rate of chalcone was low, and this was probably due to the low solubility of chalcone in water. Therefore, it was important to work with ionic liquids containing as little water as possible. Michael reactions were performed under microwaves during 45 min with a power limited to 100 W, and the ionic liquids were dried under vacuum for 4 h before use.

The reaction conducted under microwaves in DMF with the addition of K_2CO_3 as a base led to a high conversion of chalcone over 45 min (Table 4, entry 1). Next, the DMF and the base could be substituted by the ionic liquid, which played further roles (Table 4, entries 2–10). Good to very good conversions of chalcone (60 to 85%) were obtained again in 45 min, proving that the microwaves could decrease the activation energy [53,54]. According to the literature concerning the Michael addition [55], we proposed a mechanism that is described in Scheme 4.

The anion plays the role of the base towards the dimethyl malonate, the Michael's donor, which subsequently reacts with the chalcone (the Michael's acceptor) to produce a stabilized enolate. The latter will recover the proton and regenerate the anion from the ionic liquid to produce the coupling adduct. The pKa of proline (10.64) and the hydroxyproline (9.65) led to suitable basic media when ionic liquids 1–9 were used. The anions of the ionic liquids played the role of the base. The whole ionic liquids played the crucial role of phase transfer agents as explained by Ceccarelli and et al. in 2006 [56].

Concerning the enantioselectivity of the reaction, all coupling compounds were racemic. The chiral anion of the ionic liquids seemed not to influence the protonation of the enolate species. The presence of residual water in the ILs could also explain the lack of enantioselectivity. The size of our anions was also relatively small contrary to the PCT agents used by Mahajan and al., leading to good enantiomeric excesses in similar Michael's additions [57].

Table 4. Michael's reaction under microwaves.

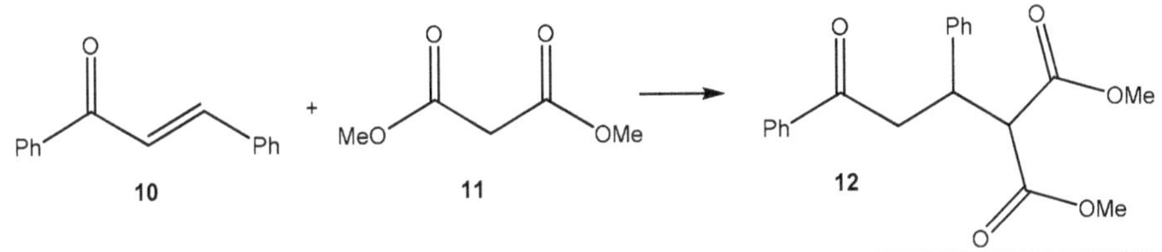

Entry	Solvent	Conversion of 10 (%)	Isolated Yield in 12 (%)
1 [a]	Dimethylformamide (DMF)	95	79
2 [b]	(S) prolinate TBP	69	48
3 [b]	(S) prolinate Ch	74	62
4 [b]	(S) prolinate TBA	90	72
5 [b]	(R) prolinate TBP	71	59
6 [b]	(R) prolinate Ch	88	65
7 [b]	(R) prolinate TBA	80	61
8 [b]	trans-4-hydroxy-S-prolinate TBP	74	50
9 [b]	trans-4-hydroxy-S-prolinate Ch	60	47
10 [b]	trans-4-hydroxy-S-prolinate TBA	95	81

[a] conditions: 1 eq. chalcone, 4 eq. dimethylmalonate, DMF (10 mL), K_2CO_3(1.5 eq.),.100W, 45 min; [b] conditions: 1 eq. chalcone, 4 eq dimethylmalonate, 1.2 eq. ionic liquid, 100W, 45 min.

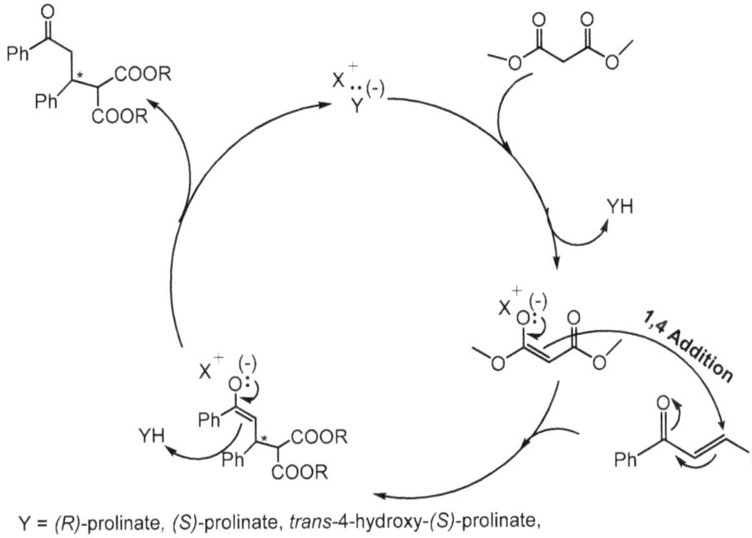

Y = (R)-prolinate, (S)-prolinate, trans-4-hydroxy-(S)-prolinate,
X = TBA, TBP, Choline

Scheme 4. A plausible mechanism for Michael's addition in ionic liquids.

The final work was to prove the possibility of recycling our ILs in the studied Michael's addition. One of the advantages of ionic liquids is the possibility to recycle them. We studied this possibility with the (S)-prolinate TBA. We chose this ionic liquid and not the *trans*-4-hydroxy-S-prolinate TBA, which furnished the best conversions because the yields of the first run were high (90%), and secondly, the price of the counter anion (S-prolinate) was lower than the *trans*-4-hydroxy-S-prolinate one. The procedure is described in Scheme 5.

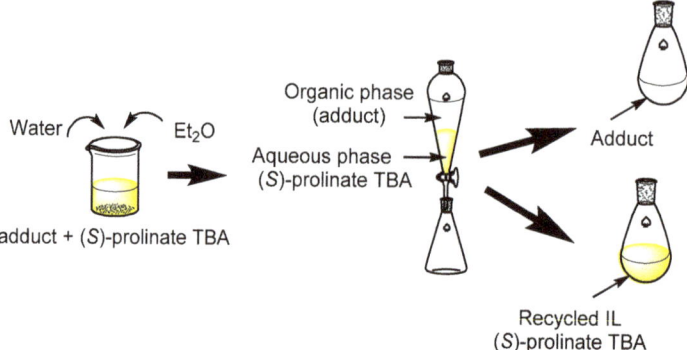

Scheme 5. Recycling procedure for the Michael addition in (S)-prolinate TBA.

The chalcone conversions for the three first cycles were good (81 to 94%). Then, we observed a loss of activity of the reacting medium, probably due to the loss of the anion, which plays a basic role in the Michael's reaction mechanism and led to the amino acid by protonation. This hypothesis has been confirmed by NMR spectrum of the regenerated IL where a slight modification of its structure has been observed (See Table 5).

Table 5. Chalcone conversion during 5 runs.

Entry	Run	Chalcone Conversion (%)
1	1	94
2	2	90
3	3	81
4	4	66
5	5	61

conditions: 1 eq. chalcone, 4 eq. dimethylmalonate, 1.2 eq. (S)-prolinate TBA, 100 W, 45 min.

3. Experimental

All reagents were commercially available and used as received. Solvents were dried and distilled under argon before use (CH_2Cl_2 over $CaCl_2$ and THF over sodium/benzophenone) and stored over molecular sieves. ^1H and ^{13}C NMR spectra were recorded on an AC 250 Bruker in CD_3OD for ^1H and ^{13}C spectra. The infrared spectra were recorded with a Spectrafile IR™ Plus MIDAC. Chromatography was carried out on an SDS Silica 60 (40–63 µm), Art 2050044 (flash-chromatography), or silica 60 F_{254} (TLC plates).

The GC analyses were recorded on a Hewlett Packard 6890 Series. The conditions used are specified in this Table 6.

Table 6. The microwaves oven is a monomode CEM DISCOVERS-CLASS.

Conditions	Columns	
	B-DM (Chiral)	TR-1
Initial Temperature (°C)	60	60
Initial time (min)	5	5
Gradient (°C/min)	10	10
Final Temperature (°C)	220	220
Final time (min)	20	20

3.1. Synthesis of Phosphonium Based Ionic Liquids-General Procedure [21d]

In a 100 mL two-neck round-bottom flask, one equivalent of a 40% wt aqueous solution of tetrabutylphosphonium hydroxide TBP$^+$OH$^-$ (30 mmol, 21 mL) and 1.2 equivalents of (R), (S)-proline or *trans*-4-hydroxy-S-proline (36 mmol) previously dissolved in 40 mL distilled water were stirred at 100 °C during 24 h. After cooling, the solvent was evaporated, and the resulting mixture was washed with ethyl acetate (4 × 80 mL) to remove the small excess of amine. Finally, the aqueous phase was evaporated.

tetrabutylphosphonium (S)-prolinate

3.1.1. General Procedure with 4.14 S-proline (36 mmol). Yield: 78%

δ_H (250.1 MHz; D$_2$O) ppm: 0.89 (t, J = 7.3 Hz, 12H, H$_4$); 1.31 (m, 16H, H$_2$ and H$_3$); 1.6 (m, 3H, H$_{2'}$ and H$_{3'a}$); 2 (m, 1H, H$_{3'b}$); 2.48 (m, 8H, H$_1$); 2.7 (m, 1H, H$_{1'a}$); 2.95 (m, 1H, H$_{1'b}$); 3.4 (m, 1H, H$_{4'}$). δ_C (62.5 MHz; D$_2$O) ppm: 13.4 (C4); 18.4 (C1); 23.6 (C2); 23.9 (C1); 25.5 (C2'); 30.9 (C3'); 46.7 (C1'); 62 (C4'); 177.9 (C = O). IR: ν (cm^{-1}) 1746 (C = O). Analysis: calculated for C$_{21}$H$_{44}$NO$_2$P: C 67.52; H 11.87; N 3.75%. Found: C 66.98; H 11.38; N 3.42%.

tetrabutylphosphonium (R)-prolinate

3.1.2. General Procedure with 4.14 g (R)-proline (36 mmol). Yield: 89%

δ_H (250.1 MHz; D$_2$O) ppm: 0.89 (t, J = 7.3 Hz, 12H, H$_4$); 1.32 (m, 16H, H$_2$ and H$_3$); 1.6 (m, 3H, H$_{2'}$ and H$_{3'a}$); 2 (m, 1H, H$_{3'b}$); 2.48 (m, 8H, H$_1$); 2.7 (m, 1H, H$_{1'a}$); 2.95 (m, 1H, H$_{1'b}$); 3.4 (m, 1H, H$_{4'}$). δ_C (62.5 MHz; D$_2$O) ppm: 13.4 (C4); 18.4 (C1); 23.6 (C2); 23.9 (C1); 25.5 (C2'); 30.9 (C3'); 46.7 (C1'); 62 (C4'); 177.9 (C=O). IR: ν (cm^{-1}) 1746 (C = O). Analysis: calculated for C$_{21}$H$_{44}$NO$_2$P: C 67.52; H 11.87; N 3.75%. Found: C 67.0; H 11.48; N 3.52%.

tetrabutylphosphonium
trans-4-hydroxy-(S)-prolinate

3.1.3. General Procedure with 4.72 g Trans-4-hydroxy-(S)-prolinate (36 mmol). Yield: 93%

δ_H (250.1 MHz; D$_2$O) ppm: 0.89 (t, J = 7.3 Hz, 12H, H$_4$); 1.31 ppm (m, 16H, H$_2$ and H$_3$); 1.50 (m, 1H, H3'b); 1.71 (m, 1H, H$_{3'a}$); 2.4 (m, 1H, H1'a); 2.92 (dd, J = 12.3; 5.1 Hz, H$_{1'b}$); 3.21 (m, 8H, H$_1$); 3.41 (m, 1H, H$_{2'}$); 4.00 (m, 1H, H$_{4'}$). δ_C (62.5 MHz; D$_2$O) ppm: 13.4 (C4); 19.5 (C3); 24.4 (C2); 41.3 (C3'); 55.4 (C1'); 58.7 (C1); 61.9 (C4'); 72.2 (C2'); 177.7 (C=O). IR: ν (cm^{-1}) 1746 (C = O). Analysis: calculated for: C$_{21}$H$_{44}$NO$_3$P: C 64.75; H 11.38; N 3.60%. Found: C 64.39; H 10.98; N 3.32%.

3.2. Synthesis of Ammonium Based Ionic Liquids-General Procedure [21c,e]

In a 100 mL two-neck round-bottom flask, one equivalent of a 40% wt aqueous solution of tetrabutylammonium hydroxide TBA$^+$OH$^-$ (30 mmol, 19.6 mL) and 1.2 equivalents of (R), (S)-proline or *trans*-4-hydroxy-S-proline (36 mmol) previously dissolved in 40 mL distilled water were stirred at 100 °C during 24 h. After cooling, the solvent was evaporated, and the resulting mixture was washed with ethyl acetate (4 × 80 mL) to remove the small excess of amine. Finally, the aqueous phase was evaporated.

Tetrabutylammonium (S)-prolinate

3.2.1. General Procedure with 4.14 S-proline (36 mmol). Yield: 85%

δ_H (250.1 MHz; D$_2$O) ppm: 0.89 (t, J = 7.3 Hz, 12H, H$_4$); 1.32 (m, 8H, H$_3$); 1.55 (m, 8H, H$_2$); 1.60 (m, 3H, H$_{2'}$ and H$_{3'b}$); 2.00 (m, 1H, H$_{3'a}$); 2.70 (m, 1H, H$_{1'a}$); 2.9 (m, 9H, H$_1$ and H$_{1'b}$); 3.40 (m, 1H, H$_{4'}$). δ_C (62.5 MHz; D$_2$O) ppm: 14.1 (C4); 20.6 (C3); 24.7 (C2); 26.7 (C2'); 32.2 (C3'); 47.6 (C1'); 59.4 (C1); 63.04 (C4'); 180.9 (C=O). IR: ν (cm^{-1}) 1743 (C = O). Analysis: calculated for: C$_{21}$H$_{44}$N$_2$O$_2$: C 70.73; H 12.44; N 7.86%. Found: C 70.29; H 11.98; N 7.37%.

Tetrabutylammonium (R)-prolinate

3.2.2. General Procedure with 4.14 R-proline (36 mmol). Yield: 78%

δ_H (250.1 MHz; D$_2$O) ppm: 0.89 (t, J = 7.3 Hz, 12H, H$_4$); 1.32 (m, 8H, H$_3$); 1.55 (m, 8H, H$_2$); 1.60 (m, 3H, H$_{2'}$ and H$_{3'b}$); 2.00 (m, 1H, H$_{3'a}$); 2.70 (m, 1H, H$_{1'a}$); 2.9 (m, 9H, H$_1$ and H$_{1'b}$); 3.40 (m, 1H, H$_{4'}$).

tetrabutylammonium trans-4-hydroxy-(S)-prolinate

3.2.3. General Procedure with 4.72 Trans-4-hydroxy-(S)-prolinate (36 mmol). Yield: 85%

δ_H (250.1 MHz; D$_2$O) ppm: 0.89 (t, J = 7.3 Hz, 12H, H$_4$); 1.31 ppm (m, 16H, H$_2$ and H$_3$); 1.5 (m, 1H, H3'b); 1.7 (m, 1H, H$_{3'a}$); 2.4 (m, 1H, H1'a); 2.9 (dd, J = 12.2; 5.1 Hz, H$_{1'b}$); 3.20 (m, 8H, H$_1$); 3.41 (m, 1H, H$_{2'}$); 4.00 (m, 1H, H$_{4'}$). δ_C (62.5 MHz; D$_2$O) ppm: 13.5 (C4); 19.9 (C3); 24.0 (C2); 41.3 (C3'); 55.4 (C1'); 58.7 (C1); 61.9 (C4'); 72.2 (C2'); 177.7 (C=O). IR: ν (cm^{-1}) 1741 (C = O). Analysis: calculated for: C$_{21}$H$_{44}$N$_2$O$_3$: C 67.70; H 11.90; N 7.52%. Found: C 67.31; H 11.48; N 7.23%.

3.3. Synthesis of Cholinium Based Ionic Liquid

In a 100 mL Bicol, a mixture of 46% wt aqueous solution of cholinium hydroxide (1 eq. 30 mmol, 7.37 mL) and proline (1.2 eq., 36 mmol, 14 g) previously dissolved in 40 mL distilled water was stirred under reflux for 24 h at 100 °C. After cooling, the mixture was evaporated, washed (4 × 80 mL) with ethyl acetate, and finally, the aqueous phase was evaporated.

cholinium-(S)-prolinate
Yield: 93%
Decomposition temperature: 230 °C

δ$_H$ (250.1 MHz; D$_2$O) ppm: 1.60 (m, 3H, H$_{2'}$ and H$_{3'a}$); 2.00 (m, 1H, H$_{3'b}$); 2.70 (m, 1H, H1'a); 2.80 (m, 1H, H$_{1'b}$); 3.11 (s, 9H, H$_1$); 3.30 (t, 2H, H$_2$); 3.45 (t, 1H, H$_{4'}$); 3.80 (t, 2H, H$_3$). δ$_C$ (62.5 MHz; D$_2$O) ppm: (62.5 MHz; D$_2$O): 22.8 (C2'); 32.3 (C3'); 46.5 (C1'); 55.7 (C1); 64.2 (C3); 67.5 (C4'); 70.2 (C2); 180.4 (C=O). IR: ν (cm^{-1}) 1741 (C = O). Analysis: calculated for: C$_{10}$H$_{22}$N$_2$O$_3$: C 55.02; H 10.16; N 12.83%. Found: C 54.78; H 9.78; N 12.47%.

cholinium-(R)-prolinate
Yield: 74%
Decomposition temperature: 230 °C

δ$_H$ (250.1 MHz; D$_2$O) ppm: 1, 60 (m, 3H, H$_{2'}$ and H$_{3'b}$); 2.00 (m, 1H, H$_{3'b}$); 2, 70 ppm (m, 1H, H$_{1'a}$); 2.80 (m, 1H, H$_{1'b}$); 3.10 (s, 9H, H$_1$); 3.30 (t, 2H, H$_2$); 3.45 (s, 1H, H$_{4'}$); 3.80 ppm (t, 2H, H$_3$). δ$_C$ (62.5 MHz; D$_2$O) ppm: 22.6 (C2'); 32.3 (C3'); 46.7 (C1'); 55.4 (C1); 64.9 (C3); 67.3 (C4'); 70.7 (C2); 180.3 (C=O). Analysis: calculated for: C$_{10}$H$_{22}$N$_2$O$_3$: C 55.02; H 10.16; N 12.83%. Found: C 54.89; H 9.78; N 12.36%.

cholinium-*trans*-4-hydroxy-(S)-prolinate
Yield: 82%
Decomposition temperature: 222 °C

δ$_H$ (250.1 MHz; D$_2$O) ppm: 1.50 (m, 1H, H$_{3'b}$); 1.70 (m, 1H, H$_{3'b}$); 2,40 ppm (m, 1H, H$_{1'a}$); 2.90 (dd, J = 12.2 Hz, 5.1 Hz,1H, H$_{1'b}$); 3.10 (s, 9H, H$_1$); 3.40 (m, 3H, H$_{2'}$); 3.45 (s, 2H, H$_3$); 3.80 ppm (t, J = 3.43 Hz, 2H, H$_2$), 4 (m, 1H, H$_{4'}$). δ$_C$ (62.5 MHz; D$_2$O) ppm: 41.3 (C3'); 54.6(C1'); 55.4 (C1); 61.9 (C4'); 64.9 (C3); 70.2 (C2); 72.2 (C2'); 177.7 (C=O). IR: ν (cm^{-1}) 1738 (C = O). Analysis: calculated for: C$_{10}$H$_{22}$N$_2$O$_4$: C 51.26; H 9.46; N 11.96%. Found: C 51.49; H 9.58; N 11.66%.

3.4. Michael Addition Procedures

3.4.1. General Procedure without Ionic Liquids

In a Schlenk tube, the chalcone (1 eq., 2 mmol, 0.42 g), the dimethyl malonate (1.2 eq or 4 eq.) and K$_2$CO$_3$ (1.2 eq.) were mixed in DMF (10 mL). The reaction mixture was stirred for 24 h at 50 °C. The reaction was stopped by adding 15 mL of ice water with the formation of a white precipitate that corresponds to the coupling product. This precipitate was dissolved in diethyl ether. Extractions with diethyl ether were carried out (3 × 60 mL). The organic phases were then dried over sodium sulfate and then evaporated. The resulting compound was purified by chromatography (silica and eluent: Petroleum ether/Ethyl acetate (7/3)).

3.4.2. General Procedure in Ionic Liquids

In a Schlenk tube, 1.2 eq. of ionic liquid (2.4 mmol) was introduced and placed under vacuum for 10 min. Then at atmospheric pressure, 1 eq. of chalcone (2 mmol, 0.42 g) and an excess of dimethyl malonate (4 eq.) were dissolved in DMF (2 mL) in order to fluidify the mixture. The reaction mixture was stirred for 24 h at 50 °C and in the presence of argon. The reaction was stopped by adding 15 mL of ice water. We observed the formation of a white precipitate that corresponded to our product. The latter was dissolved by adding diethyl ether. Extractions with diethyl ether were carried out (3 × 60 mL). The organic phases were then dried over sodium sulfate and then evaporated. The resulting compound was purified by chromatography (silica and eluent: Petroleum ether / Ethyl acetate (7/3)). Finally, the aqueous phase was also evaporated to recover the ionic liquid.

2-(3-oxo-1,3-diphenylpropyl) dimethylmalonate [58]
$C_{20}H_{20}O_5$
White Powder

δ_H (500 MHz; CDCl$_3$) ppm: 3.50 (m, 4H, H$_{1b}$ and OMe); 3.55 (dd, 1H, H$_{1a}$); 3.75 (s, 3H, OMe); 3.85 (d, 1H, H$_3$); 4.25 (m, 1H, H$_2$); 7.15 (m, 1H, H$_{arom}$); 7.25 (m, 4H, H$_{arom}$); 7.43 (t, 2H, H$_{arom}$); 7.50 (m, 1H, H$_{arom}$); 7.85 (m, 2H, H$_{arom}$). IR: ν (cm^{-1}) 1735 (C = O$_{Ester}$), 1715 (C = O$_{Ketone}$).

3.4.3. General Procedure in Ionic Liquids under Microwaves

In a balloon, 1.2 eq. of ionic liquid (2.4 mmol) was introduced and placed under vacuum for 10 min. 1 eq. of chalcone (2 mmol, 0.42 g) and an excess of dimethyl malonate (4 eq.) was added, and the mixture was stirred for 45 min under 100 W. The extraction of the Michael's adduct and the recycling of the ILs was as previously described for the reaction under thermic conditions.

4. Conclusions

Nine proline-based ionic liquids were synthesized in water with high yields and at the best sustainable chemistry processes. Michael reactions were performed using these synthesized ionic liquids as both solvent and base with improved yields by increasing the amount of dimethylmalonate. Furthermore, the use of microwaves activation reduced the reaction time drastically. Therefore, the combined use of recyclable biosourced ionic liquids and microwaves represents an alternative to achieve greener synthesis processes than commonly used methods. Eco- and cyto-toxicological analyses are in progress to support the use of such biosourced ionic solvents, and the Michael addition will be tested with more sterically enhanced compounds in order to induce good ee values.

Author Contributions: K.B. and K.A. have realized the experimental work (synthesis and analysis). J.-P.M. and S.B. have supervised this work and J.-P.M. realized also some preliminary and crucial works on the synthesis of ILs. All authors have read and agreed to the published version of the manuscript.

Funding: This research was funded by Région Champagne Ardenne Excellence Framework [Amisolver program]. The APC was funded by the ICMR.

Acknowledgments: This work was supported by the Amisolver program (Excellence Framework). We are grateful to the Public Authorities of Champagne-Ardenne and FEDER for material funds and post-doctoral fellowships to J.-P.M and to S. Hayouni and M. Mention for preliminary studies.

Conflicts of Interest: The authors declare no conflict of interest.

References

1. Prat, D.; Hayler, J.; Wells, A. A survey of solvent selection guides. *Green Chem.* **2014**, *16*, 4546–4551. [CrossRef]
2. Welton, T. Solvents and sustainable chemistry. *Proc. Math. Phys. Eng. Sci.* **2015**, *471*, 20150502. [CrossRef] [PubMed]
3. Li, C.-J.; Chan, T.-K. *Organic Reactions in Aqueous Media*; Wiley: New York, NY, USA, 1997.
4. Lindström, U.M. (Ed.) *Organic Reactions in Water: Principles, Strategies and Applications*; Blackwell: Oxford, UK, 2007.
5. Li, C.-J. (Ed.) *Handbook of Green Chemistry, Green Solvents, Vol. 5, Reactions in Water*; Wiley-VCH: Weinheim, Germany, 2010.
6. Akiya, N.; Savage, P.E. Roles of water for chemical reactions in high-temperature water. *Chem. Rev.* **2002**, *102*, 2725–2750. [CrossRef]
7. Simon, M.-O.; Lee, C.-J. Green chemistry oriented organic synthesis in water. *Chem. Soc. Rev.* **2002**, *41*, 1415–1427. [CrossRef] [PubMed]
8. Hailes, H.C. Reaction solvent selection: The potential of water as a solvent for organic transformations. *Org. Process Res. Dev.* **2007**, *11*, 114–120. [CrossRef]
9. Dallinger, D.; Kappe, C.O. Microwave-assisted synthesis in water as solvent. *Chem. Rev.* **2007**, *107*, 2563–2591. [CrossRef]
10. Hyatt, J.A. Liquid and supercritical carbon dioxide as organic solvents. *J. Org. Chem.* **1984**, *49*, 5097–5101. [CrossRef]
11. Jessop, P.G.; Leitner, W. (Eds.) *Chemical Synthesis Using Supercritical Fluids*; Wiley: New York, NY, USA, 1999.
12. Leitner, W.; Jessop, P.G. (Eds.) *Handbook of Green Chemistry, Green Solvents, Vol. 4, Supercritical Solvents*; Wiley-VCH: Weinheim, Germany, 2010.
13. Beckman, E.J. Supercritical and near-critical CO_2 in green chemical synthesis and processing. *J. Supercrit. Fluids* **2004**, *28*, 121–191. [CrossRef]
14. Rayner, C.M. The potential of carbon dioxide in synthetic organic chemistry. *Org. Process Res. Dev.* **2007**, *11*, 121–132. [CrossRef]
15. Wasserscheid, P.; Welton, T. (Eds.) *Ionic Liquids in Synthesis*, 2nd ed.; Wiley-VCH: Weinheim, Germany, 2008.
16. Wasserscheid, P.; Stark, A. (Eds.) *Handbook of Green Chemistry, Green Solvents, Vol. 6, Ionic Liquids*; Wiley-VCH: Weinheim, Germany, 2010.
17. Welton, T. Room-temperature ionic liquids. Solvents for synthesis and catalysis. *Chem. Rev.* **1999**, *99*, 2071–2084. [CrossRef]
18. Parvulescu, V.I.; Hardacre, C. Catalysis in ionic liquids. *Chem. Rev.* **2007**, *107*, 2615–2665. [CrossRef] [PubMed]
19. Plechkova, N.V.; Seddon, K.R. Applications of ionic liquids in the chemical industry. *Chem. Soc. Rev.* **2008**, *37*, 123–150. [CrossRef] [PubMed]
20. Olivier-Bourbigou, H.; Magna, L.; Morvan, D. Ionic liquids and catalysis: Recent progress from knowledge to applications. *Appl. Catal. A* **2010**, *373*, 1–56. [CrossRef]
21. Hallett, J.P.; Welton, T. Room-temperature ionic liquids: Solvents for synthesis and catalysis. *Chem. Rev.* **2011**, *111*, 3508–3576. [CrossRef]
22. Marsh, K.N.; Boxall, J.A.; Lichtenthaler, R. Room temperature ionic liquids and their mixtures—A review. *Fluid Phase Equilib.* **2004**, *219*, 93–98. [CrossRef]
23. Coleman, D.; Gathergood, N. Biodegradation studies of ionic liquids. *Chem. Soc. Rev.* **2010**, *39*, 600–637. [CrossRef]
24. Available online: https://tel.archives-ouvertes.fr/tel-01683248 (accessed on 3 January 2018).
25. Verwey, E.J.W.; Overbeek, J.T.G. Theory of the Stability of Lyophobic Colloids. *J. Phys. Chem.* **1947**, *51*, 631–636. [CrossRef]

26. Handy, S.T. Greener Solvents: Room Temperature Ionic Liquids from Biorenewable Sources. *Chem. Eur. J.* **2003**, *9*, 2938–2944. [CrossRef]
27. Webb, P.B.; Sellin, M.F.; Kunene, T.E.; Williamson, S.; Slawin, A.M.Z.; Cole-Hamilton, D.J.J. Continuous Flow Hydroformylation of Alkenes in Supercritical Fluid–Ionic Liquid Biphasic Systems. *J. Am. Chem. Soc.* **2003**, *125*, 15577–15588. [CrossRef]
28. Baker, S.N.; Baker, G.A.; Bright, F.V. Temperature-dependent microscopic solvent properties of 'dry' and 'wet' 1-butyl-3-methylimidazolium hexafluorophosphate: Correlation with ET(30) and Kamlet–Taft polarity scales. *Green Chem.* **2002**, *4*, 165–169. [CrossRef]
29. Rantwijk, F.V.; Sheldon, R.A. Biocatalysis in Ionic Liquids. *Chem. Rev.* **2007**, *107*, 2757–2785. [CrossRef] [PubMed]
30. Lafuente, L.; Diaz, G.; Bravo, R.; Ponzinibbio, A. Efficient and Selective N-, S- and O-Acetylation in TEAA Ionic Liquid as Green Solvent. Applications in Synthetic Carbohydrate Chemistry. *Lett. Org. Chem.* **2016**, *13*, 195–200. [CrossRef]
31. Hajipour, A.R.; Rafiee, F. Recent Progress in Ionic Liquids and their Applications in Organic Synthesis. *Org. Prep. Proced. Int.* **2015**, *47*, 249–308. [CrossRef]
32. Plaquevent, J.C.; Genisson, Y.; Frédéric, F. Constantes Physico-Chimiques. *Tech. l'Ing.* **2008**, *51*, K1230/1–K1230/17.
33. Wei, D.; Ivaska, A. Applications of ionic liquids in electrochemical sensors. *Anal. Chim. Acta* **2008**, *607*, 126–135. [CrossRef]
34. Karimi, M.; Dadfarnia, S.; Mohammad Haji Shabani, A.; Tamaddon, F.; Azadi, D. Deep eutectic liquid organic salt as a new solvent for liquid-phase microextraction and its application in ligand less extraction and preconcentration of lead and cadmium in edible oils. *Talanta* **2015**, *144*, 648–654. [CrossRef]
35. Makanyire, T.; Sanchez-Segado, S.; Jha, A. Separation and recovery of critical metal ions using ionic liquids. *Adv. Manuf.* **2016**, *4*, 33–46. [CrossRef]
36. Kuzmina, O.; Bordes, E.; Schmauck, J.; Hunt, P.A.; Hallett, J.P.; Welton, T. Solubility of alkali metal halides in the ionic liquid [C4C1im][OTf]. *Phys. Chem. Chem. Phys.* **2016**, *18*, 16161–16168. [CrossRef]
37. Song, C.E. Enantioselective chemo- and bio-catalysis in ionic liquids. *Chem. Commun.* **2004**, 1033–1043. [CrossRef]
38. Hayouni, S.; Ferlin, N.; Bouquillon, S. *Hydrogenation Catalysis in Biobased Ionic Liquids, "New Advances in Hydrogenation Processes—Fundamentals and Applications"*; Ravanchi, M.T., Ed.; Intech Open Science: London, UK, 2017; p. 15. [CrossRef]
39. Ghavre, M.; Morrissey, S.; Gathergood, N. *Hydrogenation in Ionic Liquids, "Ionic Liquids: Applications and Perspectives"*; Kokorin, A., Ed.; Intech Open Science: London, UK, 2011; p. 15. [CrossRef]
40. Ferlin, N.; Courty, M.; Gatard, S.; Spulak, M.; Quilty, B.; Beadham, I.; Ghavre, M.; Haiß, A.; Kümmerer, K.; Gathergood, N.; et al. Biomass derived ionic liquids: Synthesis from natural organic acids, characterization, toxicity, biodegradation and use as solvents for catalytic hydrogenation processes. *Tetrahedron* **2013**, *69*, 6150–6161. [CrossRef]
41. Hayouni, S.; Robert, A.; Ferlin, N.; Amri, H.; Bouquillon, S. New biobased tetrabutylphosphonium ionic liquids: Synthesis, characterization and use as a solvent or co-solvent for mild and greener Pd-catalyzed hydrogenation processes. *RSC Adv.* **2016**, *6*, 113583–113595. [CrossRef]
42. Ferlin, N.; Courty, M.; Nguyen Van Nhien, A.; Gatard, S.; Pour, M.; Quilty, B.; Ghavre, M.; Haiß, A.; Kümmerer, K.; Gathergood, N.; et al. Tetrabutylammonium prolinate-based ionic liquids: A combined asymmetric catalysis, antimicrobial toxicity and biodegradation assessment. *RSC Adv.* **2013**, *3*, 26241–26251. [CrossRef]
43. Chiappe, C.; Cinzia, C.; Marra, A.; Mele, A. Synthesis and Applications of Ionic Liquids Derived from Natural Sugars. *Top. Curr. Chem.* **2010**, *295*, 177–195. [PubMed]
44. Handy, S.T.; Okello, M.; Dickenson, G. Solvents from Biorenewable Sources: Ionic Liquids Based on Fructose. *Org. Lett.* **2003**, *5*, 2513–2515. [CrossRef]
45. Hayouni, S.; Ferlin, N.; Bouquillon, S. High catalytic and recyclable systems for heck reactions in biosourced ionic liquids. *Mol. Catal.* **2017**, *437*, 121–129. [CrossRef]
46. Malkar, R.S.; Jadhav, A.L.; Yadav, G.D. Innovative catalysis in Michael addition reactions for C-X bond formation. *Mol. Catal.* **2020**, *485*, 110814. [CrossRef]

47. Elhaj, E.; Wang, H.; Gu, Y. Functionalized quaternary ammonium salt ionic liquids (FQAILs) as an economic and efficient catalyst for synthesis of glycerol carbonate from glycerol and dimethyl carbonate. *Mol. Catal.* **2019**, *468*, 19–28. [CrossRef]
48. Dere, R.T.; Pal, R.R.; Patil, P.S.; Salunkhe, M.M. Influence of ionic liquids on the phase transfer-catalyzed enantioselective Michael reaction. *Tetrahedron Lett.* **2003**, *44*, 5351–5353. [CrossRef]
49. Ranu, B.C.; Banerjee, S. Ionic Liquid as Catalyst and Reaction Medium. The Dramatic Influence of a Task-Specific Ionic Liquid, [bmIm]OH, in Michael Addition of Active Methylene Compounds to Conjugated Ketones, Carboxylic Esters, and Nitriles. *Org. Lett.* **2005**, *7*, 3049–3052. [CrossRef] [PubMed]
50. Surya Prakash Rao, H.; Jothilingam, S. Solvent-free microwave-mediated Michael addition reactions. *J. Chem. Sci.* **2005**, *117*, 323–328.
51. Fukumoto, K.; Yoshizawa, M.; Ohno, H. Room Temperature Ionic Liquids from 20 Natural Amino Acids. *J. Am. Chem. Soc.* **2005**, *127*, 2398–2399. [CrossRef] [PubMed]
52. Villanueva, M.; Coronas, A.; García, J.; Salgado, J. Thermal Stability of Ionic Liquids for Their Application as New Absorbents. *Ind. Eng. Chem. Res.* **2013**, *52*, 15718–15727. [CrossRef]
53. Perreux, L.; Loupy, A. A tentative rationalization of microwave effects in organic synthesis according to the reaction medium, and mechanistic considerations. *Tetrahedron* **2001**, *57*, 9199–9223. [CrossRef]
54. Loupy, A. *Microwaves in Organic Synthesis*; Wiley-VCH: Weinheim, Germany, 2002.
55. Tsogoeva, S.B. Recent Advances in Asymmetric Organocatalytic 1,4-Conjugate Additions. *Eur. J. Org. Chem.* **2007**, *11*, 1701–1716. [CrossRef]
56. Ceccarelli, R.; Insogna, S.; Bella, M. Organocatalytic regioselective Michael additions of cyclic enones via asymmetric phase transfer catalysis. *Org. Biomol. Chem.* **2006**, *4*, 4281–4284. [CrossRef]
57. Mahajan, D.P.; Godbole, H.M.; Singh, G.P.; Shenoy, G.G. Enantioselective Michael addition of malonic esters to benzalacetophenone by using chiral phase transfer catalysts derived from proline-mandelic acid/tartaric acid. *J. Chem. Sci.* **2019**, *131*, 1642–1645. [CrossRef]
58. Dongdong, C.; Guosheng, F.; Jiaxing, Z.; Hongyu, W.; Changwu, Z.; Gang, Z. Enantioselective Michael Addition of Malonates to Chalcone Derivatives Catalyzed by Dipeptide-derived Multifunctional Phosphonium Salts. *J. Org. Chem.* **2016**, *81*, 9973–9982.

© 2020 by the authors. Licensee MDPI, Basel, Switzerland. This article is an open access article distributed under the terms and conditions of the Creative Commons Attribution (CC BY) license (http://creativecommons.org/licenses/by/4.0/).

Review

Heterogeneous Catalysis with the Participation of Ionic Liquids

Olga Bartlewicz [1], Izabela Dąbek [2], Anna Szymańska [2] and Hieronim Maciejewski [1,2,*]

[1] Faculty of Chemistry, Adam Mickiewicz University, Uniwersytetu Poznańskiego 8, 61-614 Poznań, Poland; olga.bartlewicz@amu.edu.pl
[2] Adam Mickiewicz University Foundation, Poznań Science and Technology Park, Rubież 46, 61-612 Poznań, Poland; Izabela.Dabek@ppnt.poznan.pl (I.D.); Anna.Szymanska@ppnt.poznan.pl (A.S.)
* Correspondence: hieronim.maciejewski@amu.edu.pl

Received: 30 September 2020; Accepted: 20 October 2020; Published: 22 October 2020

Abstract: This mini-review briefly describes the recent progress in the design and development of catalysts based on the presence of ionic liquids. In particular, the focus was on heterogeneous systems (supported ionic liquid (IL) phase catalysts (SILPC), solid catalysts with ILs (SCILL), porous liquids), which due to the low amounts of ionic liquids needed for their production, eliminate basic problems observed in the case of the employment of ionic liquids in homogeneous systems, such as high price, high viscosity, and efficient isolation from post-reaction mixtures.

Keywords: heterogeneous catalyst; SILPC; SCILL; porous ionic liquids

1. Introduction

Over the last few decades, ionic liquids (ILs) have undoubtedly been among the most rapidly developing and has attracted great interest as chemical compounds. Due to their special properties and a wide spectrum of applications, ionic liquids were chosen a few years ago as one of 20 materials with the highest application potential [1]. Directions of their applications are very diverse—from solvents as an alternative to volatile organic solvents, through electrochemistry (as electrolytes and conducting polymers), chemical separation, liquid and solid carriers, surfactants, and stabilizers of nanoparticles to catalysis in a broad sense (as catalysts, co-catalysts, catalyst supports, and adjuvant substances). It is no wonder that, currently, the number of publications on ionic liquids exceed 100,000 (104,310 on 23 September 2020) and, each year, about 9000 new reports appear (Figure 1) [2].

Among them, there are many excellent review papers and books on the properties and applications of ionic liquids [3–7].

Such great and continued interest in ionic liquids result mainly from their unique properties, as well as their evolution and adjustment to the new requirements and expectations. One can see that, after the first report on ionic liquids [8], as much as 40 years had passed until the appearance of ionic liquids of practical significance. These were chloroaluminate-based liquids [9] and, later, also ILs based on other metallates, so-called first-generation (Figure 2), which could be applied to electrochemistry, as well as to the electrolytic deposition of metals.

Unfortunately, chloroaluminates due to their great moisture sensitivity and instability required special conditions to use them, which significantly limited areas of their applications. After the next 40 years, the second generation of ionic liquids appeared that were stable both in air and moisture environments [10], which resulted in the rapid development of research on the properties and new directions of the applications of ionic liquids.

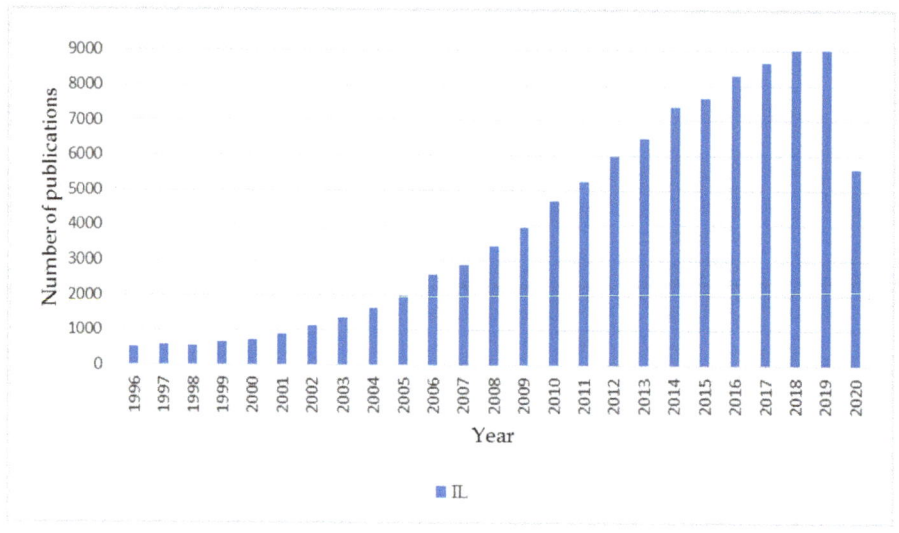

Figure 1. Number of publications on ionic liquids in the period 1996–2020.

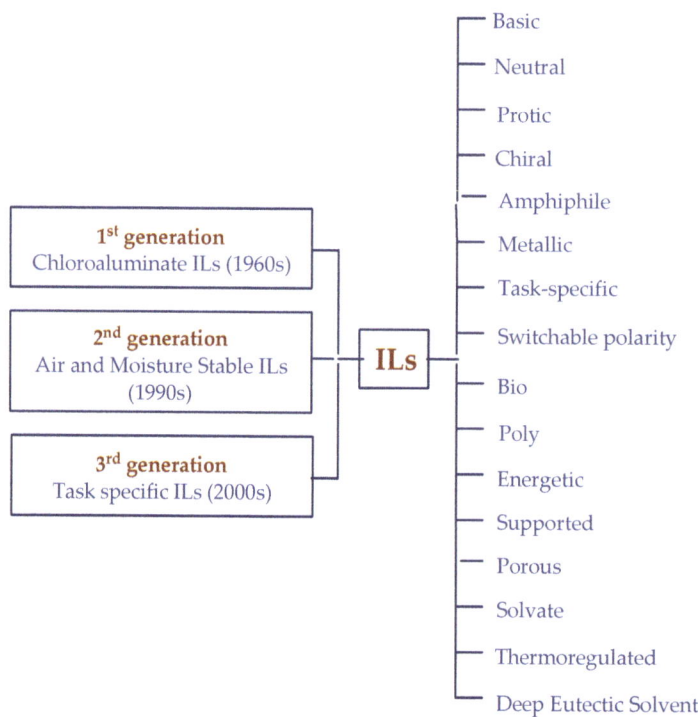

Figure 2. Structural evolution of the respective generations of ionic liquids.

The constantly developing knowledge leads to the design of new ILs that have specific properties and can be dedicated to particular tasks (task-specific ILs) [11]. This is the third generation of ionic liquids. The stages of the evolution of ILs have been very recently perfectly collected and presented by Welton [12]. However, the evolution of ionic liquids primarily finds its reflection in the permanent development and perfection of their properties, as well as in the adjustment to the required needs. At the end of the last century, the most frequently highlighted feature of ionic liquids was their low vapor pressure and high thermal stability, which should make them a green alternative to commonly used organic solvents. However, with the beginning of the new millennium, increasingly more critical opinions have been expressed that point to the toxicity, low biodegradability (or lack thereof), complicated or, in many cases, waste-generating synthesis, as well as high costs, of the manufacture of ILs [13–18]. All these voices made the term "green alternative" depreciate. However, due to the ability of ILs to evolve, after a short period of stagnation, publications of studies intended for the improvement of synthesis methods, the utilization of biodegradable and nontoxic raw materials, considerable reduction in waste, etc. began to appear [19]. One of the examples of such an approach is the application of ultrasound or microwave radiation in the synthesis of ionic liquids, which resulted in the acceleration of reactions, reduced energy inputs, and decrease in the reagent amount (compared to traditional methods), and especially, the mentioned radiation influences the selectivity of processes and waste reduction [20–23]. In addition, the approach has changed to the choice of raw materials for syntheses, which increasingly more often, are compounds originating from biorenewable sources. To this end, derivatives of choline and lignins are frequently employed as precursors of cations, whereas carboxylates of fatty acids, levulinates, and lactates are precursors of anions [24–30]. Moreover, studies are conducted on the properties of solvents and their effect on the environment and health safety, which enables their conscious and safe use [31]. Nevertheless, the most common-sense approach to the evaluation of the reasonable application of ionic liquids is to answer the question of whether they could ensure more sustainable production and cleanliness of technology compared to conventional reagents [32]. Irrespective of this, one can say that the next stage of the adjustment of ionic liquids to the expectations and pro-environmental activities resulted in a further increase in the interest in ionic liquids and pointed to new directions of their applications. Due to the diversity of combinations of anions, cations, and the properties of ionic liquids, there exist many kinds of their classification [5–7,33,34]. An example of the structural evolution of particular generations of ionic liquids is presented in Figure 2. An intensive development can be observed among ionic liquids themselves, as well as among the compounds that are not altogether regarded as ionic liquids such as deep eutectic solvents [35–37] or solvate ionic liquids [38]. Currently, the diversity of the application directions is very large, beginning from separation techniques [39] through the production and storage of energy [40,41], different types of biotransformations [42,43], organocatalysis [44], and ending with pharmaceuticals [45] and space technology [46].

2. Ionic Liquids in Catalysis

Research on ionic liquids in catalysis is one of the most interesting areas of catalytic studies [24,47–49], as evidenced by the number of publications in this field, which over the last 5 years, has been increasing by over 400 reports per year (Figure 3).

Along with the quest for new alternative methods of the synthesis of ionic liquids with the employment of biodegradable or renewable raw materials, the development of biocatalysis based on ionic liquids began. Increasingly more interesting publications have appeared on this subject [27,30,50–52].

However, one of the more important and developed wide-scale directions of applications of ionic liquids in catalysis is their use for the immobilization of homogeneous catalysts and formation of biphasic systems, where one phase is made by an ionic liquid with a catalyst dissolved in it, and the other is made by reagents. However, besides unquestionable advantages such as easier catalyst separation from the reaction mixture (e.g., by decantation and extraction) and the possibility of catalyst recycling and its multiple use, several disadvantages also exist: (i) A possibility of catalyst leaching from the ionic liquid, which is caused by the polarity of some reagents; (ii) the absorption of moisture and various contaminants by ionic liquid, which can result in catalyst deactivation; (iii) high viscosity of ionic liquids is often an obstacle to achieving maximum product yield; (iv) a relatively large amount of ionic liquid is necessary, and this is economically unfavorable because ionic liquids are expensive. Moreover, low vapor pressure, which is an asset of ionic liquids in the case of their employment as solvents, can be a problem when the separation of ionic liquid from the post-reaction mixture is necessary. Difficulties in the separation can also be caused by good solubility of ionic liquids in various solvents. Additionally, the high viscosity of most ionic liquids creates problems with mass transfer, which can be a factor determining the course of the catalytic reaction. Various approaches to reduce these problems are applied, e.g., too high a viscosity can be decreased by the employment of an additional solvent (co-solvent). Recently, interest was aroused in microemulsions based on ionic liquids and aqueous or nonaqueous solvents [53] that combine the functions of ILs with a considerably lower viscosity. Another possibility to facilitate the separation of biphasic systems is the application of thermo-regulated (temperature-responsive) ILs, which solidify (together with the catalyst contained in them) when the temperature decreases, and this enables easy separation from reagents. After rewarming, the system homogenizes and shows activity in subsequent reaction cycles. Recently, a review paper appeared that contains a very good presentation of this group of ionic liquids and their applications in catalysis [54]. Currently, there has a common trend to obtain heterogenized systems that combine advantages of homogeneous catalysis (high activity) and heterogeneous catalysis (easiness of isolation, possibility of employing fixed bed reactors), which makes the catalytic process more cost-effective.

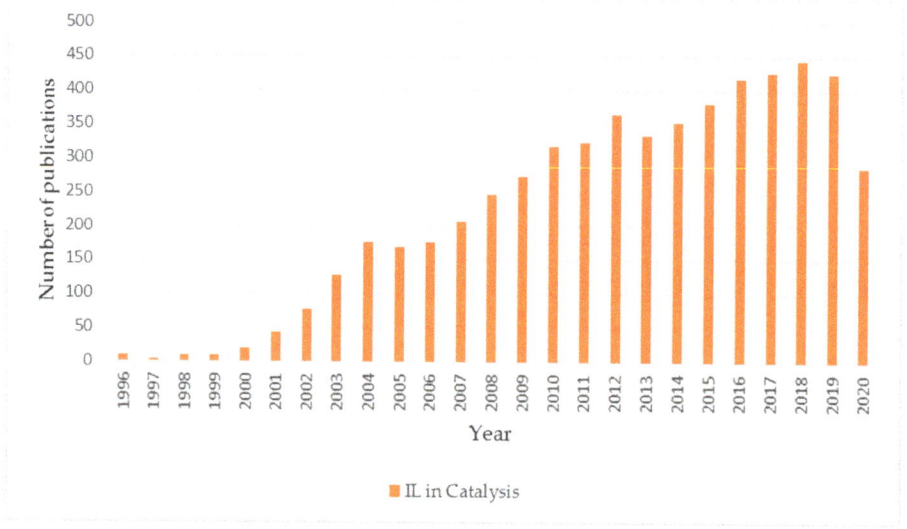

Figure 3. Number of publications on ionic liquids in catalysis in the period 1996–2020.

Such integration of homo- and heterogeneous catalysts is currently realized in several variants: Supported IL phase catalysts (SILPCs), solid catalysts with ILs (SCILLs), and supported ionic liquid catalysis (SILC). Recently, keen interest has also been seen in porous ionic liquids and metal–organic frameworks (MOFs).

2.1. Supported Ionic Liquid Phase Catalysts (SILPCs)

There is a great interest in materials of this type, which is reflected by the many publications, including review papers and books [55–58].

The SILPC materials consist of three different parts, the basis of which is a porous support, most frequently silica, alumina, or active carbon. The support surface is coated with a thin layer of an ionic liquid and the third component is a catalyst such as a nanoparticle or metal complex (Figure 4).

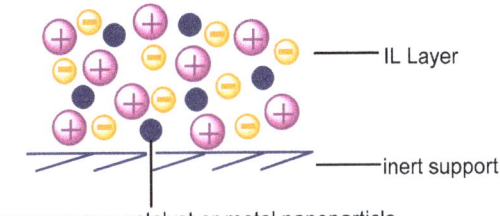

Figure 4. Model of the supported ionic liquid (IL) phase catalysts (SILPC) structure.

Different methods are used to produce materials of this type, the most important of which are the covalent anchoring method, the immersion method, and the encapsulation method [5]. The SILP systems are very popular due to the simplicity of the method of their preparation. The application of ionic liquid onto the porous support with large surface area, with or without a catalyst in the presence of a low-boiling solvent, permits the obtaining of the SILP materials [59] as a result of physical (or more rarely, chemical) adsorption. Their potential is characterized by the possibility of developing the surface with unique properties by an appropriate choice and modification of the support and ionic liquid and the addition of a catalyst. In the literature, many examples can be found on the catalytic application of the above systems; however, the vast majority of them concern the reaction of hydroformylation. The development of SILP systems applied in the reaction, as well as the development of the process itself within the last fifteen years, has been presented in the publication by Marinkovic et al. [60]. The studies carried out during the recent decades enabled the determination of conditions for conducting respective hydroformylation reactions in the liquid and gas phases, which in turn, made it possible to choose a SILP material for specific reaction conditions. One of the trends is an addition of a ligand, which together with a homogeneous catalyst, is immobilized in the SILP system, and due to this, an increase in the reaction conversion is achieved [61]. The kind and, especially, size of the employed reagent have a significant effect on the effectiveness of the SILP material used. In the case of alkenes containing less than six carbon atoms, to prevent the ionic liquid layer leaching from the surface of SILPs by the polar products formed in the system, it is recommended to employ the SILP materials in which the ionic liquid is bound to the support by a covalent bond or the application of nonpolar solvents as a mobile phase (Figure 5).

In the case of higher alkenes, a novelty is the application of supercritical carbon dioxide (scCO$_2$) as a mobile phase in the catalytic system with SILP materials [62]. Increasingly more often, particularly in reactions of hydroformylation of lower alkenes, catalytically active membranes are employed. Most likely, the reactors with SILP system-based membranes will soon be used that will enable them to conduct reactions in a continuous way and permit the direct separation of products and unreacted substrates (ROMEO approach—Reactor Optimization by Membrane Enhanced Operation) [63].

The SILP systems have also found applications in many other reactions. The reactions conducted in the presence of the supported ionic liquid phase were reviewed by Romanovsky and Tarkhanova, who have shown the applicational diversity of these systems [64]. The reactions in the liquid phase carried out with the use of SILPs include, among others, redox reactions, e.g., oxidation of sulfur-containing compounds [65], oxidation of phenols [66], reactions with haloalkynes [67,68], as well as many organic syntheses such as Heck reactions [69], Suzuki reactions [70], hydroamination of unsaturated compounds [71], Sandmayer reactions [72], and isomerization of hydrocarbons [73].

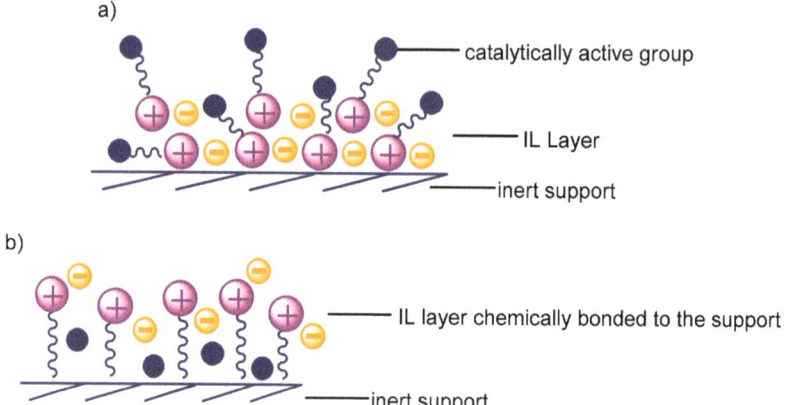

Figure 5. Methods of preventing catalyst from leaching: (**a**) By permanent bonding to ionic liquid; (**b**) by covalent bonding of ionic liquid to support.

One of the most recent novelties in the organic synthesis is the application of Pd-SILP systems with an amine group-containing imidazolium ionic liquid supported on the Merrifield resin (cross-linked chloromethyl polystyrene). The above system has been employed in the Suzuki–Miyaura coupling reaction carried out at room temperature. The reactions conducted with the use of different reagents proved that the system was stable, catalytically active, and permitted its isolation and effective reuse in five subsequent catalytic cycles [74]. Very popular are also ruthenium catalysts whose high price restricts their application to the systems that permit their recycling. An example is the preparation of SILPC systems with the use of ruthenium chloride, imidazolium ionic liquids, and activated carbon as support. The obtained system was employed in the hydrochlorination of acetylene, where the effect of the SILPC system on the improvement in conversion and selectivity was proved [75]. The imidazolium liquids are among the most popular ionic liquids applied in the SILP systems, and this brings about continuous improvement in the systems in which they are employed. An example is a new system with the use of [BMIM]PF$_6$ and pyrogenic amorphous *Aerosil* silica, which was applied as a catalyst for the synthesis of benzimidazole—the compound that has fungicidal properties [76].

Recently, a trend has been observed to employ transition metal complexes with chelating ligands that have reaction regioselectivity-improving properties. It is no wonder that these complexes have found application in SILPC systems. Brunig et al. used iron complexes of this type for the hydrogenation of aldehydes. The aforementioned catalytic system showed high chemoselectivity in reactions with aliphatic and aromatic aldehydes under mild reaction conditions [77]. When polymer-originated spherical active carbon (CARB) was employed as a support for the above catalytic system, conversions of substrates in analogous catalytic reactions were lower compared to the performance of the same catalytic system supported on silica despite an extension of the reaction time [78]. The utilization of new porous materials is one of the main trends observed in SILP materials. One of the recent examples is the employment of graphene oxide for this purpose, where graphene oxide served as a matrix for covalently bound ionic liquid (1-*N*-ferrocenylmethyl benzimidazole). The obtained system was applied with a very good result in the synthesis of 3,4-dihydro-2H-naphthol[2–e][1,3]oxazine-5,10-dione. The catalytic reaction in the presence of the SILP material proceeded fast and environmentally mild solvents were used [79]. On the other hand, phosphonium ionic liquid-modified silica with the addition of a palladium catalyst has been employed in the reaction of aminocarbonylation. Due to the utilization of the SILP system of this type, it was possible to conduct the reaction of double carbonylation in a nonpolar solvent. Moreover, the modification of the silica surface resulted in a considerable reduction in the palladium catalyst leaching from the support surface [80]. Another interesting idea seems to be the application of SILP systems to the immobilization of enzymes. The first example of the enzyme-SILP system is the immobilization of *Candida Antarctica Lipase A* with the use of 1-octyl-3-methylimidazolium tetrafluoroborate on the surface of a microglobule-forming monolith being a combination of cellulose and Teflon. By loading the reactor interior with the obtained SILP material, it was possible to conduct the reaction of transesterification in the continuous phase with very good results [81].

Reactions in the continuous phase are an essential aspect of the utilization of the SILP systems, particularly in the context of their application in the industry. The next step in this direction is the employment of multi-walled carbon nanotubes (MWCNTs) as supports in the SILP systems. Two kinds of MWCNT-based supports were obtained, namely with incorporated silica (for hydroformylation) or alumina (for water–gas shift reaction). To produce silica- or alumina-containing SILPC materials, ruthenium and rhodium catalysts, as well as imidazolium and phosphonium ionic liquids, were employed. In both cases, the reactions were carried out in a reactor and the isolation of products was performed with the use of membranes. Results obtained for 1-butene hydroformylation have shown that in the presence of the support consisting of MWCNT alone, the catalytic activity was higher than in the presence of the MWCNT-SiO_2 system. The reduction in the activity was a result of the calcination process that was carried out at a temperature at which MWCNTs are unstable. In the case of the water–gas shift reaction (WGSR), the conversion was higher when alumina was present in the SILP system. The reactions conducted in the presence of MWCNT-Al_2O_3 were characterized by a higher selectivity. It is also worth mentioning that the application of nanotube-containing SILPs as the filling of reactors enables better control of mass and heat transfer in the conducted reactions [82]. The WGSR that produces hydrogen and carbon dioxide is of great importance for several industrial processes such as Fischer–Tropsch and Haber–Bosch syntheses. The latter are conducted at high temperatures ranging from 200 to as high as 550 °C. The employment of SILPC systems in reactions of this type permits the considerable reduction in the process temperature (to 120–160 °C), which is a very promising solution. Unfortunately, a serious counter-indication to the application of the SILP materials for this purpose is the poor solubility of carbon monoxide in the ionic liquids. Wolf et al. investigated the effect of the addition of metal chlorides incorporated into the Ru-SILP system on the catalytic activity for WGSR and found that the incorporation of CuCl into the SILP material considerably improved the catalytic activity, as evidenced by the increase in the conversion by 30% [83]. The SILP materials with an ionic liquid having properties characteristic of Brønsted acids have been employed in the process of fuel desulfurization. The ionic liquid (4-(3′-ethylimidazolium)-butanesulfonate) was subjected to protonation with the use of two heteropolyacids ($H_3PMo_{12}O_{40}$ and $H_3PW_{12}O_{40}$) followed by immobilization on surfaces of silica and γ-alumina. The obtained SILP systems imparted stability to heteropolyanions, due to which the formed catalytic system became stable and enabled several oxidation reactions to be conducted with the use of the same portion of the catalyst [84]. On the other hand, the SILP system with Lewis acid centers appeared to be a good catalytic solution to selective hydrogenation of benzofuran derivatives. The SILP material consisting of silane-functionalized imidazolium ionic liquid and silica was subjected to impregnation with zinc chloride and also contained ruthenium nanoparticles. An essential aspect of the preparation of the above system was the introduction of the catalyst nanoparticles at the last stage of the synthesis, directly onto the anionic surface, due to which it became covalently bound to the support [85]. However, the utilization of SILP systems does not always require special preparation or modification of materials being its components. The relevant examples are popular hydrosilylation catalysts consisting of rhodium, phosphonium ionic liquids, and silica. The most active system ([{Rh(μ-OSiMe$_3$)(cod)}$_2$]/[P$_{66614}$][NTf$_2$]-SiO_2) enabled us to conduct as many as 20 catalytic cycles with very good yield. This made it possible to limit the amount of metal in the products, the isolation of which was also simple and less time-consuming compared to standard techniques [86]. The literature reports on the supported ionic liquid phase published in recent years indicate the application of modified supports or inorganic–organic hybrid materials as the solid phase in the SILP materials to reduce leaching of ionic liquid and catalyst from the system. Increasingly more often, the catalysts are also subject to modifications aimed at improving their adsorption on the SILP materials, among others, by employing chelating ligands or using metal nanoparticles. The enzyme-containing SILP systems, which are an entirely new solution to the enzymatic reactions, will certainly be developed in the near future. However, the greatest interest is seen in the SILP systems employed as solid beds in the reactors to enable reactions to be conducted in the continuous phase. In particular, the application of materials based on carbon nanotubes that enable better control of mass transfer in the catalytic systems will certainly find its reflection in further research works.

2.2. Solid Catalyst with Ionic Liquid Layer (SCILL)

Another way of ionic liquid application in catalysis, where the ionic liquid does not directly take part in reactions but is employed as a layer directly applied onto solid catalysts, is the so-called solid catalyst with ionic liquid layer (SCILL). Heterogeneous catalysts, despite their many advantages, are unfortunately often characterized by a lower selectivity compared to homogeneous ones. In this case, the role of IL is the modification of the surface, its homogenization, and the improvement in selectivity in some reactions, as well as the protection of the catalyst against poisoning. An appropriate selection of ionic liquid makes it possible to control the solubility of substrates and products in IL, thereby facilitating the access of desired reagents to catalytic centers and limiting the occurrence of undesired reactions. Moreover, the selection of ionic liquid in which the solubility of reactants is greater than that of the products formed enables the easy isolation of reaction products [55,87]. The essential difference between the SCILL and SILP systems consists of employing, in the SCILL materials, only heterogeneous catalysts that are covered with a thin layer of ionic liquid, whereas in the case of SILP materials, porous support is used [50]. There are two kinds of SCILLs; in the former case, a layer of the ionic liquid is placed on a solid catalyst (Figure 6a), and in the latter case, the ionic liquid covers the solid catalyst placed on an inert support (Figure 6b).

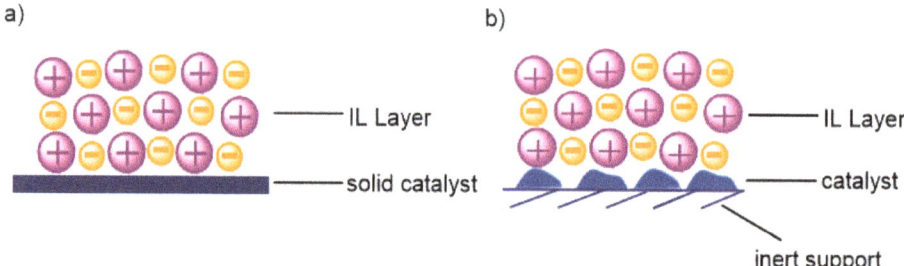

Figure 6. Two kinds of solid catalysts with ILs (SCILLs): (**a**) With catalytically active support; (**b**) with heterogeneous catalyst on an inert support.

The first reports on materials of this type were published in 2007 by Kernchen et al. [50,88,89]. They concerned the reaction of cyclooctadiene hydrogenation to cyclooctene with the use of SCILL materials based on nickel catalyst covered with ionic liquid [BMIM][n-$C_8H_{17}OSO_3$], which enabled it to achieve 70% conversion [88]. Further research widened the assortment of ionic liquids employed in the SCILL systems and proved a significant effect of the ionic liquid cation on their catalytic activity in hydrogenation reactions [90]. The latter were conducted with the use of SCILL materials, among others, for alkynes, aldehydes, and naphthalene [55]. An interesting solution for reactions of this type has been proposed by Antonels and Meijboom. In the reaction of citral hydrogenation, silica-supported nanoparticles of ruthenium catalyst with added dendrimers were employed. Then, imidazolium ionic liquids differing in anions and cations were applied onto the obtained ruthenium systems. It was shown that in some SCILL systems, the ionic liquid prevented catalyst poisoning due to the removal of formed CO from the system as a result of the competitive reaction of decarboxylation. The best catalytic results were obtained for the SCILL systems with [BMIM][NTf$_2$] [91]. The catalytic properties of the SCILL systems consisting of ruthenium dendrimers (RuDEN) and different ionic liquids were compared with the RuDEN catalytic system without ionic liquids in toluene hydrogenation. All SCILL catalytic systems have shown higher activity and better selectivity than RuDEN catalysts [92]. In the case of the reaction of 1,3-butadiene hydrogenation, the application of commercially available nickel catalyst and ionic liquid ([BMIM][BF$_4$]) made it possible to obtain butene with the yield of 95% [93]. Modified SCILL systems containing acidic ionic liquids and palladium catalyst supported on active carbon were employed for the hydrogenation of arenes, which resulted in an increase in the reaction yield, and in the case of toluene hydrogenation, it was possible to reduce the reaction temperature to 40 °C [94]. Palladium-containing SCILL systems were used to selectively obtain butene, and a 99% conversion of the substrate was achieved [95].

The ionic liquid in the SCILL system also appeared to be useful in the development of an active and selective hydrogenation catalyst, one of the components of which was the discarded fluid cracking catalyst DF3C. The latter was used as a support in the catalytic system containing 10 wt% nickel and ionic liquid (1-ethanol-3-methylimidazolium tetrafluoroborate) [96]. The system was employed for the hydrogenation of α-pinene to cis-pinane and enabled the maintaining of the conversion of α-pinene on the level of about 99% during 13 catalytic cycles, after which the performance of the catalyst deteriorated due to leaching of the ionic liquid. Both the activity and selectivity of the SCILL system depended on the amount of the IL (or in other words, on the thickness of the IL layer). At the IL loading exceeding 10 wt%, the selectivity to cis-pinane was over 98%; however, at the loading of 5 wt%, it had fallen to about 90%. Such utilization of the waste industrial catalyst as that described for the spent cracking catalyst DF3C is useful both from the viewpoint of the economy and from that of environmental protection.

The reaction of ethene hydrogenation was studied by Bauer at al. [97] in the presence of the Pd-SCILL system, who found that the above alkene reacted with the layer of ionic liquid 1-ethyl-3-methylimidazolium ethylsulfate, which resulted in the formation of a new ionic liquid 1,2-diethyl-3-methylimidazolium ethylsulfate. Their study opens the way to the synthesis of new ionic liquids or modification of their properties on the surface of the SCILL materials [98].

Not only have the SCILL systems found applications in hydrogenation reactions, but they have also been employed in the reactions of catalytic oxidation, isomerization, and Knoevenagel condensation. The selective oxidation of benzyl alcohol with air was catalyzed by platinum (5 wt%) supported on cryptomelane (KMn$_8$O$_{16}$) covered with imidazolium or pyrrolidinium ionic liquid—[BMIN][NTf$_2$] or [BMPy][NTf$_2$], respectively. In the presence of the former IL, the conversion of benzyl alcohol on the above catalytic system was 80% and the selectivity for aldehyde was 82% [99]. The oxidative coupling of aromatic thiols to disulfides in the presence of homogeneous catalysts has several shortcomings, whereas the application of heterogeneous catalytic SCILL systems to the aforementioned reactions makes it possible to limit the amount of metal in the product, to easily isolate the catalyst from the post-reaction mixture and to prevent catalyst poisoning. To take the above advantages, ruthenium- and platinum-containing SCILL systems with ionic liquids [BMIM][NTf$_2$], [EMIM][NTf$_2$], and [BMPyr][NTf$_2$] were applied in the mentioned reactions of thiols under mild conditions to result in satisfactory yields in three subsequent catalytic cycles [100]. The success of skeletal isomerization of chemical compounds largely depends on the catalyst acidity. This is why the employment of the SCILL systems with acidic ionic liquids in reactions of this type is an interesting idea. The platinum-containing SCILL system with the ionic liquid being a strong Lewis acid ([C$_4$C$_1$IM]Cl/AlCl$_3$) was applied in n-octane isomerization [101]. In one of the recent papers on the SCILL systems, their application in the Knoevenagel reaction was reported. The system employed in the above reaction made it possible to effectively conduct condensation of aldehydes with cyanoacetate and malononitrile in aqueous medium at room temperature. The catalytic system with basic properties enabled the good conversion of substrates in five subsequent catalytic cycles [102].

In the literature on the SCILL materials visible is an increase in reports on their applications in reactions other than hydrogenation. The diversity of catalytic reactions with their participation makes researchers introduce improvements in the SCILLs structure. In addition to the syntheses of specialty ionic liquids and the modification of catalyst structure, increasingly more research is often carried out on the formation of bifunctional SCILL systems. Their stability and durability, as well as applicational potential in many, often complicated, organic syntheses, will certainly be developed in further research.

2.3. Porous Ionic Liquids

The term porous ionic liquids, i.e., liquid materials that are characterized by permanent microporosity, was introduced in 2007 by James et al. [103]. The porous ionic liquids can be classified into three types. The first of them is a pure molecular unit with empty cavities that are rigid (which prevents them from collapsing) and their shape renders the intermolecular self-filling of pores of the liquid impossible. The second type, so-called liquids of "empty host" type, includes ionic liquids that are dissolved in the solvents with a steric hindrance. In addition, these liquids are characterized by high rigidity and permanence of pores, which do not collapse even when unfilled. It is also essential that solvent molecules cannot fill voids in the porous liquid structure. The third type is a dispersion of solid porous materials in the liquid matrix whose molecules are too large to enter the pores of the solid. These materials form porous phases that contain ionic liquids and additionally solid particles of microporous materials that make a skeleton of the whole system. Porous ionic liquids have properties characteristic of both ionic liquids and porous materials due to which they are characterized by large surface area, very good mechanical strength, and the capability of modifying their structure. Moreover, the presence of easy-to-exchange ions in their pores causes a change in the chemical properties of the material. It is also worth mentioning that porous ionic liquids are characterized by high polarization and high ionic density [104].

One of the methods of the preparation of porous ionic liquids is synthesis with the use of so-called templates. There exist methods based on the use of soft and hard templates. Generally, these methods consist of conducting in situ polymerization of ionic liquid monomers or the copolymerization of ionic liquid with another monomer [104–106]. An example of synthesis with the use of hard templates, which make it possible to obtain materials with a specified pore size, is the employment of colloidal silica dispersed in an ionic liquid monomer. After crosslinking followed by removal of the template (silica) from the formed structure, a mesoporous ionic liquid was obtained [105]. It is worth pointing out that the method employing hard templates enables the obtaining of meso- and macroporous materials [107], whereas that based on the use of soft templates results in the production of mesoporous materials and consists of the utilization of surfactants and formation of micelles [107,108].

Porous ionic liquids can also be obtained without the application of templates. An example of such a synthesis is the radical copolymerization of divinylbenzene and vinyl group-containing imidazolium ionic liquid [109]. In other literature reports, the possibility of obtaining porous ionic liquids via complexation of poly(ionic liquids) with the use of acids [110,111] was presented. The methods of producing porous ionic liquids have continually improved and modified. However, irrespective of the preparation method, these materials make a construction that can be successfully applied in heterogeneous catalysis. One of the first examples of the application of ionic liquids as catalysts is the Suzuki–Miyaura cross-coupling reaction, which employs palladium nanoparticles supported on a porous ionic polymer (PIP). First, the ionic liquid was obtained through the reaction of 1-vinylimidazole with bromoacetic acid, followed by its employment in the radical polymerization with vinylbenzene in the presence of azobisisobutyronitrile (AIBN) and ethanol. The system Pd-PIP was prepared in situ through the impregnation in an ethanolic solution of $Pd(OAc)_2$. The obtained catalytic material appeared to be ideally suited for conducting coupling reactions with aryl bromides and chlorides and phenylboronic acid. Moreover, it was possible to recycle and reuse it in subsequent reactions.

Due to the presence of the carboxyl group and carbene structure, palladium nanoparticles were very well-stabilized and did not aggregate [112]. On the other hand, the heteropolyanion-containing copolymer was used in the reaction of benzene hydroxylation with hydrogen peroxide. A hybrid porous catalyst was produced by polymerization of ionic liquid (3-n-butyl-1-vinylimidazolium bromide ([VBIM][Br])) with divinylbenzene (DVB) in the presence of AIBN. The copolymer obtained in this way was subjected to impregnation with heteropolyacid—$H_5PMo_{10}V_2O_{40}$, which resulted in the exchange of anions and formation of a porous cross-linked copolymer based on heteropolyacid (HPA) (Scheme 1). The produced catalytic system showed very good catalytic activity for hydroxylation, due to, among others, its large surface area and suitable pore size. Its isolation from the post-reaction mixture was easy, which made it possible to reuse it [113].

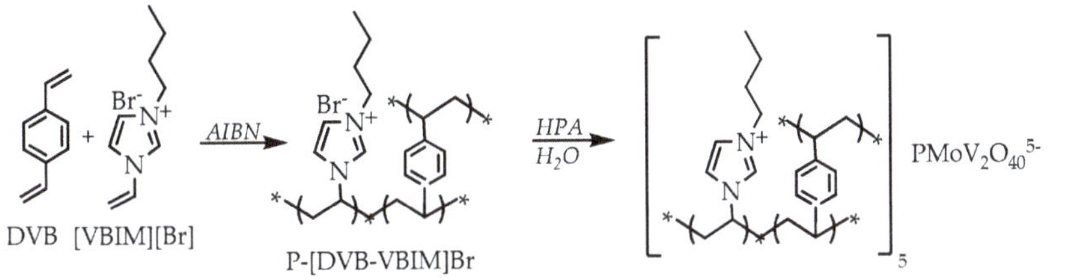

Scheme 1. Synthesis of the porous HPA-based cross-linked polymeric IL catalyst P-[DVB-VBIM]$_5$PMoV$_2$ [113].

Another example of synthesis without a template is the polymerization of 3-cyanomethyl-1-vinylimidazolium-based ionic liquid with polyacrylic acid. The resulting polymer served as a support for copper chloride, and the catalytic system produced in this way has been employed in the oxidation of cyclic compounds. The reaction was carried out in mild conditions and was characterized by very good yield and selectivity [114]. An interesting example of synthesis with the use of a soft template was the free-radical polymerization of ionic liquid 1-allyl-3-vinylimidazolium with the participation of triblock copolymer P123 (copolymer of ethylene oxide and propylene oxide). The exchange of the chloride anion for phosphotungstic heteropolyanion resulted in a very active catalyst for epoxidation of cis-cyclooctene with hydrogen peroxide [115]. A very similar procedure with the use of the same template P123 was applied in the copolymerization of 1-butyl-3-vinylimidazolium bromide and divinylbenzene followed by ion exchange for phosphotungstic heteropolyanion. The obtained material was employed in the alkylation of styrene and o-xylene [116]. Porous ternary materials, obtained by polymerization of divinylbenzene, 1-vinylimidazole, and 1-vinyl-3-butylimidazolium bromide, were applied in the same reaction. The systems of this type were characterized by a greater number of acid groups and active centers on their surface than their binary analogs. Additionally, the presence of micro- and mesopores in the structure of the porous ternary polymer increased the possibility of contact between reagents, which resulted in the high conversion of substrates (99%) [117]. The high activity of porous ionic liquids as catalytic systems can be achieved not only by the formation of certain structures but also by the generation of acid–base properties. An example of basic, hydrophobic, and mesoporous poly (ionic liquid) is the material produced by copolymerization of 1-aminoethyl-3-vinylimidazole bromide with divinylbenzene followed by the exchange of a bromide anion for a hydroxyl one. This copolymer has been employed in the solvent-free condensation of benzaldehyde with ethyl cyanoacetate (Knoevenagel reaction), and its catalytic activity significantly surpassed that of commercially available strongly basic resins, as well as NaOH, which are known for their activity for the above reaction [118]. On the other hand, a nanoporous poly (ionic liquid) with a strong acidic character was produced by copolymerization of divinylbenzene with the ionic liquid containing 1-vinylimidazolium or vinylpyridinium cation and SO_3CF_3 or SO_3H anions. The synthesized catalytic material permitted the conduction of the hydration of alkynes to ketones, and its effectiveness was higher than that of commercially available acids commonly used in reactions of this type. Moreover, this material was easy to regenerate, which enabled its employment in subsequent reactions [119].

In recent years, a strong interest has been observed in the application of porous ionic liquids to capturing CO_2 and its cycloaddition to cyclic hydrocarbons and epoxides [120]. An example of such a material is poly (divinylbenzeno-1-allyl-tetramethylguanidinium) bromide, which enables the achieving of a conversion higher than that obtained when using standard reaction systems without porous materials. In addition, the materials composed of ionic liquids and metal–organic frameworks (IL-MOF) have found application in the reactions of CO_2 cycloaddition. The use of materials of this type in the latter reactions has been recently reviewed by Wang et al. [121]. The interest in IL-MOFs has not abated, due to their structure, which captures CO_2 very effectively. Then, the latter can be utilized in the aforementioned reaction. The ordered structure, very good adsorption properties (caused, among others, by the large surface area), and metal–ligand interactions are the reason for the stability and very good catalytic properties of these materials [121,122]. The MOF materials are heterogeneous systems with very good catalytic stability. They can show, if appropriately tailored, the selectivity toward reagents of a specific size. The catalytic performance of these porous materials can be additionally upgraded by their functionalization with ionic liquids. To this end, the method of ionic liquid encapsulation within the MOF structure is employed most often. It has been shown that the ionic liquid enters the pores of the metallic structure where the metal forms a coordinate bond with a functional group (e.g., amino group) of the ionic liquid. An important aspect of the above synthesis of IL-MOF materials is the solvent choice in such a way that no affinity occurs between the solvent and the porous surface, and at the same time, the ionic liquid coordination is not restricted [107,123–125]. A relatively new method of the preparation of IL-MOF materials, developed by Khan et al. [126], is the so-called "ship-in-bottle" technique. To this purpose, they used the porous structure of chromium terephthalate, which played the role of host (bottle), inside which the ionic liquid (ship) has been synthesized. The substance of the method is that the dimensions of the ionic liquid synthesized inside the MOF host are greater than the host pores, which makes its removal impossible. The preparation of the ionic liquid occurs through the penetration of the IL precursors (N-methylimidazole and 1-bromobutane) via pores to the interior of the host structure, where the synthesis takes place. Due to the ionic liquid trapping inside the MOF structure, the problem of its eluting has been eliminated. The IL-MOF systems can also be obtained by utilizing capillary forces [122,127]. Examples of the application of the IL-MOF materials were presented in the review paper by Fujie and Kitagawa [125]. Recent reports on the IL-MOFs concern the systems containing more than one metal. The multi-metallic MOFs, as they were named by Sun at al. [128], are a real challenge. Just joining together several catalytically active systems into an integral whole that will be characterized by an ordered structure creates many problems. That is why the mechanochemical synthesis has been applied to prepare the new catalytic material. It has been shown that the obtained catalytic system, containing the binary MOF system in its structure, is characterized by the far better conversion of substrates in the cycloaddition of epoxy compounds to carbon dioxide than its single analog. The proposed method of the preparation of the IL-MOF materials consisting of two metals (and maybe, in the future, even several ones) opens the way to interesting new applications of the systems of this type. In addition, another porous organometallic structure named ZIF-8, which contains zinc ions and 2-methylimidazolium ligands, is willingly used in ionic liquid-containing systems. The addition of as small an amount as 5% of ZIF-8 material to the ionic liquid considerably improved the adsorption ability compared to the ionic liquid alone [129]. Many attempts at obtaining IL-MOF materials with just this porous structure (also in the form of colloidal solutions with permanent porosity) have confirmed their very good performance in the processes of the low-pressure separation of gases [130] and adsorption of gases in the liquid systems [131].

Porous ionic liquids provide many application possibilities. The recent synthesis of porous ionic liquids in the form of tetrahedral coordination cages proved the possibility of designing these materials in the way that enables them to fulfill specified functions [132]. A wide variety of compounds, e.g., alcohols (butanol, propanol), trichlorofluoromethane, dichlorodifluoromethane, and chlorotrifluoromethane, were adsorbed in the pores of these materials. In the future, these materials can serve for selective isolation of particular components from mixtures (e.g., by extraction), but their employment in catalytic reactions, e.g., for picking up by-products, is conceivable.

The ionic porous materials were introduced relatively recently; therefore, their increased applicational development is expected in the near future. The observation of the current research trends in the field of porous ionic liquids suggests the development of new copolymeric materials consisting of more than two monomers, as well as multimetallic IL-MOFs.

3. Conclusions

The application of ionic liquids in different fields, including catalysis, has not abated. Through the appropriate choice of new synthesis methods and the employment of renewable and biodegradable raw materials, an increasingly greater group of derivatives is being considered safe and environmentally friendly. The role of ionic liquids in catalysis is very diversified because they can be solvents in the catalytic processes, immobilizing agents for homogeneous catalysts, components of catalysts (e.g., ligands in a complex), as well as actual catalysts [5,24–27,33,47–50]. In the case of catalysis, ionic liquids make it possible to conduct reactions with higher selectivities and yields; however, their most important asset is easy catalyst isolation from the catalytic mixture. That is why we intended to present in this mini-review several variants of heterogeneous catalysts prepared with the participation of ionic liquids. We have restricted them to three groups of such catalysts, namely, supported ionic liquid phase catalyst (SILPC), solid catalyst with ionic liquid layers (SCILL), and porous ionic liquids. However, there are considerably more ionic liquid-containing heterogeneous systems that were not presented here, because of the article volume limitations. For instance, we left out nanocatalysis (in a broad sense) with the participation of dispersed metal nanoparticles stabilized by ionic liquids [133,134], ionosilicas [135,136], and poly(ionic liquids) [107,137], which can be modifying agents for polymers, copolymers, as well as play the role of catalyst in polymerization processes. Moreover, they can take part in metal–organic frameworks (MOFs) that were only very superficially mentioned when presenting porous ionic liquids [127,138], although they certainly deserve more attention. In many cases, metal-containing ionic liquids (metallate ionic liquids) also form heterogeneous systems and are successfully applied in catalysis [139,140]. Therefore, the three variants of heterogeneous systems presented in this article are only a modest, albeit one of the most common, examples of systems of this type. Undoubtedly, the SILPCs belong to the most popular and increasingly more frequently employed catalytic systems [55–58]. One of the first and best-explored processes conducted with their participation is hydroformylation [60]. However, there are currently many other SILPC-catalyzed processes [64] such as oxidation, coupling (reactions of Heck, Suzuki, etc.), hydroamination, and isomerization of hydrocarbons. Another heterogeneous system presented in this paper is SCILL, which takes advantage of the properties of ionic liquid that covers a solid catalyst with a thin layer to facilitate better access of reagents to the catalyst and thereby improves the process selectivity and yield [88]. The most prevalent process carried out with the participation of SCILL is hydrogenation, albeit recent reports appeared on the processes of oxidation, isomerization, and Knoevenagel condensation [99–102]. Finally, the third type of heterogeneous systems presented in this paper is porous ionic liquids [104,107], which also include metal–organic frameworks formed with the participation of ionic liquids [129]. The systems of this type, due to their structure and high catalytic stability, has attracted increasingly more attention [125]. Among the most interesting directions of their application is the capture of CO_2 and its cycloaddition to cyclic hydrocarbons and epoxides [120]. However, the possibilities of their applications are much wider both in chemical and enzymatic catalysis [141]. The above-presented examples of the applications of the systems are only selected ones and certainly do not exhaust all their possibilities. Our goal was to indicate their large application potential that will soon be increasingly more frequently utilized in many chemical processes. In all the above systems, the problems occurring in the case of catalysis in homogeneous systems have been eliminated. First, there is no problem with the isolation of catalyst, which is heterogeneous. Second, because ionic liquid covers the supports with a thin layer, a small amount of IL is needed, which results in a considerable cost reduction (despite the high price of ionic liquids). In addition, due to the thinness of the layer, the viscosity of IL does not influence the course of the process and the mass transfer does not create a problem. Thirdly, due to the form of catalyst, continuous flow reactors can be employed, which also influences the process economics. This is why a considerable development of heterogeneous catalysts of this type and their application in many processes can be expected in the future. Therefore, ionic liquids proved again their capability of evolving and adapting to various technological, economic, and environmental regimes.

Author Contributions: Conceptualization, H.M.; literature search, O.B. and A.S.; writing—review and editing, O.B., A.S., I.D. and H.M.; supervision, H.M.; funding acquisition, H.M. and O.B. All authors have read and agreed to the published version of the manuscript.

Funding: This research was funded by the National Science Centre (Poland), project OPUS grant number UMO-2014/15/B/ST5/04257 and project PRELUDIUM grant number UMO-2019/35/N/ST4/00494.

Conflicts of Interest: The authors declare no conflict of interest.

References

1. The New Material. The Twenty Most Potential New Materials in the Future. 2014. Available online: https://new.qq.com/omn/20191105/20191105A098C000.html (accessed on 21 October 2020).
2. ISI Web of Science Search. Available online: https://clarivate.com/webofsciencegroup/solutions/web-of-science/ (accessed on 21 October 2020).
3. Wasserscheid, P.; Welton, T. *Ionic Liquids in Synthesis*; Wiley-VCH: Weinheim, Germany, 2003.
4. Adams, D.J.; Dyson, P.J.; Tavener, S.J. *Chemistry in Alternative Reaction Media*; Wiley: Chichester, UK, 2004.
5. Vekariya, R.L. A review of ionic liquids: Applications towards catalytic transfromations. *J. Mol. Liq.* **2017**, *227*, 44–60. [CrossRef]
6. Javed, F.; Ullah, F.; Zakaria, M.R.; Akil, H.M. An approach to classification and hi-tech applications of room-temperature ionic liquids (RTILs): A review. *J. Mol. Liq.* **2018**, *271*, 403–420. [CrossRef]
7. Singh, S.K.; Savoy, A.W. Ionic liquids synthesis and applications: An overview. *J. Mol. Liq.* **2020**, *297*, 112038. [CrossRef]
8. Walden, P. Über die Molekulargrösse und elektrische Leitfähigkeit einiger geschmolzener Salze. *Bull. Acad. Imp. Sci.* **1914**, *8*, 405–422.
9. Hurley, F.H.; Weir, T.P. Electrodeposition of metals from fused quaternary ammonium salts. *J. Electrochem. Soc.* **1951**, *98*, 203–206. [CrossRef]
10. Wilkes, J.S.; Zaworotko, M.J. Air and water stable 1-ethyl-3-methylimidazolium based ionic liquids. *J. Chem. Soc. Chem. Commun.* **1992**, 965–967. [CrossRef]
11. Wierzbicki, A.; Davis, J.H., Jr. Envisioning the second generation of ionic liquid technology: Design and synthesis of Task-specific Ionic Liquids (TSILs). In Proceedings of the Symposium on Advances in Solvent Selection and Substitution for Extraction, Atlanta, Georgia, 5–9 March 2000; AIChE: New York, NY, USA, 2000.
12. Welton, T. Ionic Liquids: A brief history. *Biophys. Rev.* **2018**, *10*, 691–706. [CrossRef]
13. Earle, M.J.; Esperanca, J.M.S.S.; Gilea, M.A.; Lopes, J.N.C.; Rebelo, L.P.N.; Magee, J.W.; Seddon, K.R.; Widegren, J.A. The distillation and volatility of ionic liquids. *Nature* **2006**, *439*, 831–834. [CrossRef]
14. Clark, J.H.; Tavener, S.J. Alternative solvents: Shades of green. *Org. Proc. Res. Dev.* **2007**, *11*, 149–155. [CrossRef]
15. Jessop, P.G. Searching for green solvents. *Green Chem.* **2011**, *13*, 1391–1398. [CrossRef]
16. Cevasco, G.; Chiappe, C. Are ionic liquids a proper solution to current environmental challenges? *Green Chem.* **2014**, *16*, 2375–2385. [CrossRef]
17. Frade, R.F.; Alonso, C.A. Impact of Ionic Liquids in environment and humans: An overview. *Hum. Exp. Toxicol.* **2010**, *29*, 1038–1054. [CrossRef] [PubMed]
18. Kunz, W.; Hackl, K. The hype with ionic liquids as solvents. *Chem. Phys. Lett.* **2016**, *661*, 6–12. [CrossRef]
19. Anastas, P. (Ed.) *Green Solvents*; Wiley-VCH: Weinheim, Germany, 2010.
20. Ameta, G.; Pathak, A.K.; Ameta, C.; Punjabi, P.B. Sonochemical synthesis and imidazolium based ionic liquids: A green pathway. *J. Mol. Liq.* **2015**, *211*, 934–937. [CrossRef]

21. Naeimi, H.; Nazifi, Z.S. A facile one-pot ultrasound assisted synthesis of 1,8-dioxooctahydroxanthene derivatives catalysed by Bronsted acidic ionic liquid (BAIL) under green conditions. *J. Ind. Eng. Chem.* **2014**, *20*, 1043–1049. [CrossRef]
22. Martinez-Palou, R. Microwave-assisted synthesis using ionic liquids. *Mol. Divers.* **2010**, *14*, 3–25. [CrossRef]
23. Itoh, T.; Koo, Y.-M. (Eds.) *Application of Ionic Liquids in Biotechnology*; Springer: Cham, Switzerland, 2019.
24. Ozokwelu, D.; Zhang, S.; Okafor, O.C.; Cheng, W.; Litombe, N. *Novel Catalytic and Separation Processes Based on Ionic Liquids*; Elsevier: Amsterdam, The Netherlands, 2017.
25. Saha, B.; Fan, M.; Wang, J. (Eds.) *Sustainable Catalytic Process*; Elsevier: Amsterdam, The Netherlands, 2015.
26. Dupont, J.; Itoh, T.; Lozano, P.; Malhotra, S.V. (Eds.) *Environmentally Friendly Syntheses Using Ionic Liquids*; CRC Press: New York, NY, USA, 2015.
27. Lozano, P. *Sustainable Catalysis in Ionic Liquids*; CRC Press: New York, NY, USA, 2019.
28. Alvarez, M.S.; Zhang, Y. Sketching neoteric solvents for boosting drugs bioavailability. *J. Controll. Release* **2019**, *311*, 225–232. [CrossRef]
29. Villa, R.; Alvarez, E.; Porcar, R.; Garcia-Verdugo, E.; Luis, S.V.; Lozano, P. Ionic Liquids as an enabling tool to integrate reaction and separation processes. *Green Chem.* **2019**, *21*, 6527–6544. [CrossRef]
30. Gaida, B.; Brzęczek-Szafran, A. Insights into the Properties and Potential Applications of Renewable Carbohydrate-Based Ionic Liquids: A Review. *Molecules* **2020**, *25*, 3285. [CrossRef]
31. Gomes, J.M.; Silva, S.S.; Reis, R.L. Biocompatible ionic liquids: Fundamental behaviors and applications. *Chem. Soc. Rev.* **2019**, *48*, 4317–4335. [CrossRef]
32. Welton, T. Solvents and sustainable chemistry. *Proc. R. Soc. A Mat.* **2015**, *471*, 50502. [CrossRef] [PubMed]
33. Olivier-Bourbigou, H.; Magna, L.; Morvan, D. Ionic Liquids and Catalysis: Recent progress from knowledge to applications. *Appl. Catal. A* **2010**, *373*, 1–56. [CrossRef]
34. Hajipour, A.R.; Rafiee, F. Recent progress in ionic liquids and their applications in organic synthesis. *Org. Prep. Proced. Int.* **2015**, *47*, 249–308. [CrossRef]
35. Sahin, S. Tailor-designed deep eutectic liquids as a sustainable extraction media: An alternative to ionic liquids. *J. Pharm. Biomed. Anal.* **2019**, *174*, 324–329. [CrossRef]
36. Kalhor, P.; Ghandi, K. Deep Eutectic Solvents for Pretreatment, Extraction and Catalysis of Biomas and Food Waste. *Molecules* **2019**, *24*, 4012. [CrossRef]
37. Unli, A.E.; Arikaya, A.; Takac, S. Use of deep eutectic solvents as catalyst: A mini-review. *Green Process Synth.* **2019**, *8*, 355–372. [CrossRef]
38. Eyckens, D.J.; Henderson, L.C. A review of Solvate Ionic liquids: Physical Parameters and Synthetic Applications. *Front. Chem.* **2019**, *7*, 263. [CrossRef]
39. Salar-García, M.J.; Ortiz-Martínez, V.M.; Hernández-Fernández, F.J.; de los Ríos, A.P.; Quesada-Medina, J. Ionic liquid technology to recover volatile organic compounds (VOCs). *J. Hazard. Mater.* **2017**, *321*, 484–499. [CrossRef]
40. Watanabe, M.; Thomas, M.L.; Zhang, S.G.; Ueno, K.; Yasuda, T.; Dokko, K. Application of ionic liquids to energy storage and conversion materials and devices. *Chem. Rev.* **2017**, *117*, 7190–7239. [CrossRef]
41. Balducci, A. Ionic liquids in Lithium-ion batteries. *Top. Curr. Chem.* **2017**, *375*, 20. [CrossRef]
42. Itoh, T. Ionic liquids as tool to improve enzymatic organic synthesis. *Chem. Rev.* **2017**, *117*, 10567–10607. [CrossRef]
43. Ventura, S.P.M.; Silva, F.A.E.; Quental, M.V.; Monda, D.; Freire, M.G.; Coutinho, J.A.P. Ionic-liquid-mediated extraction and separation processes for bioactive compounds: Past, present, and future trends. *Chem. Rev.* **2017**, *117*, 6984–7052. [CrossRef] [PubMed]
44. Kristfikova, D.; Modrocka, V.; Meciarova, M.; Sebsta, R. Green Asymmetric Organocatalysis. *ChemSusChem* **2020**, *13*, 2828–2858. [CrossRef] [PubMed]
45. Egorova, K.S.; Gordeev, E.G.; Ananikov, V.P. Biological activity of ionic liquids and their application in pharmaceutics and medicine. *Chem. Rev.* **2017**, *117*, 7132–7189. [CrossRef] [PubMed]
46. Nancarrow, P.; Mohammed, H. Ionic liquids in space technology—Current and future trends. *Chembioeng. Rev.* **2017**, *4*, 106–119. [CrossRef]
47. Dyson, P.J.; Geldbach, T.J. *Metal Catalysed Reactions in Ionic Liquids*; Springer: Dordrecht, The Netherlands, 2005.
48. Dupont, J.; Kollar, L. *Ionic Liquids (ILs) in Organometallic Catalysis*; Springer: Dordrecht, The Netherlands, 2015.
49. Hardacre, C.; Parvulescu, V. Catalysis in Ionic Liquids. In *Catalysts Synthesis to Applications*; RS: Cambridge, UK, 2014.

50. Dominguez de Maria, P. (Ed.) *Ionic Liquids in Biotransformations and Organocatalysis. Solvent and Beyond*; John Wiley&Sons, Inc.: Hoboken, NJ, USA, 2012.
51. Stevens, J.C.; Shi, J. Biocatalysis in ionic liquids for lignin valorization: Opportunities and recent developments. *Biotech. Adv.* **2019**, *37*, 107418. [CrossRef]
52. Quiroz, N.R.; Norton, A.M.; Nguyen, H.; Vasileiadou, E.; Vlachos, D.G. Homogeneous metal salt solutions for biomass upgrading and other select organic reactions. *ACS Catal.* **2019**, *9*, 9923–9952. [CrossRef]
53. Hejazifar, M.; Lanaridi, O.; Bica-Schroder, K. Ionic liquid based microemulsions: A review. *J. Mol. Liq.* **2020**, *303*, 112264. [CrossRef]
54. Qiao, Y.; Ma, W.; Theyssen, N.; Chen, C.; Hou, Z. Temperature-responsive ionic liquids: Fundamental behaviors and catalytic applications. *Chem. Rev.* **2017**, *117*, 6881–6928. [CrossRef]
55. Fehrmann, R.; Riisager, A.; Haumann, M. *Supported Ionic Liquids. Fundamentals and Applications*; Wiley-VCH: Weinheim, Germany, 2014.
56. Gu, Y.; Li, G. Ionic liquids-based catalysis with solids. State of the art. *Adv. Synth. Catal.* **2009**, *351*, 817–847. [CrossRef]
57. Kaur, P.; Chopra, H. Recent Advances of Supported Ionic Liquids. *Curr. Org. Chem.* **2019**, *23*, 2881–2915. [CrossRef]
58. Feher, C.; Papp, M.; Urban, B.; Skoda-Foldes, R. Catalytic Applications of Supported Ionic Liquid Phases. In *Advances in Asymmetric Autocatalysis and Related Topics*; Academic Press: London, UK, 2017.
59. Kuhmann, E.; Haumann, M.; Jess, A.; Seeberger, A.; Wasserscheid, P. Ionic liquids in refinery desulfurization: Comparison between biphasic and supported ionic liquid phase suspension processes. *ChemSusChem* **2009**, *2*, 969–977. [CrossRef] [PubMed]
60. Marinkovic, J.M.; Riisager, A.; Franke, R.; Wasserscheid, P.; Haumann, M. Fifteen years of Supported Ionic Liquid Phase—Catalysed hydroformylation: Material and process developments. *Ind. Eng. Chem. Res.* **2019**, *58*, 2409–2420. [CrossRef]
61. Mehnert, C.P.; Cook, R.A.; Dispenziere, N.C.; Afeworki, M. Supported Ionic Liquid Catalysis—A New Concept for Homogeneous Hydroformylation Catalysis. *J. Am. Chem. Soc.* **2002**, *124*, 12932–12933. [CrossRef]
62. Hintermair, U.; Gong, Z.; Serbanovic, A.; Muldoon, M.J.; Santini, C.C.; Cole-Hamilton, D.J. Continuous flow hydroformylation using Supported Ionic Liquid Phase catalysts with carbon dioxide as a carrier. *Dlton Trans.* **2010**, *39*, 8501–8510. [CrossRef]
63. Illner, M.; Mulller, D.; Esche, E.; Pogrzeba, T.; Schmidt, M.; Schomacker, R.; Wony, G.; Repke, J.U. Hydroformylation in microemulsions: Proof of concept in a miniplant. *Ind. Eng. Chem. Res.* **2016**, *55*, 8616–8626. [CrossRef]
64. Romanovsky, B.V.; Tarhanova, I.G. Supported ionic liquids in catalysis. *Russ. Chem. Rev.* **2017**, *86*, 444–458. [CrossRef]
65. Li, M.; Zhang, M.; Wei, A.; Zhus, W.; Xun, S.; Li, Y.; Li, H.; Li, H. Facile synthesis of amphiphilic polyoxometalate-based ionic liquid supported silica induced efficient performance in oxidative desulfurization. *J. Mol. Catal. A Chem.* **2015**, *406*, 23–30. [CrossRef]
66. Navalon, S.; Alvaro, M.; Garcia, H. Heterogeneous Fenton catalysts based on clays, silicas and zeolites. *Appl. Catal. B.* **2010**, *99*, 1–26. [CrossRef]
67. Kim, D.W.; Chi, D.Y. Polymer-Supported Ionic Liquids: Imidazolium Salts as Catalysts for Nucleophilic Substitution Reactions Including Fluorinations. *Angew. Chem. Int. Ed.* **2004**, *43*, 483–485. [CrossRef]
68. Kim, D.W.; Hong, D.J.; Jang, K.S.; Chi, D.Y. Structural Modification of Polymer-Supported Ionic Liquids as Catalysts for Nucleophilic Substitution Reactions Including Fluorination. *Adv. Synth. Catal.* **2006**, *348*, 1719–1727. [CrossRef]
69. Termirbulatova, M.G.; Moskovskaya, I.F.; Romanovsky, B.V.; Yatsenko, A.V. MCM-41 mesoporous molecular sieves modified with a base or a palladium-containing ionic liquid as catalysts for certain organic synthesis reactions. *Petrol. Chem.* **2009**, *49*, 7–10. [CrossRef]
70. Trilla, M.; Borja, G.; Pleixats, R.; Man, M.W.C.; Bied, C.; Moreau, J.J.E. Recoverable Palladium Catalysts for Suzuki–Miyaura Cross- Coupling Reactions Based on Organic-Inorganic Hybrid Silica Materials Containing Imidazolium and Dihydroimidazolium Salts. *Adv. Synth. Catal.* **2008**, *350*, 2566–2574. [CrossRef]
71. Isaeva, V.I.; Prokudina, N.I.; Kozlova, L.M.; Kustov, L.M.; Glukhov, L.M.; Tarasov, D.L.; Beletskaya, I.P. Hydroamination of phenylacetylene in the presence of gold-containing catalytic systems supported on carriers modified by ionic liquids. *Russ. Chem. Bull. Int. Ed.* **2015**, *64*, 2811–2815. [CrossRef]

72. Sigeev, A.S.; Beletskaya, I.P.; Petrovskii, P.V.; Peregudov, A.S. Cu(I)/Cu(II)/TMEDA, new effective available catalyst of sandmeyer reaction. *Russ. J. Org. Chem.* **2012**, *48*, 1055–1058. [CrossRef]
73. Zavalinskaya, I.S.; Malikov, I.V.; Yas'yan, Y.P. Conversion of Straight-Run Gasoline Fraction on Combined Zeolite-Containing Catalysts. *Chem. Technol. Fuels Oils* **2015**, *51*, 154–159. [CrossRef]
74. More, S.; Jadhav, S.; Salunkhe, R.; Kumbhar, A. Palladium supported ionic liquid phase catalyst (Pd@SIPL-PS) for room temperature Suzuki-Miyaura cross-coupling reaction. *Mol. Catal.* **2017**, *442*, 126–132. [CrossRef]
75. Li, Y.; Dong, Y.; Li, W.; Han, Y.; Zhang, J. Impovement of imidazolium-based ionic liquids on the activity of ruthenium catalyst for acetylene hydrochlorination. *Mol. Catal.* **2017**, *443*, 220–227. [CrossRef]
76. Sonawane, B.D.; Rashinkar, G.S.; Sonawane, K.D.; Dhanavade, M.J.; Sonawane, V.D.; Patil, S.V. Aerosil-Supported Ionic Liquid Phase (ASILP) mediated synthesis of 2-substituted benzimidazole derivatives as AChE inhibitors. *Chem. Sel.* **2018**, *3*, 5544–5551. [CrossRef]
77. Brunig, J.; Csendes, Z.; Weber, S.; Gorgas, N.; Bittner, R.W.; Limbeck, A.; Bica, K.; Hoffmann, H.; Kirchner, K. Chemoselective Supported Ionic Liquid Phase (SILP) aldehyde hydrogenation catalysed by an Fe(II) PNP Pincer Complex. *ACS Catal.* **2018**, *8*, 1048–1051. [CrossRef]
78. Castro-Amoedo, R.; Csendes, Z.; Brunig, J.; Sauer, M.; Foelske-Schmitz, A.; Yigit, N.; Rupprechter, G.; Gupta, T.; Martins, A.M.; Bica, K.; et al. Carbon-based SILP catalysis for the selective hydrogenation of aldehydes using a well-defined Fe (II) PNP complex. *Catal. Sci. Technol.* **2018**, *8*, 4812–4820. [CrossRef]
79. Gajare, S.; Audumbar, P.; Kale, D.; Bansode, P.; Patil, P.; Rashinkar, G. Graphene oxide-supported ionic Liquid Phase catalysed synthesis of 3,4-dihydro-2H-naphtho[2,3-e][1,3]oxazine-5,10-diones. *Chem. Lett.* **2019**, *150*, 243–255.
80. Urban, B.; Skoda-Foldes, R. Development of palladium catalysts immobilized on supported phosphonium ionic liquid phases. *Phosphorus Sulfur.* **2019**, *194*, 302–306. [CrossRef]
81. Lee, C.; Sanding, B.; Buchmeiser, M.R.; Haumann, M. Supported Ionic Liquid Phase (SILP) facilitated gas-phase enzyme catalysis- CALB catalysed trensestrification of vinyl propionate. *Catal. Sci. Technol.* **2018**, *8*, 2460–2466. [CrossRef]
82. Wolf, P.; Logemann, M.; Schorner, M.; Keller, L.; Haaumann, M.; Wessling, M. Multi-walled carbo nanotube-based composite materials as catalyst support for water-gas shift and hydroformylation reactions. *RSC Adv.* **2019**, *9*, 27732–27742. [CrossRef]
83. Wolf, P.; Aubermann, M.; Wolf, M.; Bauer, T.; Blaumeiser, D.; Stepic, R.; Wick, C.R.; Smith, D.M.; Smith, A.S.; Wasserscheid, P.; et al. Improving the performance of supported ionic liquid phase (SILP) catalysts for the ultra-low-temperature water-gas shift reaction using metal salt additives. *Green Chem.* **2019**, *21*, 5008–5018. [CrossRef]
84. Bryzhin, A.A.; Gantman, M.G.; Buryak, A.K.; Tarkhanova, I.G. Brönsted acidic SILP-based catalysts with $H_3PMo_{12}O_{40}$ i $H_3PW_{12}O_{40}$ in the oxidative desulfurization of fuels. *Appl. Catal. B Env.* **2019**, *257*, 117938–117945. [CrossRef]
85. El Sayed, S.; Bordet, A.; Weidenthaler, C.; Hetaba, W.; Luska, K.L.; Leitner, W. Selective hydrogenation of benzofurans using ruthenium nanoparticles in Lewis acid-modifies Ruthenium-Supported Ionic Liquid Phase. *ACS Catal.* **2020**, *10*, 2124–2130. [CrossRef]
86. Kukawka, R.; Pawlowska-Zygarowicz, A.; Dzialkowska, J.; Pietrowski, M.; Maciejewski, H.; Bica, K.; Smiglak, M. Highly effective Supported Ionic Liquid-Phase (SILP) catalysts: Characterization and application to the hydrosilylation reaction. *Sustain. Chem. Eng.* **2019**, *7*, 4699–4706. [CrossRef]
87. Steinruck, H.P.; Wasserscheid, P. Ionic liquids in catalysis. *Catal. Lett.* **2015**, *145*, 380–397. [CrossRef]
88. Kernchen, U.; Etzold, B.; Korth, W.; Jess, A. Solid Catalyst with Ionic Liquid Layer (SCILL)- a new concept to improve the selectivity investigated for the example of hydrogenation of cyclooctadiene. *Chem. Eng. Technol.* **2007**, *79*, 807–819.
89. Werner, S.; Szesni, N.; Kaiser, M.; Haumann, M.; Wasserscheid, P. A scalable preparation method for SILP and SCILL ionic liquid thin-film materials. *Chem. Eng. Technol.* **2012**, *11*, 1962–1967. [CrossRef]
90. Miller, S.F.; Friedrich, H.B.; Holzapfel, C.W. The effects of SCILL catalyst modification on the competitive hydrogenation of 1-octyne and 1,7-octadiene versus 1-octene. *ChemCatChem* **2012**, *4*, 1337–1344. [CrossRef]
91. Antonels, N.C.; Meijboom, R. Preparation of well-defined dendrimer encapsulated ruthenium nanoparticles and their application as catalyst and enhancement of activity when utilised as SCILL catalysts in the hydrogenation of citral. *Catal. Commun.* **2014**, *57*, 148–152. [CrossRef]

92. Antonels, N.C.; Williams, M.B.; Mejboom, R.; Haumann, M. Well-defined dendrimer encapsulated ruthenium SCILL catalysts for partial hydrogenation of toluene in liquid-phase. *J. Mol. Catal. A Chem.* **2016**, *421*, 156–160. [CrossRef]
93. Jalal, A.; Uzun, A. An exceptional selectivity for partial hydrogenation on a supported nickel catalyst coated with [BMIM][BF$_4$]. *J. Catal.* **2017**, *350*, 86–96. [CrossRef]
94. Lijewski, M.; Hogg, J.M.; Swadzba-Kwasny, M.; Wasserscheid, P.; Haumann, M. Coating of PD/C catalysts with Lewis-acidic ionic liquids and liquid coordination complexes—SCILL induced activity enhancement in arene hydrogenation. *RSC Adv.* **2017**, *7*, 27558–27563. [CrossRef]
95. Bart, T.; Korth, W.; Jess, A. Selectivity-enhancing effect of a SCILL catalyst in butadiene hydrogenation. *Chem. Eng. Technol.* **2017**, *40*, 395–404. [CrossRef]
96. Hu, S.; Wang, L.; Chen, X.; Wei, X.; Tong, Z.; Yin, L. The conversion of α-pinene to cis-pinane using a nickel catalyst supported on a discarded fluid catalytic cracking catalyst with an ionic liquid layer. *RSC Adv.* **2019**, *9*, 5978–5986. [CrossRef]
97. Bauer, T.; Hager, V.; Williams, M.B.; Laurin, M.; Dopper, T.; Gorling, A.; Szesni, N.; Wasserscheid, P.; Haumann, M.; Libuda, J. Palladium-mediated ethylation of the imidazolium cation monitored in-operando on a SCILL-type catalyst. *ChemCatChem* **2017**, *9*, 109–113. [CrossRef]
98. Perdikaki, A.V.; Vangeli, O.C.; Karanikolos, G.N.; Stefanopoulos, K.L.; Beltsios, K.G.; Alexandridis, P.; Kanellopoulus, N.K.; Romanos, G.E. Ionic liquid-modified porous materials for gas separation and heterogenous catalysis. *J. Phys. Chem. C* **2012**, *31*, 16398–16411. [CrossRef]
99. Podolean, I.; Pavel, O.D.; Manyar, H.G.; Taylor, S.F.R.; Ralphs, K.; Goodrich, P.; Parvulescu, V.I.; Hardacre, C. SCILLs as selective catalysts for oxidation of aromatic alcohols. *Catal. Today* **2019**, *333*, 140–146. [CrossRef]
100. Pavel, O.D.; Podolean, I.; Parvulescu, V.I.; Taylor, S.F.R.; Manyar, H.; Ralphs, K.; Goodrich, P.; Hardacre, C. Impact of SCILL catalysts for the S-S coupling of thiols to disulfides. *Faraday Discuss.* **2018**, *206*, 535–547. [CrossRef] [PubMed]
101. Mayer, C.; Hager, V.; Schwieger, W.; Wasserscheid, P. Enhanced activity and selectivity in n-octane isomerization using a bifunctional SCILL catalyst. *J. Catal.* **2012**, *292*, 157–165. [CrossRef]
102. Li, T.; Zhang, W.; Chen, W.; Miras, H.N.; Song, Y.F. Layered double hydroxide anchored ionic liquids as amphiphilic heterogeneous catalysts for the Knoevenagel condensation reaction. *Dalton Trans.* **2018**, *47*, 3059–3067. [CrossRef]
103. O'Reilly, N.; Giri, N.; James, S.L. Porous liquids. *Chem. Eur. J.* **2007**, *13*, 3020–3025. [CrossRef]
104. Zhang, S.; Dokko, K.; Watanabe, M. Porous ionic liquids: Synthesis and application. *Chem. Sci.* **2015**, *6*, 3684–3691. [CrossRef]
105. Wilke, A.; Yuan, J.; Antonietti, M.; Weber, J. Enhanced Carbon Dioxide Adsorption by a Mesoporous Poly(ionic liquid). *ACS Macro Lett.* **2012**, *1*, 1028–1031. [CrossRef]
106. Huang, J.; Tao, C.; An, Q.; Zhang, W.; Wu, Y.; Li, X.; Shen, D.; Li, G. Visual indication of enviromental humidity by using poly(ionic liquid) photonic crystals. *Chem. Commun.* **2010**, *46*, 4103–4105. [CrossRef]
107. Eftekhari, A. Polymerized Ionic Liquids. *R. Soc. Chem.* **2018**, *2*, 23–82.
108. Jang, J.; Bae, J. Fabrication of mesoporous polymer using soft template method. *Chem. Commun.* **2005**, 1200–1202. [CrossRef] [PubMed]
109. Liu, F.; Wang, L.; Sun, Q.; Zhu, L.; Meng, X.; Xiao, F.S. Transesterification Catalysed by Ionic Liquids on Superhydrophobic Mesoporous Polymers: Heterogeneous Catalysts That Are Faster than Homogeneous Catalysts. *J. Am. Chem. Soc.* **2012**, *134*, 16948–16950. [CrossRef] [PubMed]
110. Zhao, Q.; Dunlop, J.W.C.; Qiu, X.; Huang, F.; Zhang, Z.; Heyda, J.; Dziubiella, J.; Antonielli, M.; Yuan, J. An instant multi-responsive porous polymer actuator driven by solvent molecule sorption. *Nat. Commun.* **2014**, *5*, 4293–4300. [CrossRef] [PubMed]
111. Zhao, Q.; Heyda, J.; Dziubiella, J.; Tauber, K.; Dunlop, J.W.C.; Yuan, J. Sensing solvents with ultrasensitive porous poly(ionic liquid) actuators. *Adv. Mater.* **2015**, *27*, 2913–2917. [CrossRef] [PubMed]
112. Yu, Y.; Hu, T.; Chen, X.; Xu, K.; Zhang, J.; Huang, J. Pd nanoparticles on a porous ionic copolymer: A highly active and recyclable catalyst for Suzuki-Miyaura reaction under air in water. *Chem. Commun.* **2011**, *47*, 3592–3594. [CrossRef] [PubMed]
113. Zhao, P.; Leng, Y.; Wang, J. Heteropolyanion-paired cross-linked ionic copolymer: An efficient heterogeneous catalyst for hydroxylation of benzene with hydrogen peroxide. *Chem. Eng. J.* **2012**, *204–206*, 72–78. [CrossRef]

114. Zhao, Q.; Zhang, P.; Antonielli, M.; Yuan, J. Poly(ionic liquid) complex with spontaneous micro-mesoporosity: Template-free synthesis and application as catalyst support. *J. Am. Chem. Soc.* **2012**, *134*, 11852–11855. [CrossRef]
115. Gao, C.; Chen, G.; Wang, X.; Li, J.; Zhou, Y.; Wang, J. Hierarchical meso-macroporous poly(ionic liquid) monolith derived from single soft template. *Chem. Commun.* **2015**, *51*, 4969–4972. [CrossRef]
116. Wang, B.; Sheng, X.; Zhou, Y.; Zhu, Z.; Liu, Y.; Sha, X.; Zhang, C.; Gao, H. Functional mesoporous poly(ionic liquid) derived from P123: From synthesis to catalysis and alkylation of styrene and o-xylene. *Appl. Organomet. Chem.* **2019**, *33*, 4719–4730. [CrossRef]
117. Sheng, X.; Gao, H.; Zhou, Y.; Wang, B.; Sha, X. Stable poly (ionic liquids) with unique cross-linked mesoporous-macroporous structure as efficient catalyst for alkylation of *o*-xylene and styrene. *Appl. Organomet. Chem.* **2019**, *33*, 4979–4989. [CrossRef]
118. Wang, X.; Li, J.; Chen, G.; Guo, Z.; Zhou, Y.; Wang, J. Hydrophobic mesoporous poly(ionic liquid)s towards highly efficient and contamination-resistant solid-base catalyst. *ChemCatChem* **2015**, *6*, 993–1003. [CrossRef]
119. Tao, D.J.; Liu, F.; Wang, L.; Jiang, L. A green and efficient hydration of alkynes catalysed by hierarchically porous poly(ionic liquid)s solid strong acids. *Appl. Catal. A Gen.* **2018**, *564*, 56–63. [CrossRef]
120. Hui, W.; Heb, X.M.; Xua, X.Y.; Chen, Y.M.; Zhou, Y.; Li, Z.M.; Zhang, L.; Tao, D.J. Highly efficient cycloaddition of diluted and waste CO_2 into cyclic carbonates catalyzed by porous ionic copolymers. *J. CO_2 Util.* **2020**, *36*, 169–176. [CrossRef]
121. Wang, Y.; Guo, L.; Yin, L. Progress in the Heterogeneous Catalytic Cyclization of CO_2 with Epoxides Using Immobilized Ionic Liquids. *Catal. Lett.* **2019**, *149*, 985–997. [CrossRef]
122. Fujie, K.; Yamada, T.; Ikeda, R.; Kitagawa, H. Introduction of an ionic liquid into the micropores of a metal–organic framework and its anomalous phase behavior. *Angew. Chem. Int. Ed.* **2014**, *53*, 11302–11305. [CrossRef]
123. Luo, Q.X.; An, B.W.; Ji, M.; Park, S.E.; Hao, C.; Li, Y.Q. Metal–organic frameworks HKUST-1 as porous matrix for encapsulation of basic ionic liquid catalyst: Effect of chemical behaviour of ionic liquid in solvent. *J. Porous Mater.* **2015**, *22*, 247–259. [CrossRef]
124. Luo, Q.; Song, X.; Ji, M.; Park, S.E.; Hao, C.; Li, Y. Molecular size- and shape-selective Knoevenagel condensation over microporous $Cu_3(BTC)_2$ immobilized amino-functionalized basic ionic liquid catalyst. *Appl. Catal. A* **2014**, *478*, 81–90. [CrossRef]
125. Fujie, K.; Kitagawa, H. Ionic liquid transported into metal–organic frameworks. *Coord. Chem. Rev.* **2016**, *307*, 382–390. [CrossRef]
126. Khan, N.A.; Hasan, Z.; Jhung, S.H. Ionic liquid@MIL-101 prepared *via* the ship-in-bottle technique: Remarkable adsorbents for the removal of benzothiophene from liquid fuel. *Chem. Commun.* **2016**, *52*, 2561–2564. [CrossRef]
127. Fujie, K.; Otsubo, K.; Ikeda, R.; Yamada, T.; Kitagawa, H. Low temperature ionic conductor: Ionic liquid incorporated within a metal–organic framework. *Chem. Sci.* **2015**, *6*, 4306–4310. [CrossRef]
128. Sun, Y.; Jia, X.; Huang, H.; Guo, X.; Qiao, Z.; Zhong, C. Solvent-free mechanochemical route for the construction of ionic liquid and mixed-metal MOF composites for synergistic CO_2 fixation. *J. Mater. Chem. A* **2020**, *8*, 3180–3185. [CrossRef]
129. Gomes, M.C.; Pison, L.; Cervinka, C.; Padua, A. Porous Ionic Liquids or Liquid Metal-Organic Frameworks? *Angew. Chem. Int. Ed.* **2018**, *57*, 11909–11912. [CrossRef] [PubMed]
130. Ferreira, T.J.; Ribeiro, R.P.P.L.; Mota, J.P.B.; Rebelo, L.P.N.; Esperança, J.M.S.S.; Esteves, I.A.A.C. Ionic Liquid-Impregnated Metal–Organic Frameworks for CO_2/CH_4 Separation. *ACS Appl. Nano Mater.* **2019**, *2*, 7933–7950. [CrossRef]
131. Liu, S.; Liu, J.; Hou, X.; Xu, T.; Tong, J.; Zhang, J.; Ye, B.; Liu, B. Porous Liquid: A Stable ZIF-8 Colloid in Ionic Liquid with Permanent Porosity. *Langmuir* **2018**, *34*, 3654–3660. [CrossRef]
132. Ma, L.; Haynes, C.J.E.; Grommet, A.B.; Walczak, A.; Parkins, C.C.; Doherty, C.M.; Longley, L.; Tron, A.; Stefankiewicz, A.R.; Bennett, T.D.; et al. Coordination cages as permanently porous ionic liquids. *Nat. Chem.* **2020**, *12*, 270–275. [CrossRef]
133. Prechtl, M.H.G. (Ed.) *Nanocatalysis in Ionic Liquids*; Wiley-VCH: Weinheim, Germany, 2017.
134. Verma, C.; Ebenso, E.E.; Quraishi, M.A. Transition metal nanoparticles in ionic liquids: Synthesis and stabilization. *J. Mol. Liq.* **2019**, *276*, 826–849. [CrossRef]

135. Hesemann, P. Applications of ionosilicas in heterogeneous catalysis: Opportunities for the elaboration of new functional catalytic phases. *Curr. Opin. Green Sustain. Chem.* **2018**, *10*, 21–26. [CrossRef]
136. Rajendran, A.; Rajendiran, M.; Yang, Z.F.; Fan, H.X.; Cui, T.Y.; Zhang, Y.G.; Li, W.Y. Functionalized Silicas for Metal-Free and Metal-Based Catalytic Applications: A Review in Perspective of Green Chemistry. *Chem. Rec.* **2019**, *19*, 1–29. [CrossRef]
137. Mecerreyes, D. (Ed.) *Applications of Ionic Liquids in Polymer Science and Technology*; Springer: Heidelberg, Germany, 2015.
138. Li, P.; Cheng, F.-F.; Xiong, W.-W.; Zhang, Q. New synthetic strategies to prepare metal-organic frameworks. *Inorg. Chem. Front.* **2018**, *5*, 2693–2708. [CrossRef]
139. Kore, R.; Berton, P.; Kelley, S.P.; Aduri, P.; Katti, S.S.; Rogers, R.D. Group IIIA halometallate ionic liquids: Speciation and applications in catalysis. *ACS Catal.* **2017**, *7*, 7014–7028. [CrossRef]
140. Jankowska-Wajda, M.; Bartlewicz, O.; Walczak, A.; Stefankiewicz, A.R.; Maciejewski, H. Highly efficient hydrosilylation catalysts based on chloroplatinate "ionic liquids". *J. Catal.* **2019**, *374*, 266–275. [CrossRef]
141. Zhang, H.; Xu, C.C.; Yang, S. Heterogeneously chemo/enzyme-functionalized porous polymeric catalysts of high-performance for efficient biodiesel production. *ACS Catal.* **2019**, *9*, 10990–11029. [CrossRef]

Publisher's Note: MDPI stays neutral with regard to jurisdictional claims in published maps and institutional affiliations.

© 2020 by the authors. Licensee MDPI, Basel, Switzerland. This article is an open access article distributed under the terms and conditions of the Creative Commons Attribution (CC BY) license (http://creativecommons.org/licenses/by/4.0/).

Article

Tuned Bis-Layered Supported Ionic Liquid Catalyst (SILCA) for Competitive Activity in the Heck Reaction of Iodobenzene and Butyl Acrylate

Nemanja Vucetic [1,*], Pasi Virtanen [1], Ayat Nuri [1,2], Andrey Shchukarev [3], Jyri-Pekka Mikkola [1,3] and Tapio Salmi [1]

1. Laboratory of Industrial Chemistry and Reaction Engineering, Johan Gadolin Process Chemistry Centre, Åbo Akademi University, Biskopsgatan 8, FI-20500 Turku/Åbo, Finland; pasi.virtanen@abo.fi (P.V.); ayatnuri@gmail.com (A.N.); jyri-pekka.mikkola@abo.fi (J.-P.M.); tapio.salmi@abo.fi (T.S.)
2. Department of Applied Chemistry, Faculty of Science, University of Mohaghegh Ardabili, Ardabil 56199-11367, Iran
3. Technical Chemistry, Department of Chemistry, Umeå University, SE-90187 Umeå, Sweden; andrey.shchukarev@umu.se
* Correspondence: nvucetic@abo.fi; Tel.: +358-41-703-7071

Received: 17 June 2020; Accepted: 19 August 2020; Published: 22 August 2020

Abstract: A thorough experimental optimization of supported ionic liquid catalyst (SILCA) was performed in order to obtain a stable and efficient catalyst for the Heck reaction. Out of fifteen proposed structures, propyl imidazolium bromide-tetramethylguanidinium pentanoate modified SiO_2 loaded with $PdCl_2$ appeared to be the most stable and to have a good activity in the reaction between butylacrylate and iodobezene, resulting in a complete conversion in 40 min at 100 °C, in four consecutive experiments. This study elucidated on the stability of the catalytic system with an ionic liquid layer during the catalyst synthesis but also under reaction conditions. In the bis-layered catalyst, the imidazolium moiety as a part of internal layer, brought rigidity to the structure, while in external layer pentanoic acid gave sufficiently acidic carboxylic group capable to coordinate 1,1,3,3-tetramethylguanidine (TMG) and thus, allow good dispersion of Pd nanoparticles. The catalyst was characterized by means of XPS, FT-IR, TEM, ICP-OES, ζ-potential, EDX, TGA, and ^{13}C NMR. The release and catch mechanism was observed, whereas Pd re-deposition can be hindered by catalyst poisoning and eventual loss of palladium.

Keywords: supported ionic liquid catalyst (SILCA); palladium; Heck reaction; catalyst screening; optimization

1. Introduction

Upon production of fine chemicals the palladium catalyzed coupling reactions are well known and appreciated for their capability to yield versatile intermediates with preserved functionalities [1]. Among various kinds of reactions, the Heck reaction is one of the most important ones and is used for the production of intermediates for advanced synthesis of pharmaceuticals. This reaction reported by both Heck and Mizoroki [2,3] couples olefins and acrylates with molecules possessing suitable leaving groups such as aryl or vinyl (allyl) halides, triflates, acetates, etc. Next to the transition metal catalyst, it requires a base and it benefits from high temperatures. Nowadays, the reaction is commercially applied in the production of herbicides, sunscreen agents, coatings of electronic compounds, antiasthma agents, and various drugs [4].

Initially, Heck reactions were catalyzed by homogenous palladium, but the major problem of such a process is the catalyst instability which limits the number of turnovers that can be reached with a single charge of catalysts. Due to the steady increase of the palladium price, there is a pressure on researchers to come up with recoverable catalysts. The key issue is to stabilize palladium nanoparticles, which can be achieved by introducing ligands, although, however, they can hinder the product separation and increase the total costs because the price of the tailored homogeneous ligands can be high (5000–20,000 euro/kg for chiral bisphosphine ligands) [5]. Additionally, product contamination can be an issue since the maximum limit of residual Pd acceptable in the products of pharmaceutical industry ranges from 0.5 ppm (parenteral) to 5 ppm (oral) [6]. In order to avoid expensive nanofiltration membranes or column chromatography separations (extensive and usually destructive), the emission of metal and presence in the final product should be controlled already within the chemical synthesis step.

The scientific community is now turning towards the heterogeneously catalyzed Heck reactions and avoids the use of homogeneous ligands. However, despite all the benefits that a heterogeneous catalyst can bring into the system, the major drawbacks are the price of the catalyst in correlation with the synthesis route, the lower activity and selectivity compared to homogeneous catalysts, and in some cases, the contamination of the products with leached metal. At the same time, this is a reason why a limited number of heterogeneous catalysts are used for industrial production of fine chemicals and pharmaceuticals. However, there is a lot of interest to develop inexpensive, robust, and highly active heterogeneous catalysts.

The mechanism of heterogeneously catalyzed Heck reaction is still a controversial issue. As it seems for now, the reaction path can follow a real heterogeneous route [7–10] or "so to say" pseudo-heterogeneous route where the "catalyst" is actually a precursor or reservoir for active spices often discussed as a "release and catch" mechanism or boomerang catalysis [11–14]. In the case of the pseudo-heterogeneous route, palladium is dissolved from the carrier surface under reaction conditions (mainly in the presence of an aryl halide) thus forming highly active coordinatively unsaturated Pd species that takes a part in the reaction and upon the completion redeposited back on the surface. Next to it, under the release and catch concept there were also observations of the catalysts with non-covalently immobilized catalytic moiety on the suitable support. In these cases, the whole catalytic moiety, i.e., Pd-ligand couple was identified as active spices that was leached and recaptured back on the surface at the end of the reaction [15].

Unfortunately, many truly heterogeneous studies overlook the nature of the palladium entering the reaction cycle without monitoring the reaction heterogeneity with, for example, hot filtrate tests, mercury poisoning tests, poly(vinylpyridine) polymer (PVPy) tests [16] or sometimes without appropriate characterization of used catalysts. Other reasons for not taking into account the real nature of the catalytically active species is connected to the difficulties in studying it. Operando investigations of Heck reaction requires advanced characterization techniques. Only few studies quantitatively investigate palladium cluster formation and Pd^{2+} reduction with in situ UV-visible spectroscopy [17]. Meanwhile, leaching of the palladium under reaction conditions can be monitored by quick scanning extended X-ray absorption fine structure (QEXAFS) studies, even in the sub second time scale, thus giving a deep insight into the reaction mechanism [18]. In addition to the point, if the researcher is studying the mechanism or heterogeneity, it is important to distinguish which one of these two mechanisms is actually prevailing before going deeper into the catalyst or process design. Obviously, it would be devastating to use a pseudo-heterogeneous catalyst in a continuous flow reactor.

Throughout the years, N-heterocyclic carbenes have appeared as great ligands in coordination chemistry due to their strong σ-donor and π-acceptor characteristics [19,20]. In palladium catalyzed reactions they have been proposed as a replacement for phosphine ligands as being non-toxic, less expensive, and with a higher thermal stability [21–24]. In the chemistry of Heck reactions they have been studied in their heterogenized forms as molecules anchored on the surface of the carrier [16] and more lately as supported ionic liquids [25–27] to give what is today called a supported ionic liquid catalyst (SILCA).

In our previous work, we developed a catalyst that operates through the release and catch mechanism and utilizes N-heterocyclic carbene, CO, and tetramethylguanidine ligands in an ionic liquid form [28]. The obtained results encouraged us to continue in the same direction and here we present a set of new catalysts derived from the first one. The ionic liquid layer was optimized to meet the requirements in terms of the balance between the catalyst activity and reusability in the Heck reaction of iodobenzene and butyl acrylate. The screening resulted in a $PdCl_2$ based catalyst with propyl imidazolium bromide as a first layer and tetramethylguanidinium pentanoate as a second (external) layer, which was stable in four successive experiments with a turnover frequency of 1660 h^{-1} at the temperature of 100 °C. The catalysts structures were studied and characterized with relevant techniques such as XPS, FT-IR, TEM, ICP-OES, ζ-potential, EDX, TGA, and ^{13}C NMR.

2. Results and Discussion

The new group of catalysts which were synthesized, characterized, and tested in the Heck reaction are discussed. These new catalysts were synthesized based on the modification of the previously reported procedure developed by our group [28] with a detailed experimental optimization of the ionic liquid layer structure in order to gain a higher activity and better stability in the Heck reaction. Smart adjusting of the ionic liquid layers was commenced in order to make a heterogenized ligand system capable of stabilizing the palladium nanoparticles. The anticipated structures of the synthesized SILCAs are displayed in Table 1.

Therefore, the obtained catalysts were tested for the Heck reaction of iodobenzene and butyl acrylate at 100 °C, displayed in Figure 1. On the basis of our recent studies N-Methyl-2-pyrrolidone was used as a solvent, triethanolamine as the base, and catalyst loading was 0.09 mol % of Pd [28]. The reaction was performed until completion and next to the main product—butyl cinnamate, no other products were detected. Not too high reaction temperature was chosen in order to keep the palladium cluster size as low as possible since this affects the amount of palladium that actually enrolls in the catalytic cycle. In order to compare the catalysts, the activity of the fresh catalyst in the first reaction cycle was calculated (shown as a turnover frequency, TOF) and the stability was taken into consideration, too. The stability was evaluated on the basis of the number of the reaction cycles that the catalysts were capable to catalyze until completion with no need for prolonged reaction time compared to that for the fresh catalyst.

It was noticeable that the reaction is strongly dependent on the reaction conditions, so any deviations in temperature and catalysts amount can result in a prolonged reaction time. This imposed difficulties when recycling the catalysts. Since very small amounts of material were handled here (approx. 10 mg), even loss of just 1 mg of catalyst between the cycles can have a significant impact on the outcome of the next experiment. Thus, recyclability tests were done multiple times (3–6) and the average (rounded) values are reported.

The reaction itself happens in the liquid phase while auxiliaries on the SiO_2 have a role to stabilize the palladium re-taken at the end of the reaction cycle. In the presence of aryl halide, the palladium leaches and enters the cycle through the oxidative addition step and up on halide consumption it is redeposited back on the carrier [13,29,30]. Otherwise, in the absence of supported ionic liquid or any other ligand-like molecule, there is a serious threat that the palladium can form an inactive palladium back and precipitate. Therefore, the adequate choice of the moiety will make the support capable of efficiently releasing and catching the palladium.

Table 1. Schematic presentation of assumed structures of synthetized supported ionic liquid catalysts (SILCAs) with variable auxiliaries.

SILCA	Predicted Structure	SILCA	Predicted Structure
1	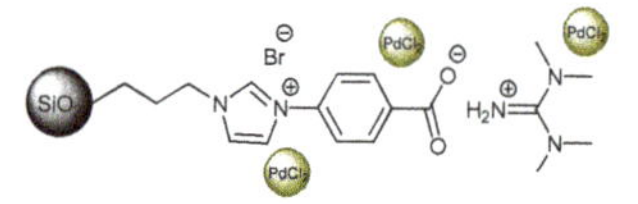	9	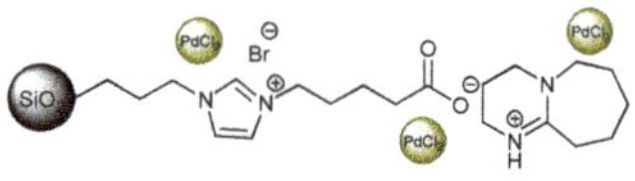
2	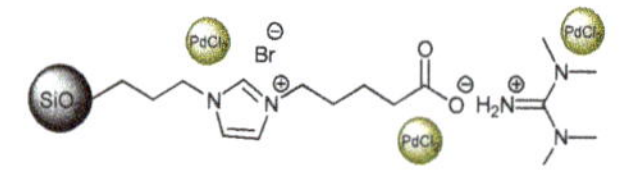	10	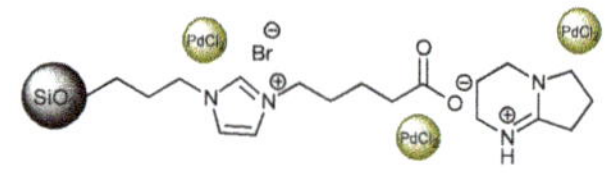
3	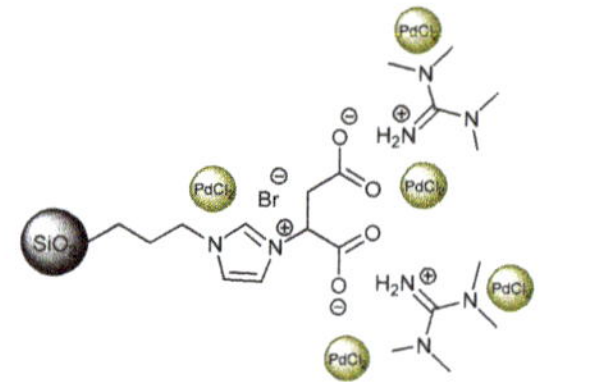	11	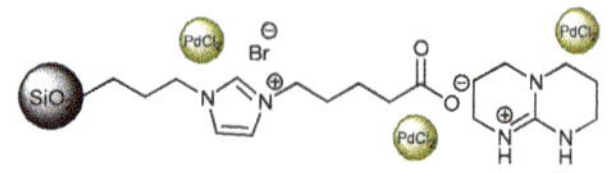
4		12	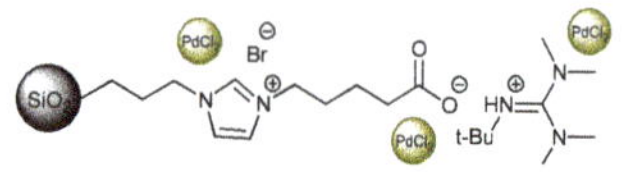
5	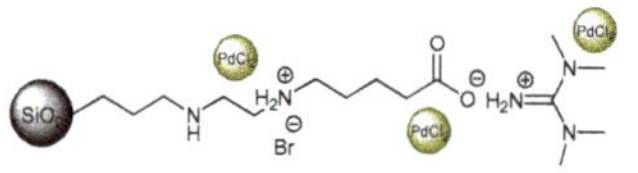	13	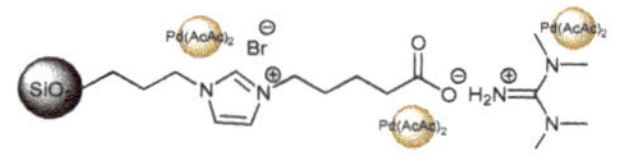
6	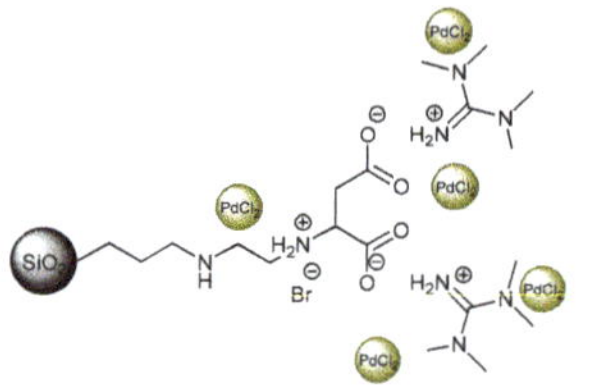	14	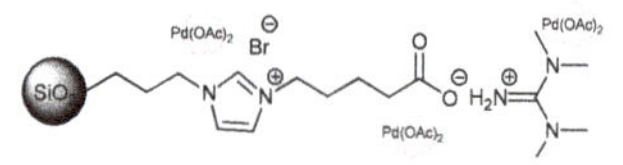
7	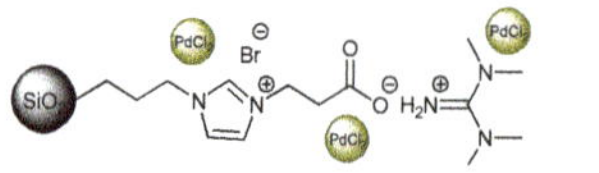	15	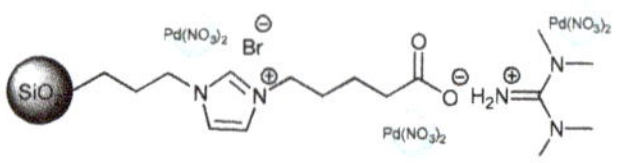
8	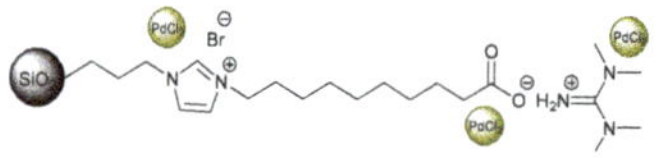		

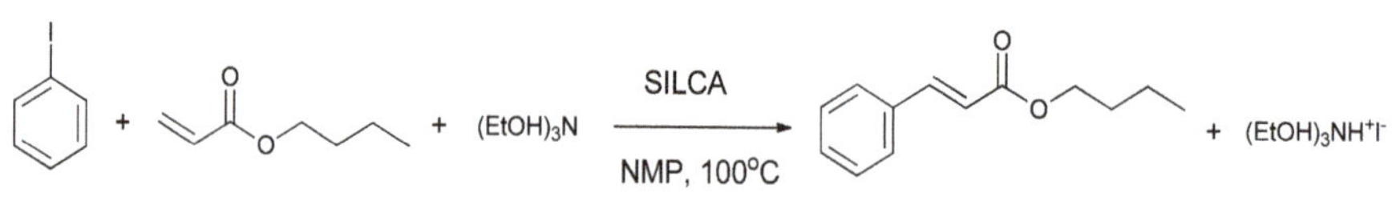

Figure 1. Heck reaction of iodobenzene and butyl acrylate in the presence of triethanolamine and different SILCAs, used to evaluate catalyst performance.

In the first set of experiments, catalysts with different acidic heads were compared (SILCAs 1–3, Figure 2). In terms of activity and stability, SILCA 2 gave the best results leading to reaction completion in 40 min and enabling the reuse of the catalyst in four consecutive cycles under identical reaction conditions. In the case of SILCA 1 and SILCA 3, a prolonged reaction time was needed (90 and 50 min) and the stability of the catalyst decreased to approximately two and three reaction cycles. To gain a deeper understanding of these reaction outcomes, the catalysts were characterized with different techniques, of which the most valuable analytics were FT-IR, XPS, TEM, and ζ-potential measurements. Firstly, by means of FT-IR and XPS, the surface composition was overviewed. The obtained FT-IR scans are shown in Figure 3. Comparing the scans of the first three catalysts (Figure 3a, SILCAs 1–3), small differences were observed in the peak at 1565 cm^{-1} which was assigned to C=N vibrations of the imidazole ring and/or TMG. This peak was most evident in the case of SILCA 2 and a potential hint of the rigidity of this C=N bond during the catalyst preparation. Unfortunately, the FT-IR spectra can just indicate the presence of various bonds and groups in a sample, but they cannot give information on how these bonds are arranged on the surface. Therefore, for the additional information on the surface composition, FT-IR was coupled with XPS. The XPS helped verify promptly if the real surface composition correspond to the expected composition, i.e., those presented in Table 1. The obtained results are presented in Figure 4 as atomic concentrations of C, N, and Pd at the surface for certain samples (Figure 4a) and percent amount of different chemical states fitted for C 1s, N 1s, and Pd 3d photoelectron lines (Figure 4b–d) whereas all the peak components are assigned to relevant functional groups on the basis of the literature data [31,32]. The presence of –CF$_2$ was noticeable in all the samples that were in contact with Teflon lined magnetic bars in DMF at high temperature, during the course of catalyst synthesis. When glass lined magnetic bars were used instead, there was no fluorine contamination (Figure 4b, SILCAs 9–12). However, it complicated the synthesis much more due to fragility of the glass bars. Due to the fluorocarbon inertness, no significant attention was given to this kind of contamination.

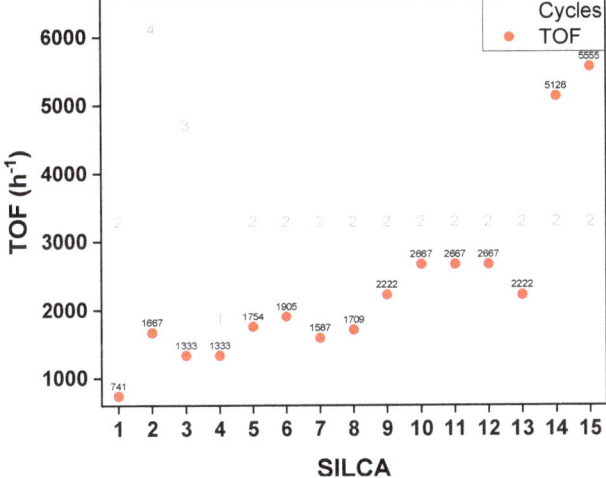

Figure 2. Catalytic performances of the SILCAs with different auxiliaries and palladium sources. Activity presented as a turnover frequency (TOF, h^{-1}) calculated at full conversion. Recyclability presented as rounded average numbers of obtained reaction cycles.

Comparing the XPS of SILCA 1–3, the first evident difference was in the nitrogen group content (Figure 4c). For SILCAs 1 and 2, the peaks assigned to amino group and quaternary ammonium cation of imidazole and/or tetramethylguanidine species were found, and thus, indicating on the ionic nature of the layer (see details in Supplementary Materials). These groups were absent in the spectra of SILCA 3, and instead, amide C=O group was observed. This was resulting from the deterioration of imidazole ring by bromosuccinate, absence of TMG species, and mostly leaching of the nitrogen compounds during the catalyst synthesis, as is also evident when looking at Figure 4a. This means that the obtained structure does not correspond to the expected structure illustrated in Table 1. Interestingly, this catalyst still turned out to be active in the test reaction for three consecutive experiments (Figure 2). The main difference between SILCA 1 and more active and stable SILCA 2 was the amount of acidic groups, the latter one having a higher amount of them: 2% vs. 5%. It was assumed that the main function of the O–C=O heads was to coordinate and keep the TMG cation which has an important role in terms of stabilizing and recapturing palladium nanoparticles [28]. Indeed, SILCA 2 turned out to not lose the activity in four consecutive cycles.

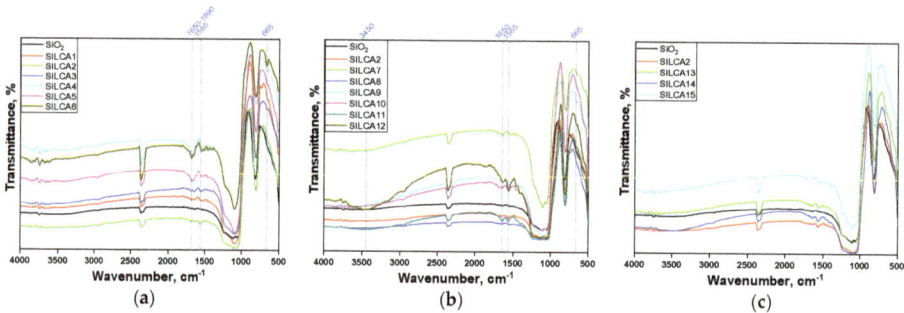

Figure 3. FT-IR spectra of reference materials and synthesized SILCAs 1-6 (**a**), SILCAs 7-12 (**b**) and SILCAs 13-15 (**c**).

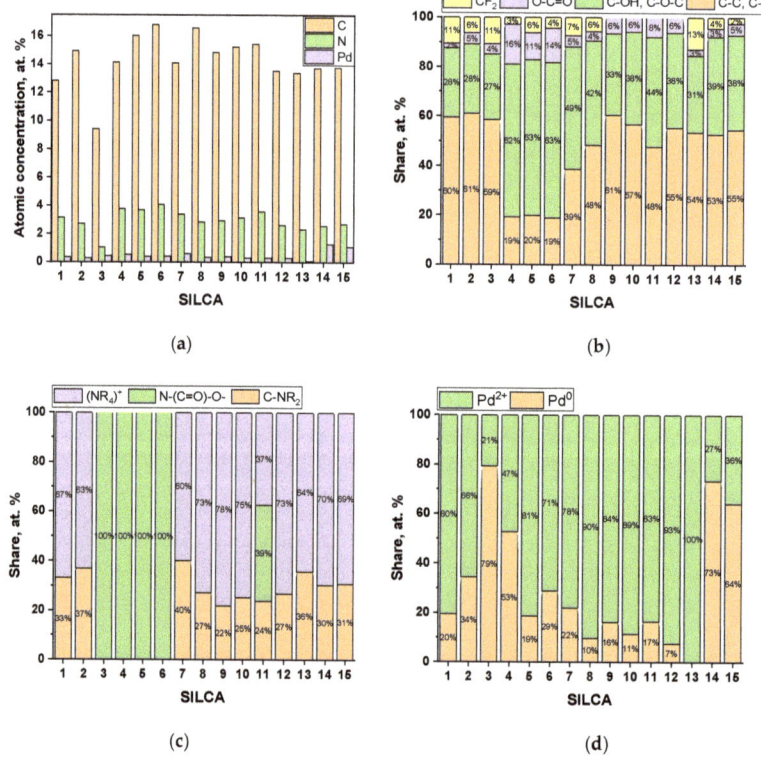

Figure 4. XPS data for synthesized catalysts. (**a**) Atomic concentrations of C, N, and Pd; (**b**) percent fractions of different carbon chemical states obtained by C 1s line fitting; (**c**) percent fractions of different nitrogen chemical states obtained by N 1s line fitting; (**d**) percent fractions of different palladium chemical states (after first 5 min of irradiation) obtained by Pd 3d line fitting.

ζ—potential measurements were carried out in the absence of more sophisticated techniques which could confirm the basicity of the modified surface that is expected to be in correlation with the stabilization of the Pd nanoparticles. Temperature programmed desorption (TPD) of CO_2 is generally used to estimate the basicity of catalyst surface; however, this should not be used for supported ionic liquids due to potential CO_2 absorption and the limited thermal stability of the material (in this case 250 °C—detected by TGA). The isoelectric point (IEP) derived from ζ—potential measurements indicates the global basicity or acidity of the molecule [33]. The molecules with a higher value of IEP can be considered more basic and more negative. Applied on heterogeneous catalytic systems, the more negative charged the surface is, the more firm the Pd^{2+} adhesion would be and, thus, the catalyst would have a better recyclability [34] and on the contrary, the more positively charged the surface is, the faster is the leaching of the cations and the reaction rate. In this context, it should be noted that if the pH value of the solution is less than the IEP, the surface charge is positive, if it is higher, then it is negative and capable of adsorbing cations. Most cross-coupling reactions are conducted under basic conditions at pH around 10, which in turn, is in most of the cases above the IEPs [10].

The IEP obtained from ζ—potential measurements of 12 catalysts (see Supplementary Materials) and with different surface auxiliaries are presented in Figure 5. The results show that among the tested catalysts, SILCA 2 gives the highest value of isoelectric point ~6.9 pH, and potentially the strongest basicity and capability to stabilize palladium, and at the same time, that those with lower value of IEP are more active in the reaction (probably due to the faster Pd leaching) such as in the case of SILCAs 9–12. However, since ζ—potential measurements were conducted in an aqueous solution of the catalyst, they should be considered with skepticism due to the possibility of TMG hydrolysis [35]. Other potential important insight gained by this analysis is the stability of the materials, namely, for many of the catalysts two additional isoelectric points were detected, above 6 pH (IEP2) and around 7 pH (IEP3), as well as lack of IEP in the case of SILCAs 7 and 8. The appearance of more than one isoelectric point can imply on leaching and instability of the analyzed materials under analysis conditions [36]. This could mean that these modifications are unstable under the reaction conditions (pH > 7) which can cause Pd leaching and catalyst deactivation. The obtained results reflected the difficulties in correlating the catalyst features with their structures and in predicting the catalyst performance.

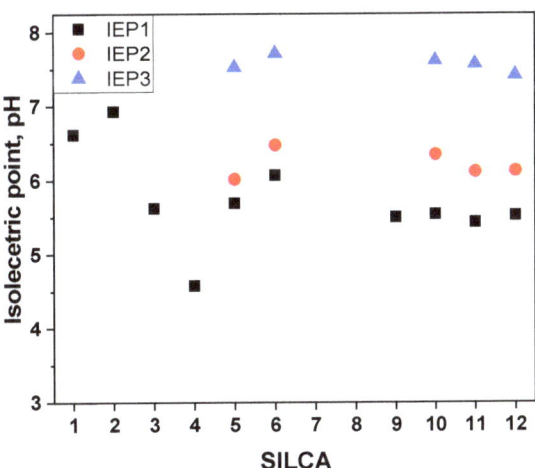

Figure 5. The values of isoelectric points (IEPs) for SILCAs 1–12 obtained with zeta potential measurements in water solution, at room temperature measured in the pH range from 2–8.

The FT-IR analysis of SILCAs 4–6 whereupon the imidazole modification was exchanged with aminoethylamino group, showed the main differences in the peaks at 1650–1690 cm^{-1} which is assumed to result from the contribution of amide C=O and acidic C=O, alongside with the difference in the peak at 665 cm^{-1} belonging to the N-H wagging of the secondary amine. Amide formation was also confirmed with XPS, showing that all the present nitrogen is within the N-(C=O)-O- group with no TMG coordinated. This disapproved the expected structures in Table 1 indicating that neither succinic acid, nor aminoethylamino group can form ionic liquid layer under the given conditions and further led to the continuation of the work with SILCA 2-like structures as the most stable and recyclable ones.

Considering the FT-IR spectra of the SILCAs 7 and 8, the catalysts with the different chain lengths compared to SILCA 2, a visible stronger peak intensity of SILCA 7 was detected. The peaks assigned to nitrogen bonds (at 665 and 1565 cm^{-1}) and to C=O bond at 1650 cm^{-1} are seen on SILCA 7 which has shorter C–C chain length, and therefore, higher N/C and O/C ratio. The dominance of nitrogen related groups is also visible from XPS results, whereas the higher relative amount of N and lower amount of C is found in SILCA 7 or more precisely higher amount of C–NR$_2$, C–OH (C–O–C), and lower amount of C–C (C–H) (see Figure 4a–c).

Shortening (SILCA 7) and prolonging (SILCA 8) the carbon chain length of SILCA 2 did not bring any significant improvement in the catalyst activity but affected negatively the catalyst recyclability (Figure 2)—for no obvious reason at this moment. Therefore, in the following work the original modification of SILCA 2 was maintained and different super bases were screened for coordination with the acidic head as shown in Table 1 for SILCAs 9–12.

Applying bicyclic amidines such as in the case of SILCAs 9 and 10 and guanidines for SILCA 11–12, respectively, brought slight differences in the FT-IR spectra. All the expected peaks were still present but more prominently, with the addition of a broad peak in the area of 3450 cm^{-1} which was assigned to CO–H stretching of acidic head implying on inefficient proton migration to the superbase. In the case of SILCA 11, the formation of amide C=O group was detected at 1650 cm^{-1} and was confirmed with XPS analysis indicating partial deterioration of the anchored modification by triazabicyclodecene superbase. New changes did not bring any improvement in terms of the catalyst recyclability. The achieved TOF was in the range of 2200–2700 h^{-1} and the activity was preserved in two consecutive cycles. Therefore, in the final study for SILCAs 13–15, the different sources of palladium were investigated with SILCA 2 structures. Instead of PdCl$_2$ the Pd(AcAc)$_2$, Pd(OAc)$_2$, and Pd(NO$_3$)$_2$ were utilized and analyzed. As expected, no significant differences were observed in the FT-IR spectra (Figure 3c). The main difference was found in the palladium amount of SILCA 13. Almost three times lower amount of palladium was impregnated in this case, which was firstly observed during the XPS analysis, and also confirmed with the ICP-OES analysis as tabulated in Table 2. The average amount of impregnated Pd was 10×10^{-2} mmol/g while in the case of SILCA 13 this was just above 3×10^{-2} mmol/g. Therefore, compared to other catalysts, it can be considered as an unsuccessful impregnation. Tested in the Heck reaction, these catalysts did not bring any improvement of the stability although the activity was enhanced to 5100–5500 h^{-1} for SILCAs 14 and 15, probably resulting from a more rapid leaching (releasing) of Pd from the surface.

Traditionally, XPS is used to identify the oxidation state of the metals. During acquisition of Pd 3d spectrum for the samples produced in this study, it was noticed that Pd is reduced under X-rays and low energy electrons during the XPS experiment (see Supplementary Materials). Therefore, for all the samples, five scans (5 min total) for Pd 3d line at the beginning of XPS measurements were included (even before survey spectrum) to gain an "original" Pd 3d spectrum. The obtained results confirmed the presence of Pd^{2+} and Pd^0 on the surface of the catalyst, with the exception of SILCA 13, where just non-metallic Pd^{2+} was found (Figure 4d). However, the amount of Pd^0 given in Figure 4d can be overestimated due to the possible formation of metallic Pd even during first 5 min of acquisition.

Table 2. Palladium amount in synthesized SILCAs detected by ICP-OES analyses and comments for the TEM images of fresh catalysts.

SILCA	Pd, ×10^{-2} mmol/g	TEM Observation
1	9.3	Low coverage of Pd, irregularly scattered with few big agglomerates
2	8.1	Good coverage of regularly dispersed nanoparticles of 8 nm average size
3	8.6	Good coverage, with raspberry-like agglomerates of average size 27 nm
4	12.7	Good coverage, with mostly individual nanoparticles of average size 13 nm
5	11.2	Low coverage, particles of irregular size and shape (round, qubic, triangle)
6	8.4	Medium coverage with nanoparticles of 8.5 nm average size
7	9.7	Few agglomerates of Pd, with no visible individual nanoparticles
8	11.4	Very few agglomerates
9	10.8	Very few big agglomerates
10	11.7	Very few agglomerated big nanoparticles
11	11.1	Almost none of Pd, few big nanoparticles
12	10.6	Very few big individual nanoparticles, with few big agglomerates
13	3.2	Almost none of Pd, just few nanoparticles
14	9.0	Densely dispersed nanoparticles of 7.5 nm average size
15	8.1	Good coverage of grape-like agglomerated nanoparticles

Visual inspection of the materials was possible with TEM analysis that provided an overall overview of the catalysts and gave the insight on how the structure of the layer affects the palladium size and disposition. The obtained images (see Supplementary Materials) are commented in Table 2. The TEM results are in relatively good agreement with the XPS analysis, i.e., the samples that initially had more metallic palladium (Pd^0) showed more visible nanoparticles and oppositely, those which had a higher percentage of Pd^{2+} (e.g., SILCAs 7–13) were hard to identify. TEM images of the most relevant catalyst—SILCA 2, being the most recyclable, are presented in Figure 6. High dispersion of the nanoparticles was observed with this catalyst, and it is assumed that the stability of this catalyst lies in the capability of the anchored layer to stabilize nanoparticles and avoid overgrowth of nanoparticles and agglomeration.

Finally, the structure of this catalyst was confirmed with energy-dispersive X-ray spectroscopy (EDX) and carbon nuclear magnet resonance (NMR). The EDX spectrum of the catalyst enabled elemental analysis and verified the presence of bromine in the ionic liquid layer. The carbon peak and its atomic percentage appears to be much higher than it is, in reality, because of the analytic shortcomings, i.e., the usage of carbon-tape to deposit the catalyst prior to exposing it to the beam. This means that the actual atomic percentage of other elements is much higher than it is presented in Figure 7. Nevertheless, a good overview of the material was obtained.

^{13}C NMR analysis of fresh and spent catalyst confirmed the layer structure and stability. The most evident peaks were those denoted to the carbons of propyl chain of modified surface at 7.3, 21.6, and 49.0 ppm (Figure 8), as well as the carbon atoms of imidazolium ring which were found at 121.0 and 134.5 ppm [26,28]. Aliphatic carbons originating from anchored acid resonated in the range of 21.6 to 34.6 ppm being partially overlapped with the peak of carbon number 2 [25], followed by ending the carboxylic group with a peak at 178.1 ppm. Tetramethylguanidinium contributed with two small peaks, the first one at 38.9 ppm belonging to methyl groups that falls in the region of aliphatic carbons and the peak at 172.7 ppm of imine carbon that is slightly shifted downfield, most probably due to the higher polarization of the ionic version of TMG compared to the molecular one [37]. The NMR analysis of the recovered catalyst after the use in one reaction cycle (Figure 8, red) showed the stability of the attached structure under the reaction conditions. All the carbon peaks were still present with an additional peak at 55.6 ppm that was assumed to belong to residual molecules of solvents and base, or more precisely NMP related structures and/or (EtOH)$_3$NH$^+$I$^-$ residual salts [37]. This indicates the potential of catalyst deactivation due to poisoning, since the adsorbed species can block the sites capable to stabilize Pd nanoparticles, therefore resulting in their leaching.

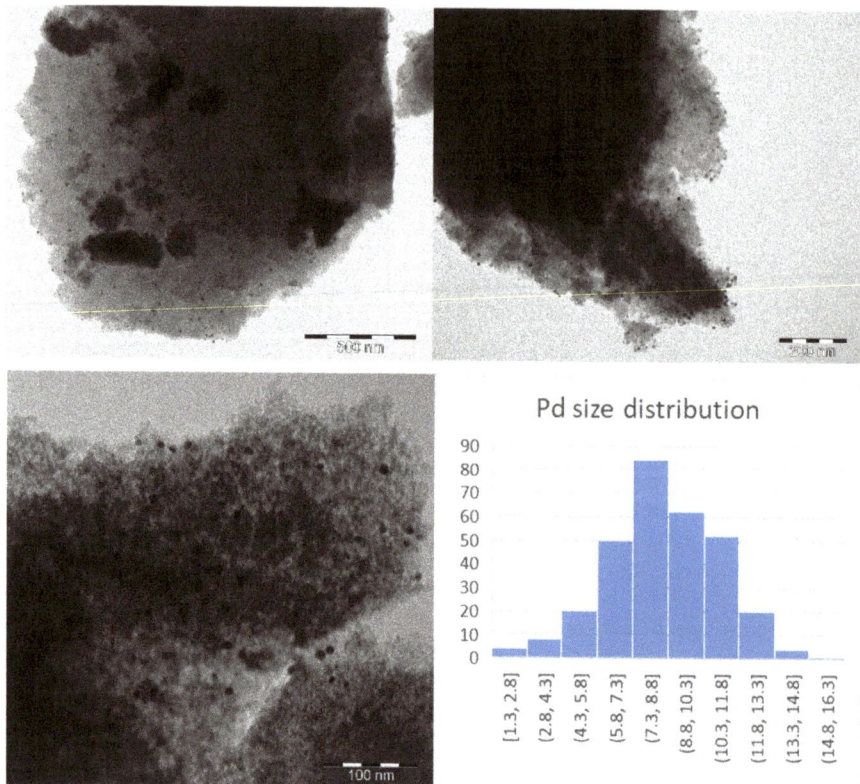

Figure 6. TEM images of SILCA 2 of a different scale and palladium size distribution.

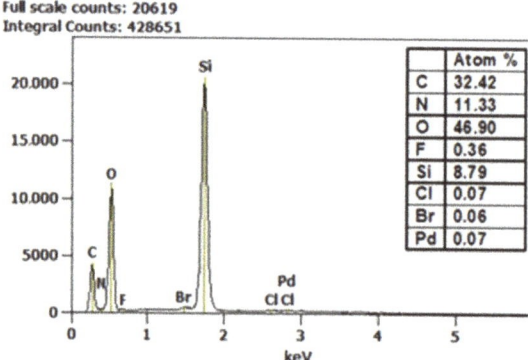

Figure 7. EDX spectrum of SILCA 2.

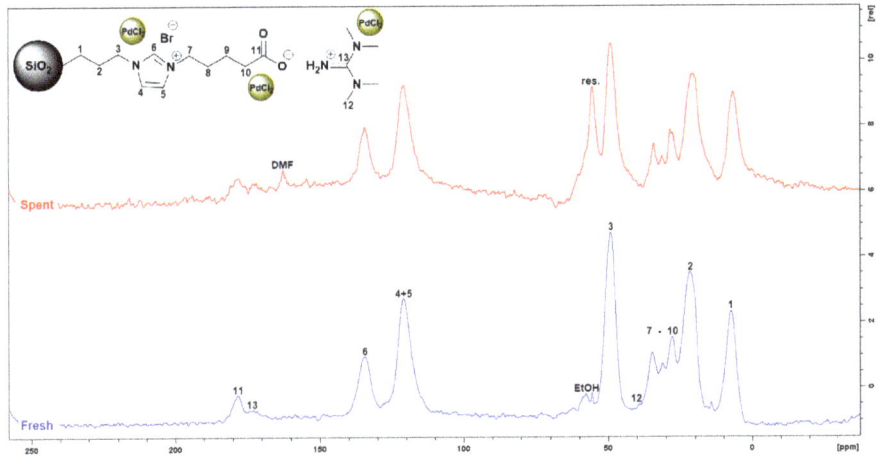

Figure 8. ^{13}C NMR spectrum of fresh and spent SILCA 2.

Thermal stability of the catalyst was analyzed with thermal gravimetric analysis (TGA). As shown in Figure 9, the ionic liquid layer is stable up to 250 °C, after which thermal decomposition of organic moieties takes place. A small mass drop of approximately 0.5 wt% at the temperatures under 250 °C was considered to belong to adsorbed water and residual solvents, while at the temperatures above this, the mass drop of 6–7 wt%, corresponds to the contribution of the ionic liquid layer. Therefore, at the reaction conditions, no loss of the ionic liquid layer should take place as a consequence of thermal destructions.

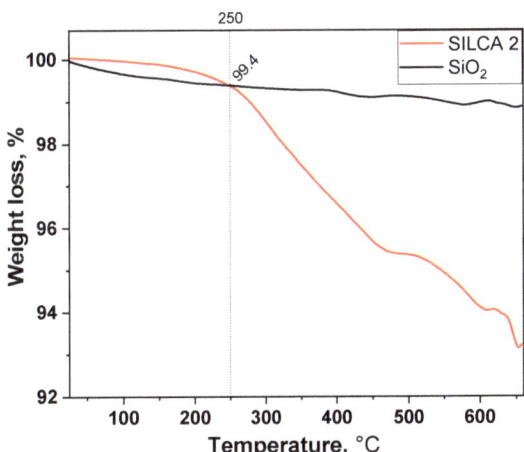

Figure 9. Thermal gravimetric analysis curve for pristine SiO_2 and synthesized catalyst.

While monitoring the reaction progress via sampling in defined time intervals (5 min), for all the catalysts, concentration curves resembling the one presented in Figure 10 for SILCA 2 were obtained. An initial induction time was observed, which in case of SILCA 2 was approximately 25 min after which the reaction ignited. To understand this induction time and to confirm the release and catch mechanism in the reaction with SILCA 2, the relative amount of leached Pd in slurry was monitored with ICP-OES during the course of the reaction and results are over-layered with concentration curves in Figure 10. The initial period corresponds to the start of slow emission of Pd from the surface, whereas the reaction completion corresponds to the highest amount of palladium in slurry. After the reaction completion (40 min), palladium needs 10 min more to redeposit back. Following the re-deposition dynamics, it was evident that for the efficient use of catalyst, the reaction plus re-deposition time should be no less than 50 min, after what the catalyst can be successfully recycled. It was not possible to fully recover the leached palladium, after the 1st reaction cycle approximately 6 ppm of palladium was detected in the solvent, which corresponds to 10% of initially loaded Pd. Together with the catalyst poisoning this can be the reason for the decline of the catalyst activity.

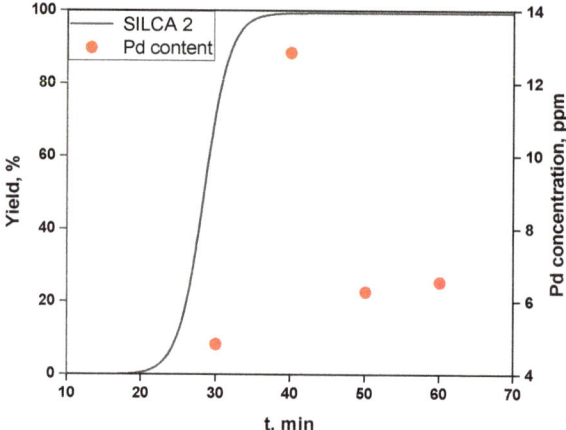

Figure 10. Progress of the reaction catalyzed by SILCA 2 correlated with Pd leaching.

3. Materials and Methods

3.1. Chemicals and Analysis

All the chemicals were of analytic grade purchased from commercial producers Sigma-Aldrich or Alfa Aesar and they were used as received.

The materials were characterized with different techniques. Fourier transform infrared spectroscopy (FT-IR) measurements of pelletized samples were performed on an Bruker IFS 66/S FT-IR (Bruker Optics, Ettlingen, Germany) in the 400–4000 cm^{-1} region.

X-ray photoelectron spectroscopy (XPS) was used to follow the changes in the oxidation state of palladium and the states of other elements. The spectra were collected with a Kratos Axis Ultra DLD electron spectrometer (Kratos Analytical Ltd., Manchester, UK) using a monochromated Al Kα source operated at 150 W. Analyzer pass energy of 160 eV was set for acquiring a wide spectra and a pass energy of 20 eV was used for individual photoelectron lines. The surface potential was stabilized by the spectrometer charge neutralization system. The binding energy (BE) scale was referenced to the Si 2p line of silica, set at 103.4 eV. Processing of the spectra was accomplished with the Kratos software. Due to the possible reduction of Pd^{2+} ions under X-rays and low energy electrons during XPS measurements, Pd 3d spectrum was acquired during first 5 min of exposure, and then again with other photoelectron lines.

The zeta potential (ζ) measurements were performed with a Zetasizer Nano ZS (Malvern Panalytical Ltd., Malvern, UK) using the light scattering technique, while for pH measurements the potentiometric method (MPT-2) was applied. The samples were made in deionized water and analyzed at ambient temperature with the use of NaOH and HCl as titrants in a pH range of 3–8. The final zeta potential vs. pH curve were averaged from three measurements and the isoelectric point was derived from it.

The analysis of metal amount in the catalyst was carried out with inductively coupled plasma optical emission spectroscopy (ICP-OES) supplied with Optima 4300 DV optical atomic emission spectrometer (Perkin Elmer, Waltham, MA, USA). The transmission electron microscopy (TEM) was performed on a JEM 1400 plus (120 kV, 0.38 nm) with OSIS Quemesa 11 Mpix digital camera (Jeol Ltd, Tokyo, Japan). The elemental analysis was performed with energy-dispersive X-ray spectroscopy (EDX) on a Zeiss Leo Gemini 1530 SEM (Zeiss Leo, Oberkochen, Germany) with a Thermo-NORAN vantage X-ray detector (Thermo Scientific, Madison, WI, USA). The thermo-gravimetric analysis (TGA) were made on CAHN D-200 instrument (Cahn Instruments, Inc., CA, USA) in the range of 20–700 °C with a heating rate of 10 °C min^{-1} and under Ar atmosphere. The Solid-State ^{13}C nuclear magnetic resonance spectroscopy (CP MAS ^{13}C NMR) spectra were recorded on a Bruker AVANCE-III spectrometer (Bruker Scientific Instruments, Billerica, MA, USA) operating at 399.75 MHz (^{1}H) and 100.5 MHz (^{13}C) equipped with a 4 mm double resonance (H/X) CP MAS probe. The powdered sample was spun at a 14 kHz spin rate. The proton 90° high-power pulse of 2.9 ms and contact time of 2 ms was used. A total of 42,000 scans (fresh catalyst) and 200,000 scans (spent catalyst) were accumulated using a recovery delay of 2 s. The solution state ^{1}H and ^{13}C NMR spectra were recorded at 300 and 298 K on a Bruker Avance-III HD 500 MHz spectrometer (Bruker Scientific Instruments, Billerica, MA, USA) equipped with a Bruker SmartProbe^{TM}. For analysis the samples were dissolved in deuterated chloroform with 0.03% tetramethylsilane (TMS) as the internal standard.

The Heck reaction yield was determined with gas chromatography, GC-HP 6890 Series with HP-5, 5% phenyl methyl siloxane capillary column (30.0 m × 320 μm, 0.25 μm) and a flame ionization detector (Agilent Technologies, Inc., Santa Clara, CA, USA) at the constant temperature of 300 °C. The temperature of the injector was 280 °C and gas flow was 9.5 mL min^{-1}. Column heating from 40 to 250 °C was carried out with a 10 °C min^{-1} heating rate. Hexadecane was used as an internal standard.

3.2. Heck Reaction Procedure and Catalyst Recovery

The reaction was run in a flat bottom vial tube (4.5 mL) sealed with screw cap equipped with a Teflon® lined septum and a magnetic stirrer, under air atmosphere and heated by an isolated oil bath. In catalyst screening tests 1 mmol of iodobenzene, 1.5 mmol of butyl acrylate, 1.5 mmol of triethanolamine, and 1 mL of N-Methyl-2-pyrrolidone were placed in a reactor together with a catalyst amount corresponding to 0.09 mol% of Pd (detected by ICP-OES). The temperature of the oil bath was kept at 100–105 °C with an assumption that this will result in a temperature of 100 °C in the reaction slurry, due to the small volume of the vials. Once the temperature reached the preset value, time measuring was initiated. In defined time intervals, samples were withdrawn with a GC syringe injector and the progress of the reaction was monitored with a GC. Ten μL of reaction mixtures were taken out from the reaction slurry and diluted with 240 μL of DMF to give 4.0% dilution samples, followed by the addition of 250 μL internal standard and final dissolution of 2.0%. The as obtained samples were analyzed with GC and the reaction was stopped after completion. For catalyst leaching tests reaction was run in multiple vial tubes in parallel and in defined time intervals (10 min) samples for ICP-OES were taken out one by one from the vials. One mL of reaction slurry was withdrawn from the vial, filtered, dissolved in aqua regia, and stirred overnight prior to the analysis.

In the catalyst recyclability studies, the reaction was conducted in a flat bottom vial tube (8.5 mL) with a screw cap and magnetic stirrer, under air atmosphere, and heated by an isolated oil bath. Vials were submerged in the oil bath just above the level of liquid, while the temperature of the oil was kept at 105–110 °C (slightly higher temperature due to the bigger vial volume) with the assumption that this will result in the temperature of 100 °C in the reaction slurry. Upon a predetermined time, the reaction was completed and the vials were cooled, and samples were withdrawn with a GC syringe injector and analyzed with a GC. The catalyst was washed and separated by adding approx. 7.0 mL of DMF (full vial) and centrifuging, followed by flushing with DMF (8.5 mL), diethyl ether (8.5 mL), and air drying for 30 min. The catalysts were reused in the next cycle without removal from the vials or stored at ambient conditions for further characterization.

Every experiment was repeated a minimum of three times and average values are reported.

3.3. Catalyst Synthesis

The new catalysts were synthesized via a modified procedure compared to that we reported recently [28]. Here, a more detailed synthesis route of the catalyst is presented with 3-propyl imidazolium bromide benzoic acid modification, 1,1,3,3-tetramethylguanidine (TMG) in the role of a cation, and $PdCl_2$ as a source of the metal. Other catalysts were prepared accordingly and all anticipated structures are displayed in Table 1.

3.3.1. Synthesis of $PdCl_2$ Supported on 3-Propyl Imidazolium Bromide-Tetramethylguanidinium Benzoate Modified SiO_2 (SILCA 1)

Imidazole Modification

In order to modify the silica surface, silanol groups were firstly reacted with (3-chloropropyl)trimethoxysilane in refluxing toluene. Under nitrogen atmosphere, 1 g of calcined SiO_2 was reacted with 5 mmol of the modifying agent for 24 h. The obtained material was filtrated, flushed with fresh toluene, and for 24 h washed with ethanol in a Soxhlet apparatus prior to drying.

Thereafter, propylated silica was reacted with imidazole. Finely crushed imidazole flakes (10 mmol) were dissolved in toluene and mixed with 1 g of silica. The slurry was stirred for 24 h under nitrogen atmosphere under refluxing conditions. As before, the obtained material was flushed with fresh toluene and for 24 h washed with ethanol in a Soxhlet apparatus prior to drying.

Acid Modification

Prior to the further modification of silica, the acidic group of 4-bromobenzoic acid had to be protected and esterified [38]. The 4-bromobenzoic acid (6 mmol) was dissolved in 20 mL of methanol and while stirred, a dropwise addition of 3 mL of 40% H_2SO_4 took place. The mixture was heated to 50 °C and reacted for 12 h. After workup, 10 mL of H_2O was added and the acidic solution was neutralized with approximately 50 mL of brine. The product was isolated in dichloromethane and dried in a rotary evaporator. The obtained white crystalline methyl 4-bromobenzoate was stored in a cold refrigerator. See the Supplementary Materials for the NMR data.

The obtained esterified product (1.5 mmol, 0.3226 g) was dissolved in DMF, followed by addition of 1 g modified silica. After reacting it for 24 h under refluxing conditions (N_2 atm), the slurry was cooled down and solid separated with centrifuge, washed with dichloromethane, ethanol, and then dried.

Finally, the recovery of the acidic group from ester modified SiO_2 was carried out in methanol and catalyzed by a strong base. To the 1 g of modified silica solution, 1.25 mL of NaOH (1 M) was added and stirred for 12 h under reflux conditions. After that, HCl (1 M) was added to acidify the solution. The solid was separated and washed with water, ethanol, and dried.

Base Modification and Palladium Loading

In the next step, the acidified surface was coordinated with a proton acceptor (TMG). To the TMG (5 mmol) ethanol solution, 1 g of 3-propyl imidazolium bromide benzoic acid modified SiO_2 was added and stirred for 6 h. The product was centrifuged, washed with diethyl ether, and dried. Finally, this material was loaded with the metal. $PdCl_2$ (0.1 mmol) was dissolved in methanol and mixed with 1 g of modified SiO_2. After 12 h stirring at room temperature, the catalyst was separated and washed with ethanol and diethyl ether. The dry material was $PdCl_2$ supported on 3-propyl imidazolium bromide-tetramethylguanidinium benzoate modified SiO_2 or more briefly, SILCA 1 with the targeted structure presented in Table 1.

3.3.2. Synthesis of PdCl$_2$ Supported on 3-Propyl Imidazolium Bromide-Tetramethylguanidinium Pentanoate Modified SiO$_2$ (SILCA 2)

SILCA 2 (see Table 1) was prepared in the same manner as SILCA 1 with a change in the acid that was used. Instead of 4-bromobenzoic acid, 5-bromo-pentanoic acid was used which preliminary gave methyl 5-bromo-pentanoate as a transparent light yellowish liquid.

3.3.3. Synthesis of PdCl$_2$ Supported on 3-Propyl Imidazolium Bromide-Tetramethylguanidinium Succinate Modified SiO$_2$ (SILCA 3)

SILCA 3 (see Table 1) was prepared analogously with SILCA 1 with a variation of the acid used. In the case of SILCA 3, bromosuccinic acid was used instead of 4-bromobenzoic acid and the reaction yielded a viscous transparent liquid dimethyl 2-bromosuccinate. In addition, in the case of SILCA 3, a double amount of TMG (10 mmol) was used due to the two acidic groups of bromosuccinic acid.

3.3.4. Synthesis of PdCl$_2$ Supported on 3-(Aminoethylamino)Propyl Bromide-Tetramethylguanidinium Benzoate (SILCA 4), Pentanoate (SILCA 5), and Succinate Modified SiO$_2$ (SILCA 6)

SILCA 4, SILCA 5, and SILCA 6 (see Table 1) were prepared via slightly modified SILCA 1, SILCA 2, and SILCA 3 synthesis routes. Instead of imidazole modification, the aminoethylamino group was utilized. To immobilize the modifying agent on the surface of the solid, 1 g of SiO$_2$ was suspended in toluene and a dropwise addition of 5 mmol of 3-(aminoethylamino)propyl trimethoxysilane was commenced. The mixture was stirred for 24 h under reflux conditions and N$_2$ atmosphere. The product was separated by filtration, flushed with toluene, and for 24 h washed with ethanol in a Soxhlet apparatus prior to drying. Further synthesis was continued as already explained with esterified 4-bromobenzoic acid, 5-bromo-pentanoic acid, and bromosuccinic acid.

3.3.5. Synthesis of PdCl$_2$ Supported on 3-Propyl Imidazolium Bromide-Tetramethylguanidinium Propionate modified SiO$_2$ (SILCA 7) and 3-Propyl Imidazolium Bromide-Tetramethylguanidinium Decanoate Modified SiO$_2$ (SILCA 8)

SILCA 7 and SILCA 8 (see Table 1) were prepared in a similar way as SILCA 1 with a change in acids that were used. For SILCA 7, instead of 4-bromobenzoic acid, 3-bromo-propionic acid was used, which gave the intermediate methyl 3-bromo-propionate as a white crystalline substance. Meanwhile, in the case of SILCA 8, 10-bromodecanoic acid was used which yielded a brownish liquid methyl 10-bromodecanoate.

3.3.6. Synthesis of PdCl$_2$ Supported on 3-Propyl Imidazolium Bromide-Diazabicycloundecenium, -Diazabicyclononenium, -Triazabicyclodecenium, and -Tert-Butyl-Tetramethylguanidinium Pentanoate Modified SiO$_2$ (SILCAs 9-12)

SILCAs 9–12 (see Table 1) were prepared based on the synthesis procedure for SILCA 2 with differentiation in terms of the base utilized. Instead of TMG the used bases were: 1,8-Diazabicyclo [5.4.0]undec-7-ene (SILCA 9), 1,5-Diazabicyclo[4.3.0]non-5-ene (SILCA 10), 1,5,7-Triazabicyclo [4.4.0]dec-5-ene (SILCA 11), and 2-Tert-Butyl-1,1,3,3-tetramethylguanidine (SILCA 12).

3.3.7. Synthesis of Pd(AcAc)$_2$, Pd(OAc)$_2$, and Pd(NO$_3$)$_2$ Supported on 3-Propyl Imidazolium Bromide-Tetramethylguanidinium Pentanoate Modified SiO$_2$ (SILCAs 13-15)

SILCAs 13-15 (see Table 1) were prepared based on the synthesis procedure for SILCA 2 with variation of the palladium source. Instead of PdCl$_2$ the used Pd was: Pd(AcAc)$_2$ (SILCA 13), Pd(OAc)$_2$ (SILCA 14), and Pd(NO$_3$)$_2$ (SILCA 15).

4. Conclusions

In this work, a new group of catalysts was synthesized and investigated in order to develop a catalyst with a high activity and reusability in the Heck reaction. While performing the study, it became clear that correlating the catalyst activity, stability, and structure is a challenging task. Catalyst performance is hardly predictable and mostly depends on the ability to successfully graft different functionalities and obtain a durable layer which can stabilize Pd nanoparticles suppressing overgrowth and agglomeration. Material containing propyl imidazolium bromide-tetramethylguanidinium pentanoate modification gave a catalyst which preserved the activity in four successive cycles in the reaction of iodobenzene and butylacrylate at 100 °C. It was demonstrated that the presence of carboxylic groups helps the coordination of tetramethylguanidine which is capable of stabilizing Pd. The imidazole ring showed to be a rigid building unit vital for the ionic liquid layer formation, on the contrary to the aminoethylamino linker. However, this was not stable in the presence of bromosuccinate. It was found out that among the examined superbases, only TMG efficiently coordinates the acidic head of the auxiliary layer and that $PdCl_2$ builds the most stable catalyst compared to other common Pd sources.

In summary, the catalyst with propyl imidazolium bromide-tetramethylguanidinium pentanoate resulted in a good metal distribution with controlled nanoparticle size and it showed a good thermal and chemical stability. This catalyst operates through the release and catch mechanism and, therefore, the use of it requires additional time for the metal re-deposition upon the end of the reaction. Slow leaching of palladium was observed which can be a consequence of catalyst poisoning by the by-products of Heck reaction and the solvents. Suppressing catalysts deactivation is an ongoing battle and further development of bis-layered catalysts have a potential to open a new field of studies to tackle this issue.

Supplementary Materials: The following are available online at http://www.mdpi.com/2073-4344/10/9/963/s1. Table S1: XPS data obtained for SILCAs 1–4; Table S2: XPS data obtained for SILCAs 5–8; Table S3: XPS data obtained for SILCAs 9–12; Table S4: XPS data obtained for SILCAs 13–15; Table S5: XPS data on Pd speciation for SILCAs 1–15; Figure S1: XPS Pd 3d spectral line of SILCA 1 after 5 min (a) and after 20 min (b) of measurements; Table S6: Representative TEM images of synthetized catalysts; Figure S2: Summarized isoelectric titration graphs for pristine SiO_2 and SILCAs 1–12. Solution state ^1H and ^{13}C NMR of the esterified acids.

Author Contributions: Conceptualization, N.V., P.V., A.N., J.-P.M., and T.S.; methodology, N.V., P.V., J.-P.M., and T.S.; validation, N.V.; formal analysis, N.V., P.V., A.S., J.-P.M., and T.S.; investigation, N.V. and A.S.; data curation, N.V.; writing—original draft preparation, N.V.; writing—review and editing, N.V., P.V., A.S., J.-P.M., and T.S.; visualization, N.V.; supervision, P.V., J.-P.M., and T.S.; funding acquisition, J.-P.M. and T.S. All authors have read and agreed to the published version of the manuscript.

Funding: This research was funded by the Graduate School in Chemical Engineering (GSCE) of Finland and the Academy of Finland grants 319002, 320115, 325186, and 316827. This work is part of activities of the Technical Chemistry, Department of Chemistry, Chemical-Biological Centre, Umeå University, Sweden as well as the Johan Gadolin Process Chemistry Centre at Åbo Akademi University in Finland.

Acknowledgments: The Swedish Bio4Energy programme, Kempe Foundations, and Wallenberg Wood Science Center under auspices of Alice and Knut Wallenberg Foundation are gratefully acknowledged.

Conflicts of Interest: The authors declare no conflict of interest.

References

1. Yin, L.; Liebscher, J. Carbon-carbon coupling reactions catalyzed by heterogeneous palladium catalysts. *Chem. Rev.* **2007**, *107*, 133–173. [CrossRef]
2. Heck, R.F.; Nolley, J.P., Jr. Palladium-catalyzed vinylic hydrogen substitution reactions with aryl, benzyl, and styryl halides. *J. Org. Chem.* **1972**, *37*, 2320–2322. [CrossRef]
3. Mizoroki, T.; Mori, K.; Ozaki, A. Arylation of Olefin with Aryl Iodide Catalyzed by Palladium. *Bull. Chem. Soc. Jpn.* **1971**, *44*, 581. [CrossRef]
4. De Vries, J.G. The Heck reaction in the production of fine chemicals. *Can. J. Chem.* **2001**, *79*, 1086–1092. [CrossRef]

5. Hübner, S.; De Vries, J.G.; Farina, V. Why Does Industry Not Use Immobilized Transition Metal Complexes as Catalysts? *Adv. Synth. Catal.* **2016**, *358*, 3–25. [CrossRef]
6. Pagliaro, M.; Pandarus, V.; Ciriminna, R.; Béland, F.; Demma Carà, P. Heterogeneous versus Homogeneous Palladium Catalysts for Cross-Coupling Reactions. *ChemCatChem* **2012**, *4*, 432–445. [CrossRef]
7. Hagiwara, H.; Shimizu, Y.; Hoshi, T.; Suzuki, T.; Ando, M.; Ohkubo, K.; Yokoyama, C. Heterogeneous Heck reaction catalyzed by Pd/C in ionic liquid. *Tetrahedron Lett.* **2001**, *42*, 4349–4351. [CrossRef]
8. Corma, A.; García, H.; Leyva, A.; Primo, A. Basic zeolites containing palladium as bifunctional heterogeneous catalysts for the Heck reaction. *Appl. Catal. A Gen.* **2003**, *247*, 41–49. [CrossRef]
9. Dams, M.; Drijkoningen, L.; De Vos, D.E.; Jacobs, P.A.; Pauwels, B.; Van Tendeloo, G. Pd-zeolites as heterogeneous catalysts in Heck chemistry. *J. Catal.* **2002**, *209*, 225–236. [CrossRef]
10. Mpungose, P.P.; Vundla, Z.P.; Maguire, G.E.M.; Friedrich, H.B. The current status of heterogeneous palladium catalysed Heck and Suzuki cross-coupling reactions. *Molecules* **2018**, *23*, 1676. [CrossRef]
11. Grasa, G.A.; Singh, R.; Stevens, E.D.; Nolan, S.P. Catalytic activity of Pd(II) and Pd(II)/DAB-R systems for the Heck arylation of olefins. *J. Organomet. Chem.* **2003**, *687*, 269–279. [CrossRef]
12. Köhler, K.; Heidenreich, R.G.; Krauter, J.G.E.; Pietsch, J. Highly active palladium/activated carbon catalysts for heck reactions: Correlation of activity, catalyst properties, and Pd leaching. *Chem.-A Eur. J.* **2002**, *8*, 622–631. [CrossRef]
13. Köhler, K.; Kleist, W.; Pröck, S.S. Genesis of coordinatively unsaturated palladium complexes dissolved from solid precursors during heck coupling reactions and their role as catalytically active species. *Inorg. Chem.* **2007**, *46*, 1876–1883. [CrossRef] [PubMed]
14. Reay, A.J.; Fairlamb, I.J.S. Catalytic C-H bond functionalisation chemistry: The case for quasi-heterogeneous catalysis. *Chem. Commun.* **2015**, *51*, 16289–16307. [CrossRef]
15. Gruttadauria, M.; Giacalone, F.; Noto, R. "Release and catch" catalytic systems. *Green Chem.* **2013**, *15*, 2608–2618. [CrossRef]
16. Aksin, Ö.; Türkmen, H.; Artok, L.; Çetinkaya, B.; Ni, C.; Büyükgüngör, O.; Özkal, E. Effect of immobilization on catalytic characteristics of saturated Pd-N-heterocyclic carbenes in Mizoroki-Heck reactions. *J. Organomet. Chem.* **2006**, *691*, 3027–3036. [CrossRef]
17. Gaikwad, A.V.; Rothenberg, G. In-situ UV-visible study of Pd nanocluster formation in solutionw. *Phys. Chem. Chem. Phys.* **2006**. [CrossRef]
18. Reimann, S.; Stötzel, J.; Frahm, R.; Kleist, W.; Grunwaldt, J.-D.; Baiker, A. Identification of the Active Species Generated from Supported Pd Catalysts in Heck Reactions: An in situ Quick Scanning EXAFS Investigation. *J. Am. Chem. Soc.* **2011**, *133*, 3921–3930. [CrossRef]
19. Wanzlick, H.-W.; Schönherr, H.-J. Direct Synthesis of a Mercury Salt-Carbene Complex. *Angew. Chemie Int. Ed. Engl.* **1968**, *7*, 141–142. [CrossRef]
20. Öfele, K. 1,3-Dimethyl-4-imidazolinyliden-(2)-pentacarbonylchrom ein neuer übergangsmetall-carben-komplex. *J. Organomet. Chem.* **1968**, *12*. [CrossRef]
21. Öfele, K.; Herrmann, W.A.; Mihalios, D.; Elison, M.; Herdtweck, E.; Scherer, W.; Mink, J. Mehrfachbindungen zwischen hauptgruppenelementen und übergangsmetallen. CXXVI. Heterocyclen-carbene als phosphananaloge liganden in metallkomplexen. *J. Organomet. Chem.* **1993**, *459*, 177–184. [CrossRef]
22. Herrmann, W.A.; Öfele, K.; Elison, M.; Kühn, F.E.; Roesky, P.W. Nucleophilic cyclocarbenes as ligands in metal halides and metal oxides. *J. Organomet. Chem.* **1994**, *480*. [CrossRef]
23. Weskamp, T.; Kohl, F.J.; Hieringer, W.; Gleich, D.; Herrmann, W.A. Highly Active Ruthenium Catalysts for Olefin Metathesis: The Synergy of N-Heterocyclic. *Angew. Chemie Int. Ed.* **1999**, *39*, 2416–2419. [CrossRef]
24. Schwarz, J.; Böhm, V.P.W.; Gardiner, M.G.; Grosche, M.; Herrmann, W.A.; Hieringer, W.; Raudaschl-Sieber, G. Polymer-Supported Carbene Complexes of Palladium: Well-Defined, Air-Stable, Recyclable Catalysts for the Heck Reaction. *Chem.-A Eur. J.* **2000**, *6*, 1773–1780. [CrossRef]
25. Gruttadauria, M.; Liotta, L.F.; Salvo, A.M.P.; Giacalone, F.; La Parola, V.; Aprile, C.; Noto, R. Multi-layered, covalently supported ionic liquid phase (mlc-SILP) as highly cross-linked support for recyclable palladium catalysts for the suzuki reaction in aqueous medium. *Adv. Synth. Catal.* **2011**, *353*, 2119–2130. [CrossRef]
26. Yang, H.; Han, X.; Li, G.; Wang, Y. N-Heterocyclic carbene palladium complex supported on ionic liquid-modified SBA-16: An efficient and highly recyclable catalyst for the Suzuki and Heck reactions. *Green Chem.* **2009**, *11*, 1184–1193. [CrossRef]

27. Jung, J.Y.; Taher, A.; Kim, H.J.; Ahn, W.S.; Jin, M.J. Heck reaction catalyzed by mesoporous SBA-15-supported ionic liquid-Pd(OAc)$_2$. *Synlett* **2009**. [CrossRef]
28. Vucetic, N.; Virtanen, P.; Nuri, A.; Mattsson, I.; Aho, A.; Mikkola, J. Preparation and characterization of a new bis-layered supported ionic liquid catalyst (SILCA) with an unprecedented activity in the Heck reaction. *J. Catal.* **2019**, *371*, 35–46. [CrossRef]
29. Ji, Y.; Jain, S.; Davis, R.J. Investigation of Pd Leaching from Supported Pd Catalysts during the Heck Reaction. *J. Phys. Chem. B* **2005**, *109*, 17232–17238. [CrossRef]
30. Zhao, F.; Bhanage, B.M.; Shirai, M.; Arai, M. Heck Reactions of Iodobenzene and Methyl Acrylate with Conventional Supported Palladium Catalysts in the Presence of Organic and/or Inorganic Bases without Ligands. *Chem.-A Eur. J.* **2000**, *6*, 843–848. [CrossRef]
31. Thermo Scientific XPS: Knowledge Base. Available online: https://xpssimplified.com/periodictable.php (accessed on 27 February 2020).
32. X-ray Photoelectron Spectroscopy (XPS) Reference Pages. Available online: http://www.xpsfitting.com/ (accessed on 27 February 2020).
33. Moldoveanu, S.C.; David, V. Properties of Analytes and Matrices Determining HPLC Selection. In *Selection of the HPLC Method in Chemical Analysis*; Elsevier: Amsterdam, The Netherlands, 2017; pp. 189–230.
34. Amoroso, F.; Colussi, S.; Del Zotto, A.; Llorca, J.; Trovarelli, A. Room-temperature Suzuki-Miyaura reaction catalyzed by Pd supported on rare earth oxides: Influence of the point of zero charge on the catalytic activity. *Catal. Lett.* **2013**, *143*, 547–554. [CrossRef]
35. Hyde, A.M.; Calabria, R.; Arvary, R.; Wang, X.; Klapars, A. Investigating the Underappreciated Hydrolytic Instability of 1,8-Diazabicyclo [5.4.0] undec-7-ene and Related Unsaturated Nitrogenous Bases. *Org. Process Res. Dev.* **2019**. [CrossRef]
36. Liu, X.; Mäki-Arvela, P.; Aho, A.; Vajglova, Z.; Gun'ko, V.M.; Heinmaa, I.; Kumar, N.; Eränen, K.; Salmi, T.; Murzin, D.Y. Zeta potential of beta zeolites: Influence of structure, acidity, pH, temperature and concentration. *Molecules* **2018**, *23*, 946. [CrossRef] [PubMed]
37. National Institute of Advanced Industrial Science and Technology (AIST). Available online: https://www.aist.go.jp/index_en.html (accessed on 23 March 2020).
38. Potter, B.V.L.; Dowden, J.; Galione, A. Therapeutics Use of Pyridinium Compounds to Modulate Naadp Activity. WO 2005054198 A2, 16 June 2005.

© 2020 by the authors. Licensee MDPI, Basel, Switzerland. This article is an open access article distributed under the terms and conditions of the Creative Commons Attribution (CC BY) license (http://creativecommons.org/licenses/by/4.0/).

Communication

SILP Materials as Effective Catalysts in Selective Monofunctionalization of 1,1,3,3-Tetramethyldisiloxane

Rafal Kukawka [1,2,*], Anna Pawlowska-Zygarowicz [2,3], Rafal Januszewski [3,4], Joanna Dzialkowska [3], Mariusz Pietrowski [3], Michal Zielinski [3], Hieronim Maciejewski [2,3] and Marcin Smiglak [1,2,*]

[1] Innosil Sp. z o.o., ul. Rubiez 46, 61-612 Poznan, Poland
[2] Poznan Science and Technology Park, Adam Mickiewicz University Foundation, ul. Rubiez 46, 61-612 Poznan, Poland; anna.pawlowska@amu.edu.pl (A.P.-Z.); hieronim.maciejewski@ppnt.poznan.pl (H.M.)
[3] Faculty of Chemistry, Adam Mickiewicz University, ul. Uniwersytetu Poznańskiego 8, 61-614 Poznan, Poland; r.janusz@amu.edu.pl (R.J.); joadzi5@st.amu.edu.pl (J.D.); mariop@amu.edu.pl (M.P.); mardok@amu.edu.pl (M.Z.)
[4] Centre for Advanced Technologies, Adam Mickiewicz University, ul. Uniwersytetu Poznańskiego 10, 61-614 Poznan, Poland
* Correspondence: kukawka.rafal@gmail.com (R.K.); marcin.smiglak@gmail.com (M.S.)

Received: 20 October 2020; Accepted: 30 November 2020; Published: 3 December 2020

Abstract: Functionalized siloxanes are one of the most important classes of organosilicon compounds, thus the enhancement of current methods of its synthesis is an important issue. Herein, we present the selective and highly effective reaction between 1,1,3,3-tetramethyldisiloxane (TMDSO) and 1-octene (1-oct), using SILP (supported ionic liquid phase) materials containing a rhodium catalyst immobilized in three phosphonium ionic liquids (ILs) differing in the structure of cation. Studies have shown high potential for using SILP materials as catalysts due to their high catalytic activity and selectivity, easy separation process, and the possibility of reusing the catalyst in subsequent reaction cycles without adding a new portion of the catalyst. Using the most active SILP material $SiO_2/[P_{66614}][NTf_2]/[\{Rh(\mu\text{-}OSiMe_3)(cod)\}_2]$ allows for reuse of the catalyst at least 50 times in an efficient and highly selective monofunctionalization of TMDSO.

Keywords: supported ionic liquid phase; ionic liquids; hydrosilylation; heterogeneous catalysis

---

1. Introduction

Functionalized organosilicon compounds, due to their unique properties, have generated much attention in a variety of organic processes [1–3]. Taking into account the importance of these compounds in industry, new methods of synthesis are being sought. One of the most interesting examples of these compounds is 1,1,3,3-tetramethyldisiloxane (TMDSO), which can be used in such applications as preparation of functionalizing agents of fluorinated polyethers [4], polysiloxane ingredients for heat conductible silicone compositions [5], and surfactants for the production of skincare cosmetics [6]. The conventional methods for obtaining unsymmetrical silicon-based compounds involve the condensation reactions and co-hydrolysis reactions of two silanols—chlorosilanes and alkoxysilanes [7]. However, the hydrosilylation reaction seems to be the most convenient procedure for the synthesis of new organosilicon derivatives [8–10]. Although hydrosilylation reactions are widely investigated and performed in homogenous single-phase systems [11–16], there is still the problem of the later separation of the catalyst from the product phase after the reaction [17]. To overcome this problem, attempts are being made to develop novel methods for the hydrosilylation reaction performed in heterogeneous systems. One possible solution to this problem is using supported or immobilized catalysts on a solid support. Such prepared catalysts can be easily recovered after the reaction, thus allowing the carrying out of many reaction cycles without an additional portion of catalyst [18]. Another approach widely described in the literature in recent years is to use biphasic reaction systems, where one phase consists of ionic liquid and catalysts dissolved in it and the second phase is formed by a mixture of substrates immiscible with the ionic liquid phase [19].

Reactions carried out in biphasic systems allow for separation of the ionic liquid/catalyst phase and its reuse in the next reaction cycles. However, the mass-transport limitations between phases in reactions carried out in biphasic liquid/liquid systems cause a need to investigate new possibilities for catalyst immobilization using ionic liquids (ILs). Adsorption of ionic liquid (with catalyst dissolved in IL phase) on a highly porous solid support may solve the problem of limited mass transfer [20]. Additionally, the adsorption of IL and the catalyst in the form of a thin layer that adsorbs the solid and highly porous solid supports, such as active carbon, mesoporous materials, or silica [21–24], allows for easy separation of the immobilized catalyst from the reaction mixture and of its reuse in the next reaction cycles, making the process more green and sustainable [22]. Such prepared catalytic systems, known as SILP (supported ionic liquid phase) materials, have found applications in hydrodeoxygenation, hydrogenation, hydrosilylation, the oxidation of alcohols, reforming of cellulose, and others [25–28]. Due to their unique properties, such as high surface area, the high thermal stability of ILs and the support, the polar character of ILs, and thus, the insolubility of ILs in many organic compounds, SILP materials could be successfully used in hydrosilylation reactions as highly efficient, selective, and reusable catalysts [24].

Until now, our group has published the results of using SILP materials with phosphonium ionic liquids as catalysts in hydrosilylation reactions between 1-octene and heptamethyltrisiloxane (HMTS) [24]. The application of SILPs to the performed reactions allowed for significant improvement in our process by decreasing the required amount of catalyst by 10 times, shortening the reaction time from 60 to 30 min, and primarily, extending the number of cycles that can be performed with the same portion of SILP catalyst. On the other hand, we have successfully developed a strategy for high-yield and selective synthesis of monofunctionalized TMDSO derivatives while performing the reaction in a biphasic system, where the catalyst is immobilized in the ionic liquid phase [29]. Moreover, we have found IL/catalyst systems that allow for easy separation of product from the IL/catalyst phase and later for its reuse up to 20 times without the need for adding a new portion of the catalyst.

The main goal of this article was to apply the concept of using efficient SILP materials (composed of a silica support, phosphonium IL, and rhodium catalyst) to obtain monofunctionalized TMDSO derivatives while maintaining easy product separation and reusability of the catalyst. Moreover, the presented approach led to the selective formation of unsymmetrical disiloxane equipped with an n-octyl group as well as the H–Si moiety, which can be used in subsequent catalytic transformations.

2. Results and Discussion

2.1. SILP Preparation

Based on our previous experiments, three ionic liquids were chosen for the study and were to be adsorbed on silica support. Selected salts, namely, tributylmethylphosphonium bis(trifluoromethane)sulfonimide $[P_{4441}][NTf_2]$, tetraoctylphosphonium bis(trifluoromethane)sulfonimide $[P_{8888}][NTf_2]$, and trihexyltetradecylphosphonium bis(trifluoromethane)sulfonimide $[P_{66614}][NTf_2]$ (Figure 1) were chosen due to their (i) proven high thermal stabilities (decomposition temperature is 371–384 °C), (ii) long thermal stability (loss of weight of ionic liquid range from 0.17 to 0.40% at 100 °C over 10 h), (iii) rhodium catalysts immobilized in ionic liquid are very efficient in the hydrosilylation reaction of 1-octene and TMDSO (yield higher than 90% even after 20 reaction cycles), and (iv) the high catalytic activity when using SILP materials with these ionic liquids (yield higher than 90% even after 20 reaction cycles) in the hydrosilylation reaction between 1-octene and HMTS. In comparison to the results published earlier, we decided to not to perform experiments with [P$_{4441}$][MeSO$_4$] due to its instability and lower activity when used in SILP material [24].

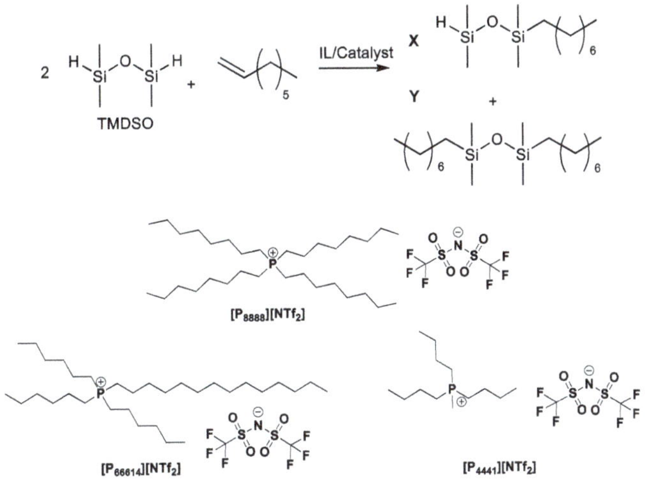

Figure 1. Top: Possible products (X—monofunctionalized; Y—bifunctionalized product synthesized in model reaction of 1,1,3,3-tetramethyldisiloxane and 1-octene. Bottom: ILs used in experiments.

Rhodium catalysts used in the experiments (Wilkinson's catalyst [RhCl(PPh$_3$)$_3$], bis(μ-trimethylsiloxy)bis(1,5-cyclooctadiene)rhodium(I) [{Rh(μ-OSiMe$_3$)(cod)}$_2$], and di(μ-chloro)bis(1,5-cyclooctadiene)dirhodium(I) [Rh(μ-Cl(cod)}$_2$]) were chosen due to their high catalytic activities reported in the literature [30].

SILP materials were prepared by impregnation of ionic liquid (Table 1) (loaded at 10% IL (w/w)) and catalyst (to be later used in the experiments at molar ratio TMDSO:1-oct:[Rh]: 2:1:2 × 10^x, where x is −5, −6, and −7) on calcined silica surface (500 °C, 24h). SILP materials were obtained with three different rhodium concentrations to examine their catalytic activity and the number of cycles over which catalyst is active [24]. The synthesis pathway of SILPs is described in details in the experimental section.

Table 1. Obtained supported ionic liquid phase (SILP) materials loaded at 10% ionic liquid (IL) (w/w).

SILP Material	IL	Complex (Catalyst)
(A1)	[P$_{4441}$][NTf$_2$]	[Rh(PPh$_3$)$_3$Cl]
(A2)	[P$_{4441}$][NTf$_2$]	[{Rh(μ-OSiMe$_3$)(cod)}$_2$]
(A3)	[P$_{4441}$][NTf$_2$]	[{Rh(μ-Cl)(cod)}]$_2$
(B1)	[P$_{8888}$][NTf$_2$]	[Rh(PPh$_3$)$_3$Cl]
(B2)	[P$_{8888}$][NTf$_2$]	[{Rh(μ-OSiMe$_3$)(cod)}$_2$]
(B3)	[P$_{8888}$][NTf$_2$]	[{Rh(μ-Cl)(cod)}]$_2$
(C1)	[P$_{66614}$][NTf$_2$]	[Rh(PPh$_3$)$_3$Cl]
(C2)	[P$_{66614}$][NTf$_2$]	[{Rh(cod)(μ-SiMe$_3$)}$_2$]
(C3)	[P$_{66614}$][NTf$_2$]	[{Rh(μ-Cl)(cod)}]$_2$

2.2. Textural Properties of SILP Materials

To characterize the obtained SILP materials and prove the successful adsorption of IL on the silica surface, an experiment of the low-temperature adsorption of nitrogen on the surface, infrared (IR) spectroscopy tests, and scanning electronic microscopy (SEM) were performed.

Low temperature nitrogen adsorption–desorption tests showed that the silica support after the impregnation of IL exhibited changed surface properties (Table 2). The Brunauer–Emmett–Teller (BET) specific surface area and the total pore volume values were reduced when comparing the SILP material to initial calcined silica from 326.5 to 220.5 m^2/g (BET surface) and from 1.10 to 0.75 cm^3/g (total pore volume), as the result of the formation of an IL film within the pores. On the other hand, the mesopores were not completely filled or blocked by the IL layer as the calculated α degree (pore filling degree of support as the ratio of IL volume to support pore volume) was within the range of 0.21 to 0.32. All obtained low temperature nitrogen adsorption/desorption isotherms belong to Type IVa (according to the International Union of Pure and Applied Chemistry (IUPAC) classification [31]) with a clearly marked hysteresis loop characteristic of mesoporous materials. Average pore diameters for calcined SiO$_2$ and SILP materials are similar (11.0–15.8 nm).

Table 2. BET characterization of calcined silica and SILP materials, pore filling degree (α) and layer thickness.

Sample	BET Surface Area [m^2/g]	Total Pore Volume [cm^3/g]	Average Pore Diameter [nm]	α ^{1}	Layer Thickness ^{2} [nm]
SiO$_2$ calcined	326.5	1.10	11.0	-	-
SILP (A1)	220.5	0.88	15.8	0.21	0.67
SILP (B1)	227.4	0.88	15.3	0.21	0.67
SILP (C1)	251.0	0.75	11.8	0.32	1.07

^{1} Pore filling degree of support as the ratio IL volume/support pore volume; ^{2} the ratio of the IL volume used for coating and the initial surface area.

The absorption of IL on the silica support was proved by comparing the IR spectra of the calcined silica with the spectra of the SILP materials (Figure 2). Characteristic bands referring to IL (2900 cm^{-1} for P-alkyl bands) were observed in all spectra of the SILP materials and not observed in the case of the spectra of calcined silica.

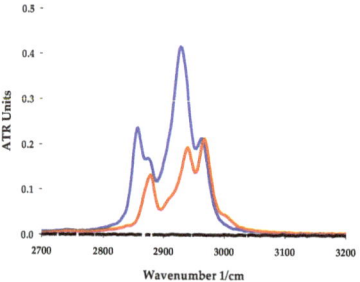

Figure 2. IR spectrum of calcined silica support (black), [P$_{4441}$][NTf$_2$] (blue) and SILP (**A1**) (red).

To examine the changes in silica surface, SEM analysis of the SILP surface prior to and after impregnation was performed on the example of SILP (**A1**). The smoother surface of the SILP material, as a consequence of filling pores of silica by a thin layer of IL, is presented in Figure 3. The analysis of SEM images is in agreement with previously described tests and confirms the successful absorption of IL on a silica support.

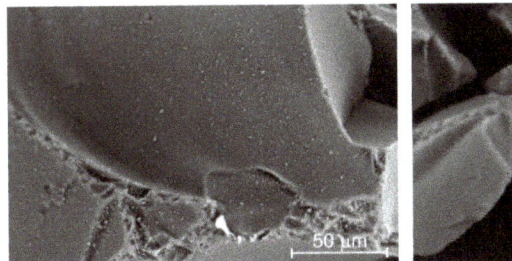 

Figure 3. SEM analysis of silica surface (**left**) and SILP (**A1**) surface (**right**) at magnification 585, Acc. Voltage; 15.0 kV.

2.3. Catalytic Activity

Catalytic activity was determined on the reaction between TMDSO and 1-octene in a molar ratio of 2:1. The reactor was charged with the SILP material in an amount corresponding to a molar ratio of TMDSO:1-oct:[Rh] to 2:1:2 × 10^x, where x is −5, −6, or −7. The mixture was stirred and heated at 50 °C for 30 min. Subsequently, the system was cooled to room temperature and the liquid phase was separated from the solid SILP material. The product phase was then analyzed by GC and GC/MS/MS to confirm conversion and selectivity. Then, the SILP material was reused in the next reaction cycle.

In general, the use of all SILP materials at a molar ratio TMDSO:1-oct:[Rh] 2:1:10^{-5} resulted in obtaining high conversion (99%) of 1-octene over at least 13 cycles (Table 3). In the case of SILPs (**A3**), (**B3**), (**C1**), (**C2**), and (**C3**), 99% conversion was maintained over 20 cycles tested with given SILP (maximum tested). The tenfold reduction in catalyst concentration (molar ratio TMDSO:1-oct:[Rh] 2:1:2 × 10^{-6}) resulted in a decrease in catalytic activity. Only in the case of SILP (**A3**), conversions above 99% were maintained to the 14th cycle, and in the case of SILP (**C3**)—to the 9th cycle. A subsequent tenfold reduction in the catalyst concentration (molar ratio TMDSO:1-oct:[Rh] 2:1:10^{-7}) resulted in a drastic decrease in activity. In most cases, only the first cycle with 99% conversion was achieved and then, a rapid drop in conversion was observed. The influence of the ionic liquid on the catalytic activity is also visible. Always, the largest number of cycles could be carried out when SILP materials contained ionic liquid [P_{666614}][NTf_2], which is in line with the published data concerning SILP activity in the hydrosilylation reaction between 1-octene and HMTS [24]. When comparing the activity of rhodium catalysts themselves, ([{Rh(μ-Cl)(cod)}]$_2$) showed the highest activity of all tested Rh catalysts. The most important advantage of the process, apart from the high catalytic activity in many cycles, is also the high selectivity of the formation of the A product. Selectivity of 95–100% is observed when using all of the tested SILPs (Tables S1–S3). This is a significant achievement in comparison to reactions carried out in conventional solvents, which allow the product to be obtained with a selectivity of 66:33% for A:B products [29]. As mentioned earlier, in the case of SILP (**A3**), (**B3**), (**C1**), (**C2**), and (**C3**), the 99% conversion was maintained for the 20 tested cycles. Therefore, in order to test catalytic activity in further cycles, reactions using SILP (**C2**) were carried out up to 50 cycles. SILP (**C2**) was used due to its high activity at catalysts concentration of 10^{-5}, 10^{-6}, and 10^{-7}. Tests have shown that after 50 reaction cycles, the conversion was still at the level of 99%, with 98% selectivity. Due to the still high catalytic activity and the absence of symptoms of poisoning and high leaching of the catalyst, subsequent cycles were not performed. Carrying out 50 cycles without a decrease in conversion and the need for adding new portion/regeneration of a catalyst makes SILP (**C2**) extremely interesting for industrial applications and is the most efficient catalyst we have obtained so far.

Table 3. Conversion of hydrosilylation reaction using SILP materials.

Catalyst	(A1)	(B1)	(C1)	(A2)	(B2)	(C2)	(A3)	(B3)	(C3)	(A1)	(B1)	(C1)	(A2)	(B2)	(C2)	(A3)	(B3)	(C3)	(A3)	(B3)	(C3)
Cycle Number	Molar Ratio TMDSO:1-oct:[Rh] 2:1:2 × 10^{-5}									Molar Ratio TMDSO:1-oct:[Rh] 2:1:2 × 10^{-6}									Molar Ratio TMDSO:1-oct:[Rh] 2:1:2 × 10^{-7}		
1	>99	>99	>99	>99	>99	>99	>99	>99	>99	>99	64	32	26	100	>99	>99	79	>99	>99	12	99
2	>99	>99	>99	>99	>99	>99	>99	>99	>99	62	52	11	0	7	56	>99	73	>99	56	0	63
3	>99	>99	>99	>99	>99	>99	>99	>99	>99	22	44	3				>99	77	>99	10		28
4	>99	>99	>99	>99	>99	>99	>99	>99	>99	11	9					>99	71	>99	0		5
5	>99	86	>99	>99	95	>99	>99	>99	>99	3						>99	65	>99			
6	>99	70	91	>99	81	>99	>99	>99	>99							>99	58	>99			
7	>99	68	80	90	37	53	>99	>99	>99							>99	42	>99			
8	>99	62	64	85	28	46	>99	>99	>99								34	99			
9	>99	54	48	77	19	32	>99	>99	>99								23	87			
10	>99	51	46	61	10	24	>99	>99	99								18	55			
11	>99		32	52	9		>99		87									39			
12	>99		58	35			>99		55									38			
13	>99			15			>99		39									20			
14	>99						>99		26												
15	>99						>99														
16	98						87														
17	98						75														
18	95						65														
19	82						34														
20	39																				
TOF [h^{-1} × 10^3]	3722	1574	3916	2414	2772	3960	3960	3960	3960	394	536	92	52	214	310	3294	1080	2312	20	24	390

2.4. Nature of Catalyst

Due to the presence of the ionic liquids and the reducing conditions ensured by the hydrosiloxane in the tested reaction conditions, the rhodium complex could be reduced to metal colloids rhodium [32], which could be a real active species. In previous studies, we performed tests to determine these active species in catalysis using SILP materials [24]. As was determined before hot filtration tests, transmission electron microscopy (TEM) measurements and mercury poisoning tests indicated that the organometallic rhodium complex, rather than metal particles, is the catalytically active species in this reaction. In these studies, we performed only TEM measurements of the SILP material after reaction and in the evaporated product phase. Tests did not show any metal particles in both examined phases.

2.5. Leaching of IL and Catalyst from SILP Material

The drop in the catalytic activity of SILP materials can be caused by leaching of the catalyst and ionic liquid/catalyst from the support. This phenomenon is known for liquid phase reactions with SILP materials and particularly problematic in continuous-flow processes, since even slight solubility of IL in the substrate, product, or solvent phase may cause removal of the thin IL film, accompanied by leaching of the catalyst [20]. To confirm the presence of IL on the SILP material after reaction (after the last reaction cycle), IR spectra of SILPs were recorded. In all cases, characteristic bands from ILs were observed, which proves that IL was still present on the support (Figure 4).

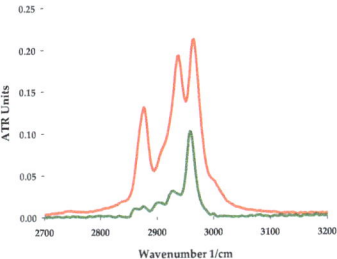

Figure 4. IR spectrum of SILP (**A1**) before reaction (red) and after the 20th cycle (green).

In order to further determine the leaching of rhodium catalysts into the product phase, inductively coupled plasma (ICP) analysis was conducted. Due to the low limit of detection (1 ppm) [19], the results of ICP analysis were inconclusive and thus, additional experiments were performed as follows: reaction was carried out following the procedure described in Section 3.4 and then, the product phase was separated from the reaction mixture. Subsequently, a new portion of substrates was added to the product phase. Then, the reaction was carried out in 50 °C over 30 min. Samples of post reaction mixture were collected and analyzed by GC. The experiments were performed only for SILP (**B1**), as for this reaction system, the yield of the reaction decreases the most significantly with every following reaction cycle and for SILP (**C2**), as for this SILP, yield of reaction was stable after the 20th cycle. Table 4 presents the conversion of 1-octene after each reaction cycle using SILP (**B1**) and (**C2**), and the conversion of 1-octene in post reaction mixtures (with a new portion of substrates added to the product phase, which was separated after each reaction cycle).

Analysis of the data presented in columns II and IV (Table 4) shows that after adding a new portion of substrates to the separated product phase, conversion of the Si–H bond is still observed (in the range of 4–10%). It is concluded that the rhodium catalyst was presented in the product phase, separated after the first and subsequent reaction cycles, which means that the catalyst had been slightly washed out from the SILP material. In the case of SILP (**C2**), where the conversion is still observed even after 50 cycles, catalyst leaching is constant, but occurs in a small amount (6% substrate conversion is observed). In the case of SILP (**B2**), where the conversion drops sharply after several cycles, it can be concluded that the decrease in activity results more from poisoning of the catalyst than from its leaching from SILP material to the product phase.

Table 4. Conversion of 1-octene obtained with reaction system containing SILP (**B1**) and (**C2**). I: Conversion using SILP (**B1**). II: Conversion after adding new portion of substrate to product phase (after separation from SILP (**B1**)). III: Conversion using SILP (**C2**). IV: Conversion after adding new portion of substrate to product phase (after separation from SILP (**C2**)).

Cycle Number	I SILP (B1)	II SILP (B1)	III SILP (C2)	IV SILP (C2)
1	>99	7	>99	10
2	>99	6	>99	10
3	>99	6	>99	9
4	>99	4	>99	7
5	86	6	>99	6
10	-	7	>99	7
20	-	-	>99	6
30	-	-	>99	7
40	-	-	>99	6
50	-	-	>99	6

2.6. Comparison of TOF

Overall, the high catalytic activity of used SILPs is confirmed by high TOF (turnover frequency) values calculated as amounts of the products expressed in moles obtained in all runs in terms of the initial catalyst used (in moles).

TOF values were compared (Table 5) with the use of two similar catalytic systems containing an ionic liquid: $[P_{66614}][NTf_2]$ or $[P_{8888}][NTf_2]$ and the $[\{Rh(\mu\text{-}Cl)(cod)\}]_2$ catalyst, when using (i) catalyst dissolved in IL and (ii) catalyst dissolved in IL absorbed on the silica surface (SILP material). The reaction carried out with the use of SILP achieves higher TOF values due to the 99% activity over 20 reaction cycles and the use of a lower catalyst concentration (10^{-5} vs. 10^{-4} in the case of the IL/catalyst system). In addition, in the case of the reaction in which the SILP catalyst was used, the reaction did not lead to a decrease in conversion, which—in the example of SILP (**C2**)—shows that the TOF value can be even several times higher. Moreover, comparison of TOF values of the catalytic system without ionic liquid and SILP (catalyst added to mixture of substrates) was performed. TOF values, when assuming the same concentration of catalyst, are significantly lower than when using biphasic IL/catalyst system or SILP systems due to the fact that the reaction can be carried out only in a single cycle (without the possibility of reusing the catalyst). This comparison shows the advantages of using SILP catalyst systems in comparison to the catalytic system without ionic liquids and the improvement over the catalytic system using the catalyst dissolved in ionic liquids.

Table 5. Comparison of TOF $[h^{-1} \times 10^3]$ without using and with using ILs as solvent for catalyst $[\{Rh(\mu\text{-}Cl)(cod)\}]_2$ in the biphasic system and supported on silica (SILP materials).

TMDSO:1-oct:[Rh] (Reactants/Catalyst Concentration)	IL	SILP System	TOF
2:1:2 × 10^{-4}	-	-	20 [1,4]
2:1:2 × 10^{-5}	-	-	1928 [2,4]
2:1:2 × 10^{-4}	$[P_{8888}][NTf_2]$	-	169 [1,4]
2:1:2 × 10^{-4}	$[P_{66614}][NTf_2]$	-	375 [1,4]
2:1:2 × 10^{-5}	$[P_{8888}][NTf_2]$	(B3)	3960 [2,3]
2:1:2 × 10^{-5}	$[P_{66614}][NTf_2]$	(C3)	3960 [2,3]

[1] t = 30 min, molar ratio TMDSO:1-oct:[Rh]:2:1:2 × 10^{-4}; [2] t = 30 min, molar ratio TMDSO:1-oct:[Rh]:2:1:2 × 10^{-5}; [3] catalytic system is still active after the 20 reaction cycles and more cycles could be performed; [4] TOF values calculated based on results published before [24,29].

3. Materials and Methods

3.1. Synthesis of Ionic Liquids

ILs: tributylmethylphosphonium bis(trifluoro-methylsulphonyl)imide [P_{4441}][NTf_2], tetraoctylphosphonium bis(trifluoro-methylsulphonyl)imide [P_{8888}][NTf_2], trihexyltetrabutylphosphonium bis(trifluoromethylsulfonyl)imide [P_{66614}][NTf_2] were synthesized by metathesis reaction from their halogen precursors as was described in the literature [33,34]. Purity of ILs was confirmed by ^{1}H NMR analysis (see ^{1}HNMR data, ESI) and IC analysis (absence of halogen ion was observed, see Figure S1). Tributylmethylphosphonium chloride [P_{4441}][Cl], tetraoctylphosphonium chloride [P_{8888}][Cl], and trihexyltetradecyl phosphonium chloride [P_{66614}][Cl] were purchased from Iolitec.

3.2. Synthesis of Catalysts

[Rh(PPh$_3$)$_3$Cl] and [{Rh(μ-Cl)(cod)}]$_2$ were purchased from Sigma Aldrich. [{Rh(μ-OSiMe$_3$)(cod)}$_2$] was prepared as described by Marciniec et. al. [35].

3.3. Preparation of SILP Materials

The silica support before impregnation was calcined (500 °C, 24 h) to decrease the number of hydroxyl groups on the silica support. In order to prepare supported ionic liquid on silica with the rhodium catalyst, ionic liquid (0.2g) was dissolved in 20 cm^3 dichloromethane. Subsequently, 10 cm^3 (to be later used in experiments at molar ratio TMDSO:1-oct:[Rh] 2:1:2 × 10^{-5}), 1 cm^3 (to be later used in experiments at molar ratio TMDSO:1-oct:[Rh] 2:1:2 × 10^{-6}), or 0.1 cm^3 (to be later used in experiments at molar ratio TMDSO:1-oct: [Rh] 2:1:2 × 10^{-7}) of solution of rhodium catalyst (([Rh(PPh$_3$)$_3$Cl]—0.68 mg, [{Rh(μ-Cl)(cod)}]$_2$—0.18 mg, and [{Rh (μ-OSiMe$_3$)(cod)}$_2$]—0.53 mg), dissolved in 10 cm^3 of dichloromethane) and mesoporous silica (high-purity grade Davisil Grade 62, pore size 150 Å, 60–200 mesh, 1.15 cm^3/g pore volume) (1.8 g) were added to a solution of IL in dichloromethane. In the next step, the solution was stirred for 30 min at room temperature and after that, the volatile solvent was very slowly evaporated, leaving a thin ionic liquid film on the surface of the support material. Then, SILP materials were dried under high vacuum at 40 °C (48 h).

3.4. Reactions Using SILP Catalysts

The reaction system was composed of a glass reactor of 2 cm^3 capacity and a side neck for substrate introduction, equipped with a magnetic stirrer and reflux condenser. The reactor was charged with 0.1 g of SILP material (in 0.1g of SILP material is the amount of catalyst corresponding to 4.6 × 10^{-5} mole of Rh (or 10^{-6} or 10^{-7})) and then, the substrates 4.6 mmol (0.62 g) of TMDSO and 2.3 mmol (0.26 g) of 1-octene (molar ratio of TMDSO:1-oct:[Rh] 2:1:2 × 10^x, where x is −5, −6 or −7) were added). Then, the entire mixture was stirred and heated at 50 °C for 30 min. Subsequently, the system was cooled to room temperature and the liquid phase was separated from the solid SILP material. The product phase was then analyzed by GC and GC/MS/MS to confirm selectivity and determine conversion and SILP material was reused in the next reaction cycle

3.5. ^{1}H NMR Spectra

^{1}H NMR spectra were recorded on a Varian XL 300 NMR (300 MHz) (Varian, Palo Alto, CA, USA) using d$_6$-DMSO as solvent with tetramethylsilane as the internal standard. Proton chemical shifts (available in the ESI file) are shown in parts per million (δ ppm).

3.6. FT-IR Spectra

FT-IR spectra were recorded in a Bruker Tensor 27 apparatus (Bruker Corporation Optik GmbH, Bremen, Germany) in the range 500–4000 1/cm of wave number. Solid samples (ca. 1 mg) were placed on a diamond window and then, covered with a sapphire tip. Using OPUS 5.5 software, baseline and peak picking were carried out in order to determine the main peaks in the product.

3.7. Determination of Surface Area, Pore Volume, and Pore Diameter

The Brunauer–Emmett–Teller (BET) surface areas were determined by N_2 adsorption at 77K using a Micromeritics ASAP 2010 sorptometer (Micromeritics Instrument Corp., Norcross, GA, USA). Total pore volume and average pore diameter were determined by applying the Barrett–Joyner–Halenda (BJH) method to the isotherm desorption branch. Prior to the measurements of adsorption/desorption isotherms, the samples were outgassed at 393K for 20 h.

3.8. GC, GC/MS/MS Analysis

The conversion and selectivity of the reactions were analyzed using a Varian CP 3800 Gas Chromatograph (Varian, Santa Clara, CA, USA) with a FactorFour VF-5ms column (15 m × 0.25 mm) with decane as internal standard. Additional experiments to confirm the selectivity of the reactions were performed using a Varian 400 GC/MS/MS (Varian, Santa Clara, CA, USA) instrument with a FactorFour VF-1ms column (15 m × 0.25 mm).

4. Conclusions

Based on previous research on the use of SILP materials in the hydrosilylation reaction and the use of ionic liquids in the selective monofunctionalization of TMDSO, it was decided to combine both concepts and apply SILP catalysts to obtain a monofunctionalized derivative of TMDSO. In general, nine SILP materials (differing in structure of cation in IL, rhodium catalyst, and its concentration in SILP material) with three different Rh concentrations were obtained and characterized by the BET, IR, and SEM methods and later used in the hydrosilylation reaction between TMDSO and 1-octene.

The obtained materials, in particular, turned out to be very efficient catalysts, and the most efficient of them (SILP **(C2)** composed of $[P_{66614}][NTf_2]/[\{Rh(\mu\text{-}OSiMe_3)(cod)\}_2]$) allowed us to obtain 50 reaction cycles with 99% conversion and 98% selectivity. Additional studies have shown that the rhodium complexes are slowly leached out of SILP materials; however, leaching does not have a decisive influence on the activity of the SILP material. Certainly, the use of SILP materials in catalysis will contribute to the reduction of energy used during the production process, will contribute to a better use of materials, and will allow for a reduction in the amount of post reaction waste.

Supplementary Materials: The following are available online at http://www.mdpi.com/2073-4344/10/12/1414/s1, ^{1}H NMR data for obtained ionic liquids; IC analysis of Ionic Liquids; GC/MS/MS chromatogram and analysis; Figure S1. IC chromatograms of synthesized ILs with $[NTf_2]^-$ anion from their chloride precursors. No peak observance at ~4.9 min means that $[Cl]^-$ content in ionic liquid is below limit of detection; Figure S2. GC chromatogram of post reaction mixture obtained from reaction using SILP **(A1)** after 1st reaction cycle; retention times 0.9 min—acetone, 1.5–2.0 min—substrates (1-octene, TMDSO), 6.7 min—decane (internal standard), 10.1 min—product A (octylotetramethyldisiloxane), Figure S3. GC chromatogram of post reaction mixture obtained from reaction using SILP **(B1)** after 1st reaction cycle; retention times 0.9 min—acetone, 1.5–2.0 min—substrates (1-octene, TMDSO), 6.7 min—decane (internal standard), 10.1 min—product A (octylotetramethyldisiloxane), Figure S4. GC chromatogram of post reaction mixture obtained from reaction using SILP **(C1)** after 1st reaction cycle; retention times 0.9 min—acetone, 1.5–2.0 min—substrates (1-octene, TMDSO), 6.7 min—decane (internal standard), 10.1 min—product A (octylotetramethyldisiloxane), Figure S5. Example of GC chromatogram of post reaction mixture obtained from reaction using SILP **(C1)** after 4st reaction cycle; retention times 0.9 min—acetone, 1.5–2.0 min—substrates (1-octene, TMDSO), 6.7 min—decane (internal standard), 10.1 min—product A (octylotetramethyldisiloxane), 14.7 min—product B (1,3-dioctylo-1,1,3,3-tetramethyldisiloxane), Figure S6. GC-MS mass spectrum of octylotetramethyldisiloxane obtained from reaction using SILP **(A1)** recorded at retention time 8.842 min., Table S1. Conversion and selectivity of hydrosilylation reaction using SILP materials ($[RhCl(PPh_3)_3]$), Table S2. Conversion and selectivity of hydrosilylation reaction using SILP materials ($[\{Rh(\mu\text{-}OSiMe_3)(cod)\}_2]$), Table S3. Conversion and selectivity of hydrosilylation reaction using SILP materials $[\{Rh(\mu\text{-}Cl)(cod)\}]_2$.

Author Contributions: Conceptualization, H.M., M.S., R.K. and A.P.-Z.; methodology, R.K. and A.P.-Z.; synthesis of ILs, A.P.-Z.; analysis of purity of ILs, R.K., A.P.-Z. and J.D.; synthesis of catalysts, R.J.; synthesis of SILP materials, A.P.-Z.; IR analysis of SILP materials, R.K.; SEM analysis of SILP materials, R.K.; textural analysis of SILP materials, M.P. and M.Z.; catalysis with using SILP materials, A.P.-Z., J.D. and R.K.; GC, GC/MS analysis, R.K. and A.P.-Z.;

HNMR analysis, R.K.; examination of nature of catalyst, R.K.; leaching studies, R.K.; writing—original draft preparation, R.K.; writing—review and editing, R.K., M.S., R.J., A.P.-Z., M.P. and H.M.; supervision, H.M. and M.S.; project administration, M.S.; funding acquisition, M.S. All authors have read and agreed to the published version of the manuscript.

Funding: M.S. greatly acknowledges funding from the National Science Centre (Poland), project SONATA BIS; grant number UMO-2017/26/E/ST8/01059.

Conflicts of Interest: The authors declare no conflict of interest

References

1. Lee, V.Y. (Ed.) *Organosilicon Compounds. Theory and Experiment (Synthesis)*; Elsevier: London, UK, 2017. [CrossRef]
2. Januszewski, R.; Kownacki, I.; Maciejewski, H.; Marciniec, B.; Szymańska, A. An Efficient Catalytic Route for the Synthesis of Silane Coupling Agents Based on the 1,1,3,3-Tetramethyldisiloxane Core. *Eur. J. Inorg. Chem.* **2017**, *2017*, 851–856. [CrossRef]
3. Hiyama, T.; Shirakawa, E. Organosilicon Compounds. *Chemin* **2003**, *34*, 169–218. [CrossRef]
4. Yamane, Y.; Koike, N.; Yamaguchi, K.; Kishita, H. Fluorine-Containing Organopolysiloxane, a Surface Treatment Composition Comprising the Same and an Article Treated with the Composition. European Patent EP1813640B1, 24 February 2010.
5. Sakurai, I.; Matsumoto, N.; Miyoshi, K.; Yamada, K. Heat-Conductive Silicone Composition and Cured Product Thereof. U.S. Patent US8119758B2, 21 February 2012.
6. Leatherman, M.D.; Policello, G.A.; Rajaraman, K. Extreme Environment Surfactant Compositions Comprising Hydrolysis Resistant Organomodified Disiloxane Surfactants. U.S. Patent US7645720B2, 12 January 2010.
7. Brook, M.A. *Silicon in Organic, Organometallic and Polymer Chemistry*; Wiley-VCH: Weinheim, Germany, 2000.
8. Marciniec, B.; Maciejewski, H.; Pietraszuk, C.; Pawluć, P. *Hydrosilylation. A Comprehensive Review on Recent Advances*; Springer: Dordrecht, The Netherlands, 2009. [CrossRef]
9. Troegel, D.; Stohrer, J. Recent advances and actual challenges in late transition metal catalyzed hydrosilylation of olefins from an industrial point of view. *Co-Ord. Chem. Rev.* **2011**, *255*, 1440–1459. [CrossRef]
10. Marciniec, B.; Maciejewski, H.; Pietraszuk, C.; Pawluc, P. *Applied Homogeneous Catalysis with Organometallic Compounds*; Cornils, B., Hermmann, W.A., Belier, M., Pawelo, R., Eds.; Wiley-VCH: Weinheim, Germany, 2017; pp. 569–620. [CrossRef]
11. Zhao, Z.-Y.; Nie, Y.-X.; Tang, R.-H.; Yin, G.-W.; Cao, J.; Xu, Z.; Cui, Y.-M.; Zheng, Z.-J.; Xu, L.-W. Enantioselective Rhodium-Catalyzed Desymmetric Hydrosilylation of Cyclopropenes. *ACS Catal.* **2019**, *9*, 9110–9116. [CrossRef]
12. Wen, H.; Wang, K.; Zhang, Y.; Liu, G.; Huang, Z. Cobalt-Catalyzed Regio- and Enantioselective Markovnikov 1,2-Hydrosilylation of Conjugated Dienes. *ACS Catal.* **2019**, *9*, 1612–1618. [CrossRef]
13. Schuhknecht, D.; Spaniol, T.P.; Maron, L.; Okuda, J. Regioselective Hydrosilylation of Olefins Catalyzed by a Molecular Calcium Hydride Cation. *Angew. Chem. Int. Ed.* **2020**, *59*, 310–314. [CrossRef]
14. Raya-Barón, Á.; Oña-Burgos, P.; Fernández, I. Iron-Catalyzed Homogeneous Hydrosilylation of Ketones and Aldehydes: Advances and Mechanistic Perspective. *ACS Catal.* **2019**, *9*, 5400–5417. [CrossRef]
15. Jankowska-Wajda, M.; Bartlewicz, O.; Pietras, P.; Maciejewski, H. Piperidinium and Pyrrolidinium Ionic Liquids as Precursors in the Synthesis of New Platinum Catalysts for Hydrosilylation. *Catalysts* **2020**, *10*, 919. [CrossRef]
16. Wu, C.; Peng, J.; Li, J.; Bai, Y.; Hu, Y.; Lai, G. Synthesis of poly(ethylene glycol) (PEG) functionalized ionic liquids and the application to hydrosilylation. *Catal. Commun.* **2008**, *10*, 248–250. [CrossRef]
17. Pawlowska-Zygarowicz, A.; Kukawka, R.; Maciejewski, H.; Smiglak, M. Optimization and intensification of hydrosilylation reactions using a microreactor system. *New J. Chem.* **2018**, *42*. [CrossRef]
18. Cano, R.; Yus, M.; Ramón, D.J. Impregnated Platinum on Magnetite as an Efficient, Fast, and Recyclable Catalyst for the Hydrosilylation of Alkynes. *ACS Catal.* **2012**, *2*, 1070–1078. [CrossRef]
19. Jankowska-Wajda, M.; Kukawka, R.; Smiglak, M.; Maciejewski, H. The effect of the catalyst and the type of ionic liquid on the hydrosilylation process under batch and continuous reaction conditions. *New J. Chem.* **2018**, *42*, 5229–5236. [CrossRef]

20. Fehrmann, R.; Riisager, A.; Haumann, M. (Eds.) *Supported Ionic Liquids: Fundamentals and Applications*; Wiley-VCH: Weinheim, Germany, 2014. [CrossRef]
21. Van Doorslaer, C.; Wahlen, J.; Mertens, P.; Binnemans, K.; De Vos, D. Immobilization of molecular catalysts in supported ionic liquid phases. *Dalton Trans.* **2010**, *39*, 8377–8390. [CrossRef] [PubMed]
22. Riisager, A.; Fehrmann, R.; Haumann, M.; Wasserscheid, P. Supported Ionic Liquid Phase (SILP) Catalysis: An Innovative Concept for Homogeneous Catalysis in Continuous Fixed-Bed Reactors. *Eur. J. Inorg. Chem.* **2006**, *2006*, 695–706. [CrossRef]
23. Sowińska, A.; Maciejewska, M.; Guo, L.; Delebecq, E. Effect of SILPs on the Vulcanization and Properties of Ethylene–Propylene–Diene Elastomer. *Polymers* **2020**, *12*, 1220. [CrossRef] [PubMed]
24. Kukawka, R.; Pawlowska-Zygarowicz, A.; Dzialkowska, J.; Pietrowski, M.; Maciejewski, H.; Bica, K.; Smiglak, M. Highly Effective Supported Ionic Liquid-Phase (SILP) Catalysts: Characterization and Application to the Hydrosilylation Reaction. *ACS Sustain. Chem. Eng.* **2019**, *7*, 4699–4706. [CrossRef]
25. Geier, D.; Schmitz, P.; Walkowiak, J.; Leitner, W.; Franciò, G. Continuous Flow Asymmetric Hydrogenation with Supported Ionic Liquid Phase Catalysts Using Modified CO_2 as the Mobile Phase: From Model Substrate to an Active Pharmaceutical Ingredient. *ACS Catal.* **2018**, *8*, 3297–3303. [CrossRef]
26. Giacalone, F.; Gruttadauria, M. Covalently Supported Ionic Liquid Phases: An Advanced Class of Recyclable Catalytic Systems. *ChemCatChem* **2016**, *8*, 664–684. [CrossRef]
27. Kudo, S.; Goto, N.; Sperry, J.; Norinaga, K.; Hayashi, J.-I. Production of Levoglucosenone and Dihydrolevoglucosenone by Catalytic Reforming of Volatiles from Cellulose Pyrolysis Using Supported Ionic Liquid Phase. *ACS Sustain. Chem. Eng.* **2017**, *5*, 1132–1140. [CrossRef]
28. Bartlewicz, O.; Dąbek, I.; Szymańska, A.; Maciejewski, H. Heterogeneous Catalysis with the Participation of Ionic Liquids. *Catalysts* **2020**, *10*, 1227. [CrossRef]
29. Kukawka, R.; Januszewski, R.; Kownacki, I.; Smiglak, M.; Maciejewski, H. An efficient method for synthesizing monofunctionalized derivatives of 1,1,3,3-tetramethyldisiloxane in ionic liquids as recoverable solvents for rhodium catalyst. *Catal. Commun.* **2018**, *108*, 59–63. [CrossRef]
30. Maciejewski, H.; Wawrzynczak, A.; Dutkiewicz, M.; Fiedorow, R. Silicone waxes—synthesis via hydrosilylation in homo- and heterogeneous systems. *J. Mol. Catal. A Chem.* **2006**, *257*, 141–148. [CrossRef]
31. Thommes, M.; Kaneko, K.; Neimark, A.V.; Olivier, J.P.; Rodriguez-Reinoso, F.; Rouquerol, J.; Sing, K.S. Physisorption of gases, with special reference to the evaluation of surface area and pore size distribution (IUPAC Technical Report). *Pure Appl. Chem.* **2015**, *87*, 1051–1069. [CrossRef]
32. Lewis, L.N.; Uriarte, R.J. Hydrosilylation catalyzed by metal colloids: A relative activity study. *Organometallics* **1990**, *9*, 621–625. [CrossRef]
33. Wasserscheid, P.; Welton, T. (Eds.) *Ionic Liquids in Synthesis*; Wiley-VCH: Weinheim, Germany, 2008. [CrossRef]
34. Zieliński, W.; Kukawka, R.; Maciejewski, H.; Smiglak, M. Ionic Liquids as Solvents for Rhodium and Platinum Catalysts Used in Hydrosilylation Reaction. *Molecules* **2016**, *21*, 1115. [CrossRef]
35. Marciniec, B.; Krzyżanowski, P. Synthesis, characterization and some reactions of [(diene)Rh(μ-OSiMe3)]2. *J. Organomet. Chem.* **1995**, *493*, 261–266. [CrossRef]

Publisher's Note: MDPI stays neutral with regard to jurisdictional claims in published maps and institutional affiliations.

 © 2020 by the authors. Licensee MDPI, Basel, Switzerland. This article is an open access article distributed under the terms and conditions of the Creative Commons Attribution (CC BY) license (http://creativecommons.org/licenses/by/4.0/).

Article

N-Hydroxyphthalimide Supported on Silica Coated with Ionic Liquids Containing CoCl₂ (SCILLs) as New Catalytic System for Solvent-Free Ethylbenzene Oxidation

Gabriela Dobras [1], Kornela Kasperczyk [1], Sebastian Jurczyk [2] and Beata Orlińska [1,*]

[1] Department of Organic Chemical Technology and Petrochemistry, Silesian University of Technology, Krzywoustego 4, 44-100 Gliwice, Poland; gabriela.dobras@polsl.pl (G.D.); kornela.kasperczyk@gmail.com (K.K.)
[2] Łukasiewicz Research Network - Institute for Engineering of Polymer Materials and Dyes, Skłodowskiej-Curie 55, 87-100 Toruń, Poland; s.jurczyk@impib.pl
* Correspondence: beata.orlinska@polsl.pl

Received: 30 January 2020; Accepted: 15 February 2020; Published: 19 February 2020

Abstract: N-Hydroxyphthalimide was immobilized via ester bond on commercially available silica gel (SiOCONHPI) and then coated with various ionic liquids containing dissolved CoCl₂ (SiOCONHPI@CoCl₂@IL). New catalysts were characterized by means of FT IR spectroscopy, elemental analysis, SEM and TGA analysis and used in ethylbenzene oxidation with oxygen under mild solvent-free conditions (80 °C, 0.1 MPa). High catalytic activity of SiOCONHPI was proved. In comparison to a non-catalytic reaction, a two-fold increase in conversion of ethylbenzene was observed (from 4.7% to 8.6%). Coating of SiOCONHPI with [bmim][OcOSO₃], [bmim][Cl] and [bmim][CF₃SO₃] containing CoCl₂ enabled to increase the catalytic activity in relation to systems in which IL and CoCl₂ were added directly to reaction mixture. The highest conversion of ethylbenzene was obtained while SiOCONHPI@CoCl₂@[bmim][OcOSO₃] were used (12.1%). Catalysts recovery and reuse was also studied.

Keywords: oxidation; N-hydroxyphthalimide; immobilization; ionic liquids; SCILL

1. Introduction

Hydrocarbon oxidation processes play an important role in organic synthesis and industrial processes. Processes which use oxygen or air as an oxidizing factor often require the use of suitable catalytic systems. The most commonly used ones include compounds of transition metals, such as Co and Mn [1,2]. High activity of N-hydroxyphthalimide (NHPI) has also been proved in reactions of this kind [3–5]. Its catalytic activity is attributed to a PINO radical generated in the system, which abstracts a hydrogen atom from oxidized hydrocarbon, from 2 to 20 times faster than peroxyl radicals in an autocatalytic process [3] (Figure 1). PINO is generated by various additives, e.g., azo compounds, peroxides, transition metals salts, aldehydes, and quinones and its derivatives [6].

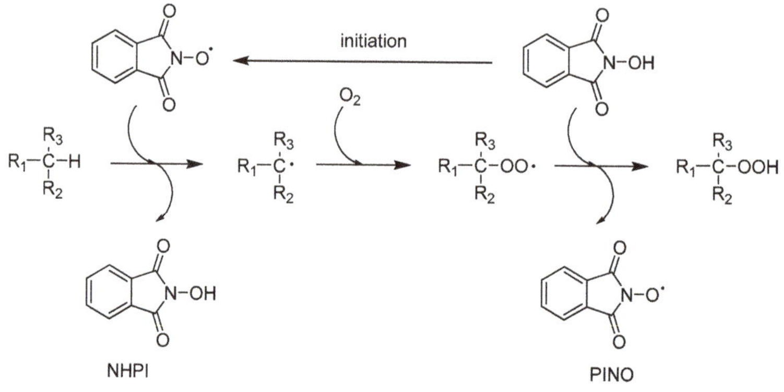

Figure 1. Mechanism of hydrocarbons oxidation using *N*-hydroxyphthalimide (NHPI).

The advantage of NHPI is its availability, low price, and a simple method of its synthesis from phthalic anhydride and hydroxylamine [3]. Because of its poor solubility in hydrocarbons, the processes are usually carried out in polar solvents, e.g., in acetic acid [7], acetonitrile [8], or benzonitrile [9]. The use of compressed CO_2 [10–12] and ionic liquids [11,13–18] has also been reported. In order to use NHPI in industrial processes, it is necessary to develop efficient methods of its recovery and reuse. NHPI separation by evaporation of organic solvent [8,19] or CO_2 has been described in literature [10,11]. However, this simple method may prove to be ineffective, if the product of the reaction will be characterized by high polarity and related good solubility of NHPI. Therefore, there were attempts to immobilize NHPI on solid carriers: Silicas [20,21], polymers [22–25], and zeolites [26]. As an example, Hermans [20] et al., in the process of oxidation of cyclohexane, used NHPI immobilized by simple impregnation on silica gel. Attempts of heterogeneous NHPI reuse, proved that the activity of catalyst in the first cycle drops down twofold, but in the following two cycles it is maintained at a constant level. In the work [21], in the reactions of toluene oxidation in acetic acid as a solvent, NHPI immobilized on silica via amide bond was used. It was proven that it is possible to reuse heterogeneous NHPI with only a slight decrease in activity. In our work [22] the results of oxidation of *p*-methoxytoluene with oxygen, Co(II) salt, and NHPI immobilized via amide bond on aminomethyl polystyrene or via ester bond on a chloromethyl polystyrene, were discussed. We proved that two-fold recovery and reuse of catalysts is possible. Examples of use of NHPI immobilized in ionic liquids was also described [17,18]. NHPI was incorporated into the structure of imidazolium [17] cation and pirydinium [18] cation. The catalysts belong to the so called task-specific ionic liquids, compounds built of ionic liquid with an introduced functional group with specific properties [27], in this case—NHPI (Figure 2). They were used in the processes of oxidation of 1-phenylethanol [17], *N*-alkylamides, and various benzylic hydrocarbons [18]. The authors of [17] claim that IL-NHPI has higher activity than NHPI. For example, acetophenone was obtained in 98% yield in the presence of IL-NHPI, after two hours, and in 93% yield with NHPI, after six hours. It is also possible to use IL-NHPI five times without decrease in activity.

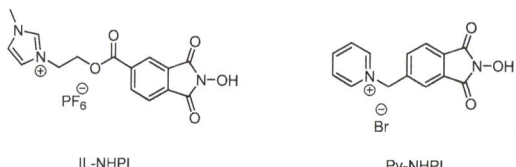

Figure 2. NHPI immobilized in ionic liquids.

Heterogenization of catalysts, is often related to the decrease in their activity. Recently, in order to minimize or even to eliminate this problem, studies on supported catalysts with ionic liquid layer (SCILL) have been undertaken [28,29]. In this case, heterogeneous catalysts, not inert support as in the case of the supported ionic liquid phase (SILP) catalysis [30–32], are coated with ionic liquids. It was confirmed that ionic liquid can positively influence properties of catalyst (cocatalytic effect) and may cause an increase in concentration of substrates or intermediate products on its surface, because of ionic properties of ionic liquid phase (solvent effect) [29]. However, it is important that the organic phase and ionic liquid compose biphasic system, otherwise IL film will be washed out from a catalyst [28,29]. SCILL method has been already used in hydrogenation [33–35] and oxidation [36,37] processes. Karimi and Badreh [36], in the oxidation reactions of a great variety of alcohols, used a system composed of TEMPO chemically bonded to silica SBA-15 coated with ionic liquid [bmim][Br]. Catalyst IL@SBA-15-TEMPO showed higher activity than not modified by ionic liquid SBA-15-TEMPO. In addition, it was possible to obtain 11-fold recycling values of this catalyst without loss in activity. In a similar system, a catalyst composed of TEMPO immobilized on a polymer support coated with ionic liquid [bmim][PF_6], containing $CuCl_2$, was used in oxidation reactions of 4-methoxybenzyl alcohol [37]. In oxidation process carried out in the presence of this catalyst, a significantly higher conversion of alcohol was acquired (85%) than in the presence of the same catalyst without ionic liquid (58%). However, catalyst recycling attempts showed that the activity of this catalyst drops down insignificantly, together with every reaction cycle, what results from washing out [bmim][PF_6] from the support surface.

Herein, NHPI was immobilized via ester bond on silica (SiOCONHPI), which was coated with a layer of ionic liquid containing dissolved $CoCl_2$ salt (SiOCONHPI@$CoCl_2$@IL). To the best of our knowledge, both catalysts have not been previously described. Consequently, catalytic activity of SiOCONHPI as well as SCILL system was examined for the first time, in oxidation reaction using oxygen. The influence of the type of ionic liquid on activity of the system and a possibility of recovery and reuse was determined. In the previous work [38] it was proved that selected ionic liquid, used in catalytic amounts, can positively influence the hydrocarbons oxidation process using NHPI/Co(II) system without solvent additive.

2. Results and Discussion

2.1. Preparation of Catalyst

2.1.1. SiOCONHPI Synthesis

Immobilization of NHPI was carried out on commercially available silica. For that purpose, in accordance with the procedure described for polymer support [22], in the first stage, trimellitic anhydride was immobilized on chloropropyl-functionalized silica gel via ester bond and consequently it underwent the reaction with hydroxylamine hydrochloride (Figure 3).

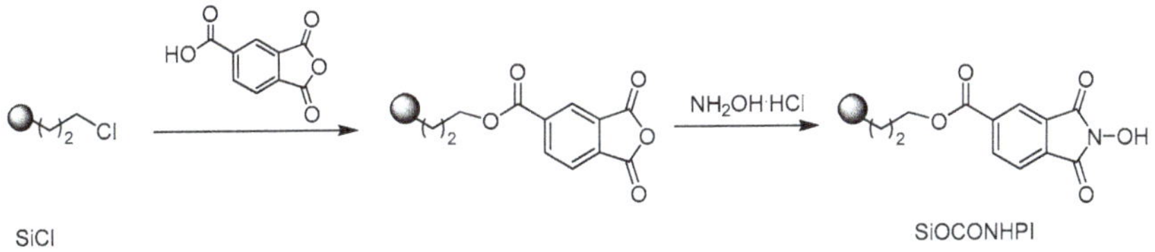

Figure 3. Synthesis of NHPI immobilized on chloropropyl silica gel via ester bond.

On the basis of elemental analysis, it was confirmed that the content of immobilized NHPI was equal to 0.45 mmol NHPI/g.

Comparing FT-IR spectra of chloropropyl functionalized silica SiCl (Figure 4I) and of NHPI immobilized on silica SiOCONHPI (Figure 4II) it is possible to unambiguously identify only signals of carbonyl groups at around 1650 cm^{-1}. N-hydroxyl group signals at around 3300 cm^{-1} overlap with vibration band of SiO–H hydroxyl groups of silica support. Because of intensive vibration bands characteristic for silica gel, this is silicon-oxygen bond peaks 1007 cm^{-1} and hydroxyl group bonded with silicon atom 793 cm^{-1}, it was not possible to identify a signal originating from C–O in ester bond.

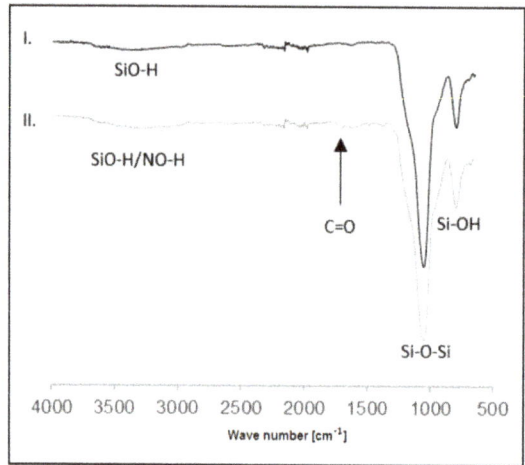

Figure 4. FT-IR spectra of I. SiCl and II. SiOCONHPI.

During the synthesis irregular chloropropyl silica granules were mechanically degraded to smaller, irregular granules with size equal to 1–20 μm (Figure 5).

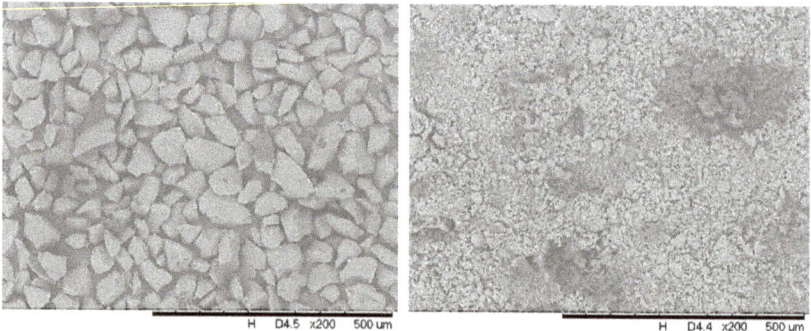

Figure 5. SEM analysis of support (SiCl) and NHPI immobilized on chloropropyl silica gel (SiOCONHPI).

2.1.2. SiOCONHPI@CoCl$_2$@IL Preparation

Obtained SiOCONHPI was coated with a layer of ionic liquid containing dissolved cobalt(II) chloride (SiOCONHPI@CoCl$_2$@IL). Ionic liquids with the following cations were selected for the study: 1-ethyl-3-methylimidazolium [emim], 1-butyl-3-methylimidazolium [bmim], 1-hexyl-3-methylimidazolium [hmim], 1-octyl-3-methylimidazolium [omim], 1-decyl-3-methylimidazolium [C$_{10}$mim], and anions: [CH$_3$OSO$_3$], octylsulfate ([OcOSO$_3$]),

[$C_{12}H_{25}OSO_3$], [Br], [Cl], [CF_3SO_3], [BF_4], [PF_6], and [($CF_3SO_2)_2N$] ([NTf_2]). Used ionic liquids differ in polarity, solubility of oxygen, and ethylbenzene.

2.2. Study on Ethylbenzene Oxidation Reaction

The catalytic activity of SiOCONHPI and SiOCONHPI@CoCl$_2$@IL was studied in oxidation reaction of ethylbenzene, used as a model alkylaromatic hydrocarbon, with the use of oxygen. The selection of CoCl$_2$ as cocatalyst resulted from known activity of cobalt(II) salt in oxidation reactions using NHPI as well as from limited solubility of CoCl$_2$ in the hydrocarbon in order to limit its leaching from the surface of catalyst. It is known that the cobalt(II) salts in NHPI catalyzed oxidation reactions of hydrocarbons with oxygen not only influence the generation of PINO radical but also accelerate decomposition of hydroperoxides to stable products. As a result of ethylbenzene oxidation mixtures composed of unreacted ethylbenzene, ethylbenzene hydroperoxide (EBOOH), acetophenone (AP), and 1-phenylethanol (PEOH) were obtained (Figure 6).

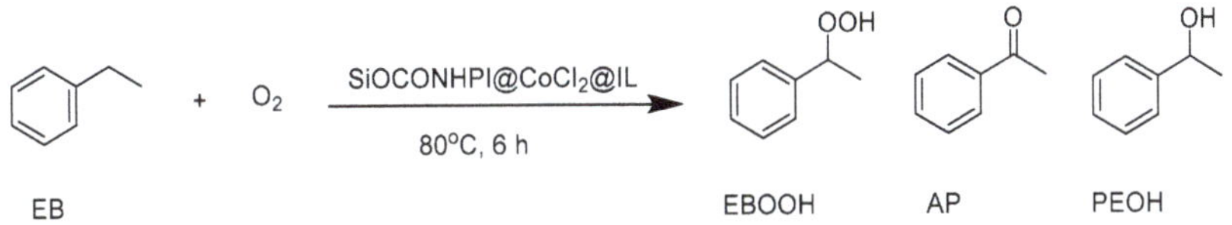

Figure 6. Ethylbenzene oxidation with oxygen in the presence of SiOCONHPI@CoCl$_2$@IL.

2.2.1. Ethylbenzene Oxidation with Oxygen using SiOCONHPI or SiOCONHPI/CoCl$_2$ System

High catalytic activity of NHPI immobilized on silica was proved in the studied oxidation process. In comparison to a non-catalytic reaction carried out only with the use of azo initiator a two-fold increase in conversion was observed (Table 1, entries 1, 3). At the same time a minimal decrease in selectivity to EBOOH was observed, what may result from the influence of silica support on its stability.

Table 1. Ethylbenzene oxidation with oxygen in the presence of SiOCONHPI@CoCl$_2$@[bmim][X].

Entry	Catalyst	α (%)	S_{EBOOH} (%)	S_{AP} (%)	S_{PEOH} (%)
1	-	4.7	67.1	12.3	20.6
2 [a]	CoCl$_2$	2.8	81.4	-	-
3 [b]	SiOCONHPI	8.6	51.0	19.4	29.6
4 [c]	SiOCONHPI/CoCl$_2$	8.2	20.0	52.9	26.2
5 [d]	SiOCONHPI/[bmim][OcOSO$_3$]	6.4	71.2	19.0	9.8
6 [e]	SiOCONHPI/CoCl$_2$/[bmim][OcOSO$_3$]	11.3	25.4	50.0	24.6
7 [f]	SiOCONHPI@[bmim][OcOSO$_3$]	6.1	70.0	19.1	10.9
8 [g]	SiOCONHPI@CoCl$_2$@[bmim][OcOSO$_3$]	12.1	23.0	50.6	26.3
9 [g]	SiOCONHPI@CoCl$_2$@[bmim][Cl]	10.2	26.8	55.0	17.9
10 [g]	SiOCONHPI@CoCl$_2$@[bmim][CF$_3$SO$_3$]	9.0	19.8	54.3	25.9
11 [g]	SiOCONHPI@CoCl$_2$@[bmim][BF$_4$]	8.4	36.5	49.1	14.4
12 [g]	SiOCONHPI@CoCl$_2$@[bmim][CH$_3$COO]	8.4	21.1	55.6	23.3
13 [g]	SiOCONHPI@CoCl$_2$@[bmim][PF$_6$]	7.3	23.0	69.7	7.7
14 [g]	SiOCONHPI@CoCl$_2$@[bmim][Br]	6.9	29.9	51.7	18.4
15 [g]	SiOCONHPI@CoCl$_2$@[bmim][NTf$_2$]	6.6	23.5	50.2	26.3

Ethylbenzene 2 mL, AIBN 1.0 mol%, 80 °C, 0.1 MPa, 6 h, 1200 rpm [a] only CoCl$_2$ 0.1 mol% (0.0021 g), [b] SiOCONHPI 0.033 g [c] SiOCONHPI 0.033 g, CoCl$_2$ 0.0021 g, [d] SiOCONHPI 0.033 g, [bmim][OcOSO$_3$] 0.017 g, [e] SiOCONHPI 0.033 g, [bmim][OcOSO$_3$] 0.017 g, CoCl$_2$ 0.0021 g, and [f] SiOCONHPI 0.033 g, [bmim][OcOSO$_3$] 0.017 g was dissolved in 3 mL of acetone and stirred for 3 h, then the acetone was evaporated [g] SiOCONHPI (0.033 g), [bmim][X] (0.017 g), CoCl$_2$ (0.0021 g) was dissolved in 3 mL of acetone and stirred for 3 h, then the acetone was evaporated.

It was confirmed that the addition of CoCl$_2$ directly to a reaction mixture does not influence the conversion of ethylbenzene, in comparison to non-catalytic reaction, as well as to the one carried out with the use of SiOCONHPI (Table 1, entries 3, 4). In the second case, however, it was observed that the selectivity to EBOOH dropped down from 51.0% to 20.0%. The obtained results are probably influenced by poor solubility of CoCl$_2$ in hydrocarbon, which can, however, increase together with the progress of the reaction and increase in polarity of the mixture.

2.2.2. Ethylbenzene Oxidation with Oxygen using SiOCONHPI@CoCl$_2$@[bmim][X] System

Catalytic activity of SCILL systems SiOCONHPI@CoCl$_2$@IL was studied. In those systems, ionic liquids with [bmim] cation and following anions: [OcOSO$_3$], [Br], [Cl], [CF$_3$SO$_3$], [BF$_4$], [PF$_6$], [CH$_3$COO], and [NTf$_2$] containing dissolved CoCl$_2$ were applied. The results are presented in the Table 1. For comparison, reactions in which [bmim][OcOSO$_3$] and/or CoCl$_2$ were added directly to reaction mixture containing SiOCONHPI were also carried out (Table 1, entry 5, 6).

It was confirmed that coating of SiOCONHPI with a layer of [bmim][OcOSO$_3$] influences the decrease in conversion from 8.6% to 6.1% (Table 1, entries 3, 7) and at the same time the increase in selectivity to EBOOH, from 51% to 70%. It indicates that the IL layer can limit the contact between hydrocarbon and active center of the catalyst (N–OH), as well as between EBOOH and support (silica may accelerate EBOOH decomposition). On the other hand, coating heterogeneous catalyst SiOCONHPI with ionic liquids [bmim][OcOSO$_3$], [bmim][Cl], and [bmim][CF$_3$SO$_3$] containing dissolved CoCl$_2$ (Table 1, entries 8–10) enabled to increase the catalytic activity in relation to systems in which IL and CoCl$_2$ were added directly to reaction mixture (Table 1, entries 4–6). The highest conversion of ethylbenzene was obtained while SiOCONHPI@CoCl$_2$@[bmim][OcOSO$_3$] were used (12.1%). Our previous study [38] on the influence of catalytic amounts of various ILs on ethylbenzene oxidation catalyzed by Co(II) and NHPI/Co(II) system, has proved a positive effect of [bmim][OcOSO$_3$].

The effect observed while using SiOCONHPI@CoCl$_2$@IL, may be influenced by such factors as: Interaction between N-OH and IL, better contact between CoCl$_2$ dissolved in a layer of IL and N-OH, as well as between reagents and N-OH groups or NO radicals and CoCl$_2$, and also undesired washing out of ionic liquid containing CoCl$_2$ from the surface of silica. Among all of the studied ionic liquids, [bmim][OcOSO$_3$] is characterized by the highest solubility of oxygen as well as used hydrocarbon. It was observed that after the reaction catalyzed by SiOCONHPI@CoCl$_2$@[bmim][OcOSO$_3$] post-reaction mixture had a pale-green shade, what suggests that a part of Co(II) salt was leaching from the surface of the catalyst during the reaction. Similar selectivities to EBOOH, obtained while using SiOCONHPI@CoCl$_2$@[bmim][OcOSO$_3$] and SiOCONHPI/CoCl$_2$, respectively equal to 23.0% and 20.0% (Table 1, entries 4, 8), may indicate similar concentration of CoCl$_2$ in the reaction mixture. Amount of [bmim][OcOSO$_3$] introduced together with the catalyst, this is 0.017 g, can dissolve in ethylbenzene as showed in [38]. In this case, it can cause an increase in polarity of reaction mixture and solubility of CoCl$_2$. Among studied ILs it is the [bmim][OcOSO$_3$] which is characterized by the highest solubility in ethylbenzene.

In the reactions, in which [bmim][OcOSO$_3$] and CoCl$_2$ was introduced directly into the reaction mixture containing heterogeneous SiOCONHPI (Table 1, entry 6), also it was possible to observe a positive influence of [bmim][OcOSO$_3$]. However, during the reaction, the polar ionic liquid can also physically interact with the layer of silica. Therefore, it is difficult to explicitly determine, whether the observed effect results from the presence of IL dissolved in the system or supported onto the surface of silica.

As opposed to [bmim][OcOSO$_3$], [bmim][Cl] and [bmim][CF$_3$SO$_3$] poorly dissolve in ethylbenzene. Additionally, in the previous work [38] we ascertained that the addition of catalytic amounts of [bmim][CF$_3$SO$_3$] to the process of ethylbenzene oxidation catalyzed by Co(II) or NHPI/Co(II), in both cases influenced the decrease in conversion of substrate. However, in the studies presented in this paper, it was observed that there was a slight increase in conversion, which was related to the process of coating SiOCONHPI with a layer of [bmim][CF$_3$SO$_3$] containing CoCl$_2$, what would indicate a positive influence of IL bonded with silica on the whole process.

The main product of the oxidation of ethylbenzene using SiOCONHPI@CoCl$_2$@[bmim][X] was AP obtained with selectivities equal to ca. 50% (only in the case of using [bmim][PF$_6$] additive, it was obtained with selectivity equal to 70%), furthermore, selectivities to EBOOH and PEOH were varying respectively between 21.1% and 36.5% and between 7.7 and 26.3%.

2.2.3. Influence of the Structure of Alkyl Substituent in Cation or Anion of Ionic Liquid on Catalytic Activity of SiOCONHPI@CoCl$_2$@IL

The study on the influence of the length of alkyl chain in 1-alkyl-3-methylimidazolium cation or alkyl sulfate anion on catalytic activity of SiOCONHPI@CoCl$_2$@IL, in the processes of oxidation of ethylbenzene with oxygen, was carried out. Results are presented in Table 2.

Table 2. Influence of the ionic liquid structure on catalytic activity of SiOCONHPI@CoCl$_2$@IL.

Entry	Catalyst	α (%)	S$_{EBOOH}$ (%)	S$_{AP}$ (%)	S$_{PEOH}$ (%)
1 [a]	SiOCONHPI/CoCl$_2$	8.2	20.0	52.9	26.2
2	SiOCONHPI@CoCl$_2$@[bmim][Cl]	10.2	26.8	55.0	17.9
3	SiOCONHPI@CoCl$_2$@[hmim][Cl]	11.1	21.8	52.6	25.4
4	SiOCONHPI@CoCl$_2$@[omim][Cl]	11.3	17.5	57.0	25.2
5	SiOCONHPI@CoCl$_2$@[C$_{10}$mim][Cl]	11.9	20.0	57.1	22.6
6	SiOCONHPI@CoCl$_2$@[bmim][CH$_3$OSO$_3$]	10.2	36.2	38.7	25.1
7	SiOCONHPI@CoCl$_2$@[bmim][OcOSO$_3$]	12.1	23.0	50.6	26.3
8	SiOCONHPI@CoCl$_2$@[bmim][C$_{12}$H$_{25}$OSO$_3$]	14.3	15.8	58.3	25.9
9	SiOCONHPI@CoCl$_2$@[emim][OcOSO$_3$]	12.8	28.8	44.6	26.6

Ethylbenzene 2 mL, AIBN 1.0 mol%, 80 °C, 0.1 MPa, 6 h, 1200 rpm, SiOCONHPI@CoCl$_2$@IL: SiOCONHPI (0.033 g), IL (0.017 g), CoCl$_2$ (0.0021 g, 0.1 mol%) was dissolved in 3 mL of acetone and stirred for 3 h, then the acetone was evaporated [a] SiOCONHPI 0.033 g, CoCl$_2$ 0.1 mol%.

It was proved that together with the increase of the number of carbon atoms in the alkyl group, both in anion and cation, conversion of ethylbenzene is increased. At the same time selectivity to EBOOH slightly decreases. Lipophilicity increase of IL influences the increase of oxygen and hydrocarbon solubility in IL supported on silica, as well as solubility of IL in non-polar starting material and related greater solubility of CoCl$_2$ in ethylbenzene. The decrease in selectivity to EBOOH may indicate that IL containing CoCl$_2$ is washed out from the surface of the catalyst.

2.2.4. Attempts to Reuse SiOCONHPI@CoCl$_2$@[bmim][X] System.

The advantage of heterogeneous catalysts is the possibility to recover them in a simple way. Therefore, attempts were made to reuse studied catalytic systems. After the reaction (first cycle) the mixture was decanted, and the catalyst was washed with hexane and subsequently the hexane was also decanted. Before the reuse, the catalyst was dried under vacuum. Table 3 compares the results of conversion of ethylbenzene in four reaction cycles, obtained with the use of the most active catalytic systems, i.e., SiOCONHPI@CoCl$_2$@[bmim][X] containing layers of [bmim][OcOSO$_3$], [bmim][Cl], and [bmim][CF$_3$SO$_3$]. For comparison also SiOCONHPI was recycled 3 times from the reaction carried out in the presence of CoCl$_2$ introduced directly to the reaction mixture. Portions of fresh CoCl$_2$ were not introduced in the following cycles.

The study on recovery and recycling of SiOCONHPI catalyst from reaction mixture containing only an additive of CoCl$_2$, in the first cycle, proved that the catalyst maintains its activity. In the successive cycles conversion between 8.9% and 8.1% was obtained. These conversion values are higher than the one obtained in non-catalytic process (4.7%) (Table 1, entry 1) and are comparable to the one obtained using SiOCONHPI only (8.6%) (Table 1, entry 3). Selectivity to EBOOH gradually increased in every recycling from 20% to 43% (Table 3, entry 1–4) and got close to the selectivity obtained while using SiOCONHPI without CoCl$_2$, this is 51% (Table 1, entry 3). It indicates binding of CoCl$_2$, introduced to the reaction mixture, in the first cycle, on the surface of silica and consequently its gradual wash out in the following cycles two to four.

Table 3. Recovery and recycling of SiOCONHPI@CoCl$_2$@[bmim][X] systems in ethylbenzene oxidation reaction.

Entry	Cycle	Catalyst	α (%)	S$_{EBOOH}$ (%)	S$_{AP}$ (%)	S$_{PEOH}$ (%)
1 [a]	1	SiOCONHPI/CoCl$_2$	8.2	20.0	52.9	26.2
2	2		8.9	28.5	48.1	23.3
3	3		8.4	30.7	46.3	23.0
4	4		8.1	43.4	34.2	22.3
5 [b]	1	SiOCONHPI@CoCl$_2$@[bmim][OcOSO$_3$]	12.1	23.0	50.6	26.3
6	2		12.4	24.3	43.8	31.8
7	3		11.3	30.5	36.2	33.3
8	4		8.0	44.1	37.1	18.6
9 [b]	1	SiOCONHPI@CoCl$_2$@[bmim][Cl]	10.2	26.8	55.0	17.9
10	2		10.1	42.5	40.2	17.3
11	3		7.4	59.0	25.2	15.8
12	4		7.7	61.6	24.0	14.4
13 [b]	1	SiOCONHPI@CoCl$_2$@[bmim][CF$_3$SO$_3$]	9.0	19.8	54.3	25.9
14	2		9.5	26.8	48.4	24.5
15	3		9.0	38.3	40.6	20.7
16	4		7.4	57.1	32.8	10.1

Ethylbenzene 2 mL, AIBN 1.0 mol%, 80 °C, 0.1 MPa, 6 h, 1200 rpm [a] SiOCONHPI 0.033 g, CoCl$_2$ 0.1 mol% [b] SiOCONHPI@CoCl$_2$@IL: SiOCONHPI (0.033 g), IL (0.017 g), CoCl$_2$ (0.0021 g, 0.1 mol%) was dissolved in 3 mL of acetone and stirred for 3 h, then the acetone was evaporated.

In the reactions carried out with the use of SiOCONHPI@CoCl$_2$@[bmim][OcOSO$_3$] and SiOCONHPI@CoCl$_2$@[bmim][CF$_3$SO$_3$] systems, it is possible to recycle the catalyst two times, and a significant decrease in conversion and an increase in selectivity to EBOOH related to washing out of CoCl$_2$ was observed no sooner than in the fourth cycle. In case of SiOCONHPI@CoCl$_2$@[bmim][Cl] the decrease in activity occurred already in the third cycle.

Figure 7 compares the FTIR spectra of SiOCONHPI@CoCl$_2$@[bmim][OcOSO$_3$] before the reaction and after the fourth cycle. Based on that, it is possible to determine that after the fourth cycle the IL was washed out from the surface of silica (Figure 7). Thermogravimetric analysis of fresh and recovered SiOCONHPI@CoCl$_2$@[bmim][OcOSO$_3$] also demonstrates that after the fourth cycle the IL was partly washed out from the surface of silica (Figure 8). The weight loss of 34% was observed between 200 and 500 °C for fresh SiOCONHPI@CoCl$_2$@[bmim][OcOSO$_3$] and 13% for recovered catalyst (Supplementary Materials; Figure S1–S3).

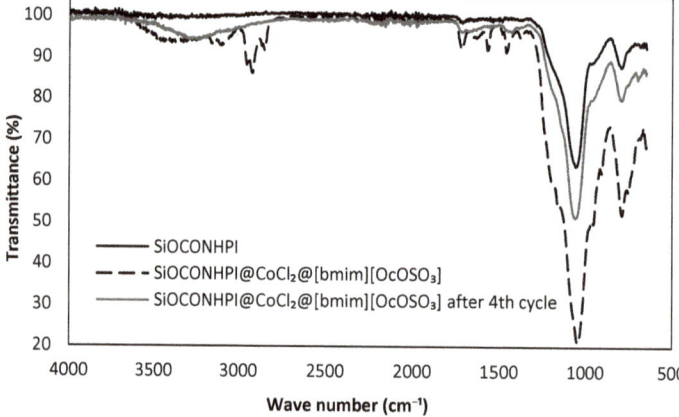

Figure 7. FT-IR spectra of SiOCONHPI, SiOCONHPI@CoCl$_2$@[bmim][OcOSO$_3$] and SiOCONHPI@CoCl$_2$@[bmim][OcOSO$_3$] after the fourth cycle.

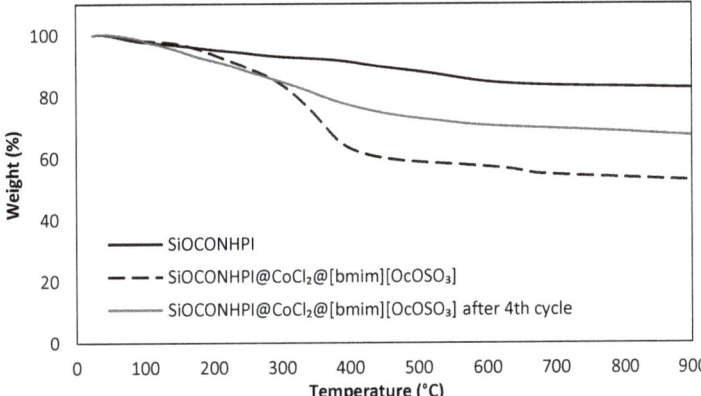

Figure 8. TGA curves obtained for SiOCONHPI, SiOCONHPI@CoCl$_2$@[bmim][OcOSO$_3$] and SiOCONHPI@CoCl$_2$@[bmim][OcOSO$_3$] after the fourth cycle.

The influence of the length of carbon chain in alkyl substituent in the alkyl sulfate anion and 1-alkyl-3-methylimidazolium cation, on the possibility of SiOCONHPI@CoCl$_2$@IL system recycling (Table 4), was also studied.

Table 4. Influence of ionic liquid structure on possibility to recycle SiOCONHPI@CoCl$_2$@IL.

Entry	Cycle	Catalyst	α (%)	S$_{EBOOH}$ (%)	S$_{AP}$ (%)	S$_{PEOH}$ (%)
1 [a]	1		10.2	36.2	38.7	25.1
2	2	SiOCONHPI@CoCl$_2$@[bmim][CH$_3$OSO$_3$]	9.3	47.7	30.0	22.3
3	3		8.2	58.8	25.5	15.7
4	4		6.8	61.4	23.2	14.5
5 [a]	1		12.1	23.0	50.6	26.3
6	2	SiOCONHPI@CoCl$_2$@[bmim][OcOSO$_3$]	12.4	24.3	43.8	31.8
7	3		11.3	30.5	36.2	33.3
8	4		8.0	44.1	37.1	18.6
9 [a]	1		14.3	15.8	58.3	25.9
10	2	SiOCONHPI@CoCl$_2$@[bmim][C$_{12}$H$_{25}$OSO$_3$]	11.6	26.3	50.5	23.2
11	3		9.8	37.5	36.8	25.7
12	4		8.8	47.8	30.2	22.0
13 [a]	1		12.8	28.8	44.6	26.6
14	2	SiOCONHPI@CoCl$_2$@[emim][OcOSO$_3$]	10.0	36.0	38.1	25.9
15	3		9.9	36.4	34.6	29.0
16	4		9.5	44.9	27.1	28.0

Ethylbenzene 2 mL, AIBN 1.0 mol%, 80 °C, 0.1 MPa, 6 h, 1200 rpm [a] SiOCONHPI@CoCl$_2$@IL: SiOCONHPI (0.033 g), IL (0.017 g), CoCl$_2$ (0.0021 g, 0.1 mol%) was dissolved in 3 mL of acetone and stirred for 3 h, then the acetone was evaporated.

The results presented in Table 4, indicate that the greatest decrease in conversion and increase in selectivity to EBOOH in subsequent cycles is observed while using SiOCONHPI@CoCl$_2$@[bmim][CH$_3$OSO$_3$]. In the fourth cycle, the higher conversion, in comparison to the reaction carried out with the use of SiOCONHPI (8.6%) (Table 1, entry 3), was obtained with the use of SiOCONHPI@CoCl$_2$@[emim][OcOSO$_3$] (Table 4, entry 16).

3. Materials and Methods

3.1. Materials

Ethylbenzene (Acros 99.8%, Geel, BE) was purified by washing with H$_2$SO$_4$ and vacuum distillation. Cobalt(II) chloride, N-hydroxyphthalimide (NHPI) were purchased from commercial sources and used without purification. 3-Chloropropyl-functionalized silica gel were purchased from Sigma Aldrich (St. Louis, MO, USA) (Si–Cl): Loading 1.0 mmol Cl/g, 230–400 mesh.

Ionic liquids: 1-butyl-3-methylimidazolium chloride ([bmim][Cl]), 1-butyl-3-methylimidazolium bromide ([bmim][Br]), 1-butyl-3-methylimidazolium methyl sulfate ([bmim][CH_3OSO_3]), 1-butyl-3-methylimidazolium octyl sulfate ([bmim][$OcOSO_3$]), 1-butyl-3-methylimidazolium acetate ([bmim][CH_3COO]), 1-butyl-3-methylimidazolium trifluoromethanesulfonate ([bmim][CF_3SO_3]), 1-butyl-3-methylimidazolium hexafluorophosphate ([bmim][PF_6]), 1-butyl-3-methylimidazolium tetrafluoroborate ([bmim][BF_4]), 1-butyl-3-methylimidazolium bis(trifluoromethylsulfonyl)imide ([bmim][NTf_2]), 1-hexyl-3-methylimidazolium chloride ([hmim][Cl]), 1-octyl-3-methylimidazolium chloride ([omim][Cl]), 1-decyl-3-methylimidazolium chloride (C_{10}mim][Cl]), 1-ethyl-3-methylimidazolium octyl sulfate ([emim][$OcOSO_3$]) were commercial materials and were dried under vacuum before use (50 °C, 0.1 bar). 1-Butyl-3-methylimidazolium lauryl sulfate ([bmim][$C_{12}H_{25}OSO_3$]) was prepared according procedure described in literature [38].

3.2. SiOCONHPI Synthesis

Chloropropyl silica gel SiCl (2.0 g, 1.0 mmol Cl/g), trimellitic anhydride (8 mmol) and triethylamine (20 mmol) were added to 20 mL of dioxane. The mixture was stirred for 48 h under reflux. The product was filtered and washed twice with H_2O (2 times 25 mL) at RT and of each of the following solvents at 50 °C: MeOH, THF and CH_2Cl_2. The solid was dried under vacuum and then was added to a 40 mL mixture of pyridine:1,2-dichloroethane (3:1, v/v) followed by the addition of hydroxylamine hydrochloride (20 mmol). The mixture was stirred for 24 h at 75 °C, then filtered. The filtrate was washed with H_2O (2 times 25 mL) and H_2O:MeOH (1:1, v/v, 25 mL) at RT and MeOH, DMF, THF, and CH_2Cl_2 (2 times 25 mL) at 50 °C. The product was died under vacuum to afford 1.26 g of immobilized NHPI (SiOCONHPI). SiOCONHPI was analyzed using FT-IR spectroscopy (Figure 4), SEM (Figure 5), TGA (Figure 8 and Figure S1) and elemental analysis (%C: 9.26; %H: 1.34; %N: 0.64). Based on the elemental analysis the content of immobilized NHPI was 0.45 mmol/g.

3.3. SiOCONHPI@Co(II)@IL Preparation

Catalysts were prepared in accordance with the described procedure [36]. Ionic liquid (0.017 g), NHPI immobilized on silica SiOCONHPI (0.033 g) and $CoCl_2$ (0.0021 g; 0.1 mol% in relation to etylbenzene) were introduced into a two-neck flask. Consequently, about 3 mL of acetone was added, and the flask was placed on a magnetic stirrer and stirred for 3 hours, the next step was evaporation of the acetone with the use of rotary evaporator.

3.4. Ethylbenzene Oxidation

The oxidation reactions were performed in a gasometric apparatus as described in [39]. Ethylbenzene, heterogeneous NHPI, AIBN and $CoCl_2$ were placed in a 10 mL flask connected to a gas burette filled with oxygen under atmospheric pressure. The reaction was conducted at 80 °C for 6 h with magnetic stirring at 1200 rpm. The oxygen uptake (n_{O2}) was measured, recalculated for normal conditions (273 K, 1 atm), and this value was used to calculate the ethylbenzene conversion (α). The amount of ethylbenzene hydroperoxide (EBOOH) was determined iodometrically according to the described method [40], and the result was used for the calculation of the reaction's selectivity (S_{EBOOH}).

$$n_{O2} = (V_{O2} \cdot 273 \cdot p)/(101325 \cdot 22.415 \cdot T) \text{ (mol)}$$

$$\alpha = n_{O2}/n \cdot 100\%$$

$$S_{EBOOH} = n_{OOH}/n_{O2} \cdot 100\%$$

EBOOH is thermally unstable and easily decomposes to 1-phenylethanol (PEOH) and acetophenone (AP). Thus, before GC analysis, the hydroperoxide was quantitatively reduced to

1-phenylethanol by addition of triethylphosphite (EtO)$_3$P, which is oxidized to triethyl phosphate (EtO)$_3$PO [41]. The amount of PEOH determined by GC was the sum of the alcohol and hydroperoxide formed. In order to calculate the PEOH selectivity, the amount of hydroperoxide determined by iodometric analysis was subtracted. The amount of acetophenone was determined based on the GC analysis and used for the calculation of the reaction's selectivity.

3.5. Analytic Methods

FT-IR analysis was carried out in Nicolet 6700 FT-IR Spectrometer (Thermo Electron Scientific Instruments Corporation, Madison, WI, USA), with the use of ATR technique (attenuated total reflection technique). Elemental analysis was carried out with the use of CHNS Vario Micro Cube equipment (Elementar Analysensysteme GmbH, Langenselbold, DE).

Morphology of the surface of solid carriers and immobilized NHPI was carried out with the use of Scanning Electron Microscope (SEM) (TM3000 TableTop, Hitachi High-Technologies Corporation brand, USA).

GC analysis was performed using a gas chromatograph (Agilent Technologies 7890C, Santa Clara, CA, USA) (Zebron ZB-5HT capillary column 30 m by 0.25 mm by 0.25 μm) with an FID detector. The injection port temperature was 200 °C. The detector temperature was 300 °C. The temperature program was: Hold at 80 °C for 10 min, then ramp at 3 °C/min to 110 °C, hold at 110 °C for 4 min, then ramp at 20 °C/min to 210 °C, hold for 4 min. Calibration curves for acetophenone and 1-phenylethanol were determined using *p*-methoxytoluene as the internal standard.

Thermogravimetric analysis (TGA) was performed using a TGA851e thermobalance (Mettler Toledo, Greifensee, CHE). Samples of approximately 10 mg were heated from 25 °C to 800 °C at a rate of 10 °C/min in standard 70 μL Al$_2$O$_3$ crucibles under a dynamic nitrogen flow of 60 mL/min (99.9992%).

4. Conclusions

For the first time, the catalytic activity of NHPI immobilized on silica via ester bond SiOCONHPI in oxidation of model ethylbenzene with oxygen without any solvent, as well as possibility of its (at least) three-fold recycling without any loss in activity, was reported.

It was also proved that coating the surface of SiOCONHPI with a layer of IL such as [bmim][OcOSO$_3$], [bmim][Cl] and [bmim][CF$_3$SO$_3$] containing dissolved CoCl$_2$, enables to obtain higher conversions of ethylbenzene. The highest conversion was obtained while using SCILL system SiOCONHPI@CoCl$_2$@[bmim][OcOSO$_3$].

Study on recovery and reuse of reported new SCILLs systems SiOCONHPI@CoCl$_2$@IL demonstrated that a significant decrease in conversion was observed no sooner than in the fourth cycle using [bmim][OcOSO$_3$] and [bmim][CF$_3$SO$_3$] or in the third cycle using [bmim][Cl]. Fortunately, after IL and CoCl$_2$ are washed out, SiOCONHPI preserves its catalytic activity.

Supplementary Materials: The following are available online at http://www.mdpi.com/2073-4344/10/2/252/s1, Figure S1: TGA curves obtained for SiOCONHPI, Figure S2: TGA curves obtained for SiOCONHPI@CoCl$_2$@[bmim][OcOSO$_3$], Figure S3: TGA curves obtained for SiOCONHPI@CoCl$_2$@[bmim][OcOSO$_3$] after the fourth cycle.

Author Contributions: Conceptualization, B.O.; methodology, G.D. and K.K. and B.O.; investigation, G.D.; TGA analysis, S.J.; writing—original draft preparation, G.D. and B.O.; supervision, B.O. All authors have read and agreed to the published version of the manuscript.

Funding: This research was funded by Rector of the Silesian University of Technology grant No. 04/050/RGJ19/0076.

Conflicts of Interest: The authors declare no conflict of interest.

References

1. Franz, G.; Sheldon, R.A. Ullmann's Encyclopedia of Industrial Chemistry. In *Ullmann's Encyclopedia of Industrial Chemistry*; Wiley-VCH: Weinheim, Germany, 2012; pp. 1–28.
2. Weissermel, K.; Arpe, H.J. *Industrial Organic Chemistry*; VCH: Weinheim, Germany, 1997.

3. Recupero, F.; Punta, C. Free Radical Functionalization of Organic Compounds Catalyzed by N-Hydroxyphthalimide. *Chem. Rev.* **2007**, *107*, 3800–3842. [CrossRef] [PubMed]
4. Chen, K.; Zhang, P.; Wang, Y.; Li, H. Metal-Free Allylic/Benzylic Oxidation Strategies with Molecular Oxygen: Recent Advances and Future Prospects. *Green Chem.* **2014**, *16*, 2344–2374. [CrossRef]
5. Coseri, S. Phthalimide-N-Oxyl (PINO) Radical, a Powerful Catalytic Agent: Its Generation and Versatility towards Various Organic Substrates. *Catal. Rev. Sci. Eng.* **2009**, *51*, 218–292. [CrossRef]
6. Melone, L.; Punta, C. Metal-Free Aerobic Oxidations Mediated by N-Hydroxyphthalimide: A Concise Review. *Beilstein J. Org. Chem.* **2013**, *9*, 1296–1310. [CrossRef] [PubMed]
7. Ishii, Y.; Iwahama, T.; Sakaguchi, S.; Nakayama, K.; Nishiyama, Y. Alkane Oxidation with Molecular Oxygen Using a New Efficient Catalytic System: N-Hydroxyphthalimide (NHPI) Combined with Co(acac)$_n$ (n = 2 or 3). *J. Org. Chem.* **1996**, *61*, 4520–4526. [CrossRef]
8. Orlińska, B.; Zawadiak, J. Aerobic Oxidation of Isopropylaromatic Hydrocarbons to Hydroperoxides Catalyzed by N-Hydroxyphthalimide. *React. Kinet. Mech. Catal.* **2013**, *110*, 15–30. [CrossRef]
9. Ishii, Y.; Nakayama, K.; Takeno, M.; Sakaguchi, S.; Iwahama, T.; Nishiyama, Y. A Novel Catalysis of N-Hydroxyphthalimide in the Oxidation of Organic Substrates by Molecular Oxygen. *J. Org. Chem.* **1995**, *60*, 3934–3935. [CrossRef]
10. Lisicki, D.; Orlińska, B. Oxidation of Cycloalkanes Catalysed by N-Hydroxyimides in Supercritical Carbon Dioxide. *Chem. Pap.* **2020**, *74*, 711–716. [CrossRef]
11. Dobras, G.; Lisicki, D.; Pyszny, D.; Orlińska, B. Badania Reakcji Utleniającego Rozszczepienia α-Metylostyrenu Tlenem Wobec N-Hydroksyftalimidu w Alternatywnych Rozpuszczalnikach. *Przem. Chem.* **2019**, *1*, 124–129. [CrossRef]
12. Yu, K.M.K.; Abutaki, A.; Zhou, Y.; Yue, B.; He, H.Y.; Tsang, S.C. Selective Oxidation of Cyclohexane in Supercritical Carbon Dioxide. *Catal. Lett.* **2007**, *113*, 115–119.
13. Yavari, I.; Karimi, E. N-Hydroxyphthalimide-Catalyzed Oxidative Production of Phthalic Acids from Xylenes Using O$_2$/HNO$_3$ in an Ionic Liquid. *Synth. Commun.* **2009**, *39*, 3420–3427. [CrossRef]
14. Dobras, G.; Orlińska, B. Aerobic Oxidation of Alkylaromatic Hydrocarbons to Hydroperoxides Catalysed by N-Hydroxyimides in Ionic Liquids as Solvents. *Appl. Catal. A Gen.* **2018**, *561*, 59–67. [CrossRef]
15. Wang, Y.; Lu, T. PEG1000-DAIL Enhanced Catalysis Activity: Oxidation of Ethylbenzene and Its Derivatives by N-Hydroxyphthalimide and Oxime in 1000-Based Dicationic Acidic Ionic Liquid. *Chiang Mai J. Sci.* **2014**, *41*, 138–147.
16. Lu, T.; Lu, M.; Yu, W.; Liu, Z. Remarkable Effect of PEG-1000-Based Dicationic Ionic Liquid for N-Hydroxyphthalimide-Catalyzed Aerobic Selective Oxidation of Alkylaromatics. *Croat. Chem. Acta* **2012**, *85*, 277–282. [CrossRef]
17. Koguchi, S.; Kitazume, T. Synthetic Utilities of Ionic Liquid-Supported NHPI Complex. *Tetrahedron Lett.* **2006**, *47*, 2797–2801. [CrossRef]
18. Wang, J.R.; Liu, L.; Wang, Y.F.; Zhang, Y.; Deng, W.; Guo, Q.X. Aerobic Oxidation with N-Hydroxyphthalimide Catalysts in Ionic Liquid. *Tetrahedron Lett.* **2005**, *46*, 4647–4651. [CrossRef]
19. Melone, L.; Prosperini, S.; Ercole, G.; Pastori, N.; Punta, C. Is It Possible to Implement N-Hydroxyphthalimide Homogeneous Catalysis for Industrial Applications? A Case Study of Cumene Aerobic Oxidation. *J. Chem. Technol. Biotechnol.* **2014**, *89*, 1370–1378. [CrossRef]
20. Hermans, I.; Van Deun, J.; Houthoofd, K.; Peeters, J.; Jacobs, P.A. Silica-Immobilized N-Hydroxyphthalimide: An Efficient Heterogeneous Autoxidation Catalyst. *J. Catal.* **2007**, *251*, 204–212. [CrossRef]
21. Rajabi, F.; Clark, J.H.; Karimi, B.; Macquarrie, D.J. The Selective Aerobic Oxidation of Methylaromatics to Benzaldehydes Using a Unique Combination of Two Heterogeneous Catalysts. *Org. Biomol. Chem.* **2005**, *3*, 725–726. [CrossRef]
22. Kasperczyk, K.; Orlinska, B.; Witek, E.; Łątka, P.; Zawadiak, J.; Proniewicz, L. Polymer-Supported N-Hydroxyphthalimide as Catalyst for Toluene and p-Methoxytoluene Aerobic Oxidation. *Catal. Lett.* **2015**, *145*, 1856–1867. [CrossRef]
23. Łątka, P.; Kasperczyk, K.; Orlińska, B.; Drozdek, M.; Skorupska, B.; Witek, E. N-Hydroxyphthalimide Immobilized on Poly(HEA-Co-DVB) as Catalyst for Aerobic Oxidation Processes. *Catal. Lett.* **2016**, *146*, 1991–2000. [CrossRef]

24. Łątka, P.; Berniak, T.; Drozdek, M.; Witek, E.; Kuśtrowski, P. Formation of N-Hydroxyphthalimide Species in Poly(Vinyl-Diisopropyl-Phtalate Ester-Co-Styrene-Co-Divinylbenzene) and Its Application in Aerobic Oxidation of p-Methoxytoluene. *Catal. Commun.* **2018**, *115*, 73–77. [CrossRef]
25. Gao, B.; Meng, S.; Yang, X. Synchronously Synthesizing and Immobilizing N-Hydroxyphthalimide on Polymer Microspheres and Catalytic Performance of Solid Catalyst in Oxidation of Ethylbenzene by Molecular Oxygen. *Org. Process Res. Dev.* **2015**, *19*, 1374–1382. [CrossRef]
26. Hosseinzadeh, R.; Mavvaji, M.; Tajbakhsh, M.; Lasemi, Z. Synthesis and Characterization of N-Hydroxyphthalimide Immobilized on NaY Nano-Zeolite as a Novel and Efficient Catalyst for the Selective Oxidation of Hydrocarbons and Benzyl Alcohols. *React. Kinet. Mech. Catal.* **2018**, *124*, 839–855. [CrossRef]
27. Sawant, A.D.; Raut, D.G.; Darvatkar, N.B.; Salunkhe, M.M. Recent Developments of Task-Specific Ionic Liquids in Organic Synthesis. *Green Chem. Lett. Rev.* **2011**, *4*, 41–54. [CrossRef]
28. Korth, W.; Jess, A. Solid Catalysts with Ionic Liquid Layer (SCILL). In *Supported Ionic Liquids: Fundamentals and Applications*; Wiley-VCH: Weinheim, Germany, 2014; pp. 279–306.
29. Gu, Y.; Li, G. Ionic Liquids-Based Catalysis with Solids: State of the Art. *Adv. Synth. Catal.* **2009**, *351*, 817–847. [CrossRef]
30. Mehnert, C.P.; Cook, R.A.; Dispenziere, N.C.; Afeworki, M. Supported Ionic Liquid Catalysis—A New Concept for Homogeneous Hydroformylation Catalysis. *J. Am. Chem. Soc.* **2002**, *124*, 12932–12933. [CrossRef]
31. Riisager, A.; Wasserscheid, P.; Van Hal, R.; Fehrmann, R. Continuous Fixed-Bed Gas-Phase Hydroformylation Using Supported Ionic Liquid-Phase (SILP) Rh Catalysts. *J. Catal.* **2003**, *219*, 452–455. [CrossRef]
32. Chrobok, A.; Baj, S.; Pudło, W.; Jarzębski, A. Supported Hydrogensulfate Ionic Liquid Catalysis in Baeyer-Villiger Reaction. *Appl. Catal. A Gen.* **2009**, *366*, 22–28. [CrossRef]
33. Arras, J.; Steffan, M.; Shayeghi, Y.; Claus, P. The Promoting Effect of a Dicyanamide Based Ionic Liquid in the Selective Hydrogenation of Citral. *Chem. Commun.* **2008**, 4058–4060. [CrossRef]
34. Arras, J.; Steffan, M.; Shayeghi, Y.; Ruppert, D.; Claus, P. Regioselective Catalytic Hydrogenation of Citral with Ionic Liquids as Reaction Modifiers. *Green Chem.* **2009**, *11*, 716–723. [CrossRef]
35. Kernchen, U.; Etzold, B.; Korth, W.; Jess, A. Solid Catalyst Ionic Liquid Layer (SCILL)—A New Concept to Improve Selectivity Illustrated by Hydrogenation of Cyclooctadiene. *Chem. Eng. Technol.* **2007**, *30*, 985–994. [CrossRef]
36. Karimi, B.; Badreh, E. SBA-15-Functionalized TEMPO Confined Ionic Liquid: An Efficient Catalyst System for Transition-Metal-Free Aerobic Oxidation of Alcohols with Improved Selectivity. *Org. Biomol. Chem.* **2011**, *9*, 4194–4198. [CrossRef] [PubMed]
37. Liu, L.; Liu, D.; Xia, Z.; Gao, J.; Zhang, T.; Ma, J.; Zhang, D.; Tong, Z. Supported Ionic-Liquid Layer on Polystyrene-TEMPO Resin: A Highly Efficient Catalyst for Selective Oxidation of Activated Alcohols with Molecular Oxygen. *Mon. Chem.* **2013**, *144*, 251–254. [CrossRef]
38. Dobras, G.; Sitko, M.; Petroselli, M.; Caruso, M.; Cametti, M.; Punta, C.; Orlińska, B. Solvent-Free Aerobic Oxidation of Ethylbenzene Promoted by NHPI/Co(II) Catalytic System: The Key Role of Ionic Liquids. *ChemCatChem* **2020**, *12*, 259–266. [CrossRef]
39. Kasperczyk, K.; Orlińska, B.; Zawadiak, J. Aerobic Oxidation of Cumene Catalysed by 4-Alkyloxycarbonyl-N-Hydroxyphthalimide. *Cent. Eur. J. Chem.* **2014**, *12*, 1176–1182. [CrossRef]
40. Zawadiak, J.; Gilner, D.; Kulicki, Z.; Baj, S. Concurrent Iodimetric Determination of Cumene Hydroperoxide and Dicumenyl Peroxide Used for Reaction Control in Dicumenyl Peroxide Synthesis. *Analyst* **1993**, *118*, 1081–1083. [CrossRef]
41. Denney, D.B.; Goodyear, W.F.; Goldstein, B. Concerning the Mechanism of the Reduction of Hydroperoxides by Trisubstituted Phosphines and Trisubstituted Phosphites. *J. Am. Chem. Soc.* **1960**, *82*, 1393–1395. [CrossRef]

© 2020 by the authors. Licensee MDPI, Basel, Switzerland. This article is an open access article distributed under the terms and conditions of the Creative Commons Attribution (CC BY) license (http://creativecommons.org/licenses/by/4.0/).

Article

Highly Efficient and Reusable Alkyne Hydrosilylation Catalysts Based on Rhodium Complexes Ligated by Imidazolium-Substituted Phosphine

Olga Bartlewicz [1],*, Magdalena Jankowska-Wajda [1] and Hieronim Maciejewski [1,2]

1. Faculty of Chemistry, Adam Mickiewicz University in Poznań, Uniwersytetu Poznańskiego 8, 61-614 Poznań, Poland; magdalena.jankowska-wajda@amu.edu.pl (M.J.-W.); hieronim.maciejewski@amu.edu.pl (H.M.)
2. Adam Mickiewicz University Foundation, Poznań Science and Technology Park, Rubież 46, 61-612 Poznań, Poland
* Correspondence: olga.bartlewicz@amu.edu.pl; Tel.: +48-61-829-1702

Received: 15 April 2020; Accepted: 28 May 2020; Published: 1 June 2020

Abstract: Rhodium complexes ligated by imidazolium-substituted phosphine were used as catalysts in the hydrosilylation of alkynes (1-heptyne, 1-octyne, and phenylacetylene) with 1,1,1,3,5,5,5-heptamethyltrisiloxane (HMTS) or triethylsilane (TES). In all cases, the above complexes showed higher activity and selectivity compared to their precursors ([Rh(PPh$_3$)$_3$Cl] and [{Rh(μ-Cl)(cod)}$_2$]). In the reactions with aliphatic alkynes (both when HMTS and TES were used as hydrosilylating agents), β(Z) isomer was mainly formed, but, in the reaction of phenylacetylene with TES, the β(E) product was formed. The catalysts are very durable, stable in air and first and foremost insoluble in the reactants which facilitated their isolation and permitted their multiple use in subsequent catalytic runs. They make a very good alternative to the commonly used homogeneous catalysts.

Keywords: hydrosilylation; alkynes; heterogeneous catalysis; rhodium catalysts; ionic liquids

1. Introduction

Vinylsilanes and siloxanes due to a relatively low cost of their synthesis, low toxicity, and good chemical stability are valuable raw materials applied in many organic syntheses such a nucleophilic substitution, alkylation [1,2], and coupling [3,4]. However, the course of the above syntheses and formation of desired products are influenced by the kind and purity of isomer of vinyl organosilicon derivative employed. This is why various methods are developed for the synthesis of the aforementioned derivative to ensure high regio- and stereoselectivity of vinylsilanes and siloxanes formed. One of the most popular and commonly applied (also on a commercial scale) methods of synthesis of organofunctional silicon compounds is hydrosilylation [5,6], which enables obtaining vinyl derivatives in the reaction with alkynes. However, the reaction course and the kind of products formed depend on many factors such as the type of alkyne (terminal or internal) and hydrogen silane (siloxane) as well as reaction conditions (temperature, time, the kind of solvent) and particularly the kind of catalyst used [7]. In the case of hydrosilylation of terminal alkynes, there is a possibility of the formation of three isomers β-(E), β-(Z), and α as shown in Scheme 1:

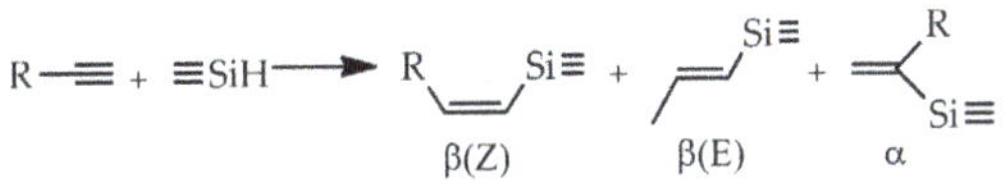

Scheme 1. The example hydrosilylation reaction of terminal alkynes with hydrosilanes.

The kind of isomer formed depends on the way of ≡Si–H addition to the C≡C triple bond. In the former two cases, the silicon atom bonds to the terminal carbon atom which results in β-(E) isomer (if *cis*-addition occurs) or β-(Z) isomer (if *trans*-addition occurs). On the other hand, the reverse addition leads to the geminal vinyl silane (α isomer) as an internal adduct.

As mentioned above, the catalyst has a significant influence on the reaction course. The catalysts used for alkyne hydrosilylation are diverse simple compounds and complexes of transition metals such as Rh [8], Pt [9,10], Ru [11], Ni [12], Co [13], and Ir [13]. Their activity in many cases is high, but they differ in selectivity [7].

For instance, platinum catalysts (including Karstedt's catalyst), which are among the most popular catalysts for hydrosilylation, in the case of the reaction with terminal alkynes preferentially form β(E) isomer as the main product and, to a small extent, also α isomer. On the other hand, β(Z) isomer is not formed (or formed only in negligible amounts) in the reactions catalyzed by platinum catalysts [14], whereas rhodium catalysts favor mainly the formation of β(Z) product [8], except for cationic rhodium complexes, e.g., [{Rh(cod)$_2$}BF$_4$/PPh$_3$], in the presence of which the hydrosilylation reactions result in the formation of predominant amounts of the β(E) product [15,16]. In the literature, there are reports on the effect of different factors on the stereoselectivity of the reaction of hydrosilylation conducted in the presence of rhodium complexes. They include the kind of the structure of starting compounds [17], the type of solvent [16,18] as well as the way and sequence of adding reactants [19–21]. On the basis of the above data, some conclusions can be drawn that will enable directing the reaction to the desired isomer. For example, the reactions of silanes with electron-donor substituents (e.g. trialkylsilanes) and terminal alkynes not containing bulky substituents (with a large steric hindrance) result mainly in the formation of β(Z) isomer, whereas the reactions of silanes with electron-withdrawing substituents (e.g. alkoxy- or chlorosilanes) lead preferentially to the formation of β(E) isomer [20,22]. This effect is enhanced when an alkyne contains large substituents. Quite a large group of catalysts used in the hydrosilylation of terminal alkynes consists of various rhodium(I)-NHC complexes which in most cases catalyze the selective formation of β(Z) product [23–25]. However, if the steric hindrance caused by substituents both in NHC ligand and alkyne is large, then the formation of mostly β(E) product is possible [26]. The use of complexes with NHC ligands, as well as the addition of other ligands (e.g., phosphines) that are sensitive to various contaminants and are unstable in the presence of oxygen and moisture, results in the necessity of performing the reaction in a closed system in the atmosphere of inert gases while maintaining an appropriate level of purity of the reagents [12,23,27]. This is why researchers keep searching for new catalysts that are characterized by high activity and selectivity as well as stability and resistance to contaminants. An alternative solution can be heterogeneous catalysts.

Our research group has been involved for several decades in the development of new catalysts for hydrosilylation. The application of ionic liquids as immobilizing agents for transition metal complexes is an interesting aspect of the above research [28–34]. In all the cases, ionic liquids played the role of a solvent and immobilizing agent for metal complexes which provided a biphasic system with reagents. All the above systems were very effective in the hydrosilylation of olefins. However, ionic liquids can also be employed in another way, i.a. as structural components of the complexing ions [35,36]. Very recently, we have obtained rhodium and platinum complexes with phosphine ligands that were functionalized with imidazolium ionic liquids [37]. As precursors of the above complexes, we have employed Wilkinson's catalyst or *bis*[chloro(1,5-cyclooctadiene)rhodium(I)]. Catalytic studies of olefin hydrosilylation proved very high activity, durability, and stability of the obtained catalysts as well as the possibility of their multiple use [35].

In this paper, we present the results of the study on the catalytic activity of rhodium complexes with phosphine ligands functionalized with ionic liquids for reactions of hydrosilylation of terminal alkynes. Our study was aimed at determining the effect of the kind of catalyst, olefin, and hydrosilylating agent on the yield and selectivity of the hydrosilylation process as well on multiple use of the same portion of the catalyst.

2. Results and Discussion

The reaction of alkyne hydrosilylation was studied according to Scheme 2, using three kinds of alkynes.

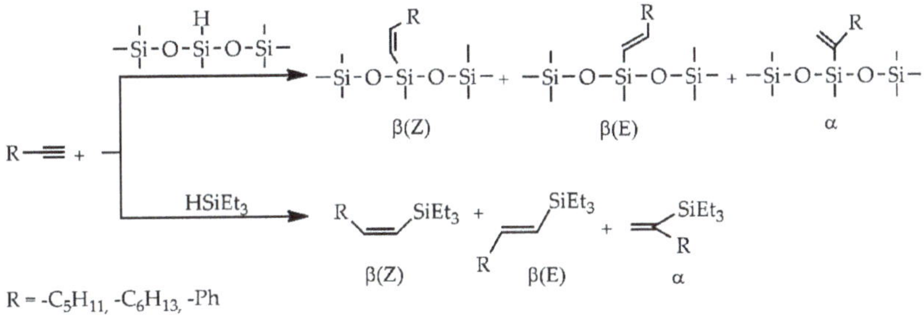

Scheme 2. The reactions of the hydrosilylation of alkynes studied.

We have employed aliphatic alkynes differing in their chain length, 1-heptyne and 1-octyne, as well as an alkyne that contains an aromatic ring in its structure, namely phenylacetylene. Hydrosilylating agents were 1,1,1,3,5,5,5-heptamethyltrisiloxane (HMTS) and triethylsilane (TES). The former of the compounds, due to its high stability, is frequently employed in the synthesis of organosilicon compounds. Moreover, it is a good model compound for obtaining vinylpolysiloxanes, whereas triethylsilane, due to its electron donor properties is a popular reducing agent and a starting material for obtaining vinylsilanes. As catalysts for the above reactions, we have employed Rh(I) complexes that have been recently obtained by our research group. They contain phosphine ligands in which imidazolium ionic liquid is a substituent. Synthesis of the ligands was described in our earlier paper [37], whereas syntheses of rhodium complexes (1 and 2) ligated by these ligands are presented in Scheme 3.

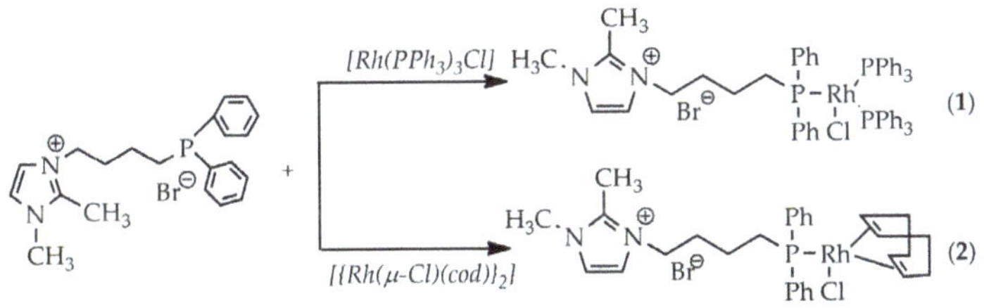

Scheme 3. Syntheses of rhodium complexes ligated by imidazolium-substituted phosphine.

For the sake of comparison, the catalytic study also included both precursors of rhodium catalysts, i.e., Wilkinson's catalyst [Rh(PPh$_3$)$_3$Cl] and cyclooctadiene rhodium complex [{Rh(cod)(μ-Cl)}$_2$]. The complexes were highly active for hydrosilylation of alkenes and, for this reason, we tested their effectiveness in the reactions with alkynes. Phosphine complexes (1 and 2), contrary to the starting complexes (precursors), are insoluble in reactants. Moreover, they are characterized by high thermal stability [38] and resistance to oxidation and moisture. This is why they can be used without the necessity of reagent purification and under normal conditions (in the presence of air in open systems).

At the initial stage, when optimizing hydrosilylation reaction conditions, we performed test reactions in the presence of Wilkinson's catalyst and determined that the optimal stoichiometric ratio of reactants is [HSi≡]:[RC≡CH] = 1.3:1. The excess of HSi≡ enabled obtaining the considerably higher conversion of alkyne. Moreover, test reactions were conducted at different temperatures and the optimum reaction temperature appeared to be 90 °C. For example, when the reaction of 1-octyne with HMTS was carried out at 60 °C for four hours, the alkyne conversion was 11%, whereas at 90 °C the conversion reached 100% already after one hour. Such results were obtained at the catalyst concentration of 1×10^{-3} mol/L mol HSi≡. For this reason, to compare the activity of all the catalysts studied, the ratio [HSi≡]:[RC≡CH]:[cat] = 1.3:1:2 × 10^{-3} was used. Results of our earlier studies of the reaction of hydrosilylation conducted in the presence of commercially available rhodium catalysts have shown that the optimal reaction time is one hour. However, to verify these results in the presence of the catalysts tested in the present study, we carried out FT-IR in situ analysis for two catalytic systems (the complex **1** and the Wilkinson's catalyst) that enabled us to monitor the reaction course in real time. As test reactions, we have chosen reactions of HTMS and 1-octyne. During the experiments, we kept track of the decay of the HSi≡ band originated from HMTS (913 cm^{-1}) and the increase in the band originated from the C=C bond (1600 cm^{-1}) that shows the formation of the hydrosilylation reaction product (Figure S41 and Figure S42 in Supplementary Materials). To determine the reaction profiles, we have used the HSi≡ conversion calculated as a change in the area of the HSi≡ band (Figure 1). Non-stoichiometric amount of the above reactant, which is also shown by the presence of the remainder of the HSi≡ band (Figure S41 and Figure S42), was taken into account during the determination of the reaction profiles and converted to the stoichiometric amount.

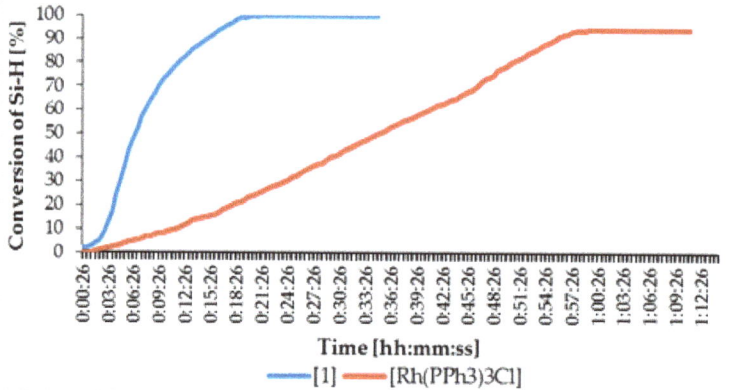

Figure 1. The dependence of the conversion of Si–H on time for the reaction of 1-octyne with 1,1,1,3,5,5,5-heptamethyltrisiloxane in the presence of catalyst **1** and Wilkinson's catalyst.

The measurements showed that the new complex **1** makes it possible to reach full conversion considerably faster than the Wilkinson's catalyst. In the case of the complex **1**, full conversion was reached as early as after 15 min, whereas, in that of Wilkinson's catalyst, it took 55 min. Therefore, to compare the activity of all complexes in the same conditions and to achieve the highest conversion, we decided to conduct all the further reactions at 90 °C for 1 h in air in the open system without a solvent. The reactions of alkynes with HMTS were performed first. After the reaction completion, the catalyst was in the form of a fine suspension; therefore, its separation from the post-reaction mixture was carried out by centrifugation. After the separation, the post-reaction mixture was analyzed chromatographically and the obtained results are presented in Table 1.

Chromatographic and spectroscopic analyses of the post-reaction mixture showed the presence of isomers β(Z), β(E) and α. The obtained results confirmed the high catalytic activity of rhodium phosphine complexes (**1, 2**) in all the reaction systems (Table 1). The conversion in the reaction of hydrosilylation of aliphatic alkynes has been considerably higher (100%) than in the reaction with phenylacetylene (87%). Some surprise is relatively low activity and selectivity (and its lack in the case of the reaction with phenylacetylene) of the catalyst [{Rh(μ-Cl)(cod)}$_2$] since this complex belongs to the most active catalysts for hydrosilylation of alkenes [6,7]. New rhodium complexes (**1** and **2**) do not lead to the formation of α isomer, but only to β isomers. For the majority of the complexes studied (except for the catalyst [{Rh(μ-Cl)(cod)}$_2$]), *cis* isomer is overwhelmingly formed, which is in agreement with results obtained for other rhodium complexes. However, selectivities obtained in reactions conducted in the presence of the catalysts **1** and **2** are higher than those observed in the case of reactions catalyzed by precursors of the rhodium catalysts. One can also notice that, in the case of aliphatic alkynes, the selectivity towards β(Z) isomer increases with the chain growth. It is modest growth, albeit the chain lengths of both alkynes differ only by one carbon atom. When comparing selectivities of the catalysts **1** and **2**, one can observe that in each case the catalyst **1** is characterized by a higher selectivity. To a large extent, it can be explained by differences in the structure and steric hindrance of both complexes. Taking into consideration the mechanism of catalysis occurring in analogous phosphine and cyclooctadiene complexes [7], one can assume that, in the case of the complex **1**, the activation of the catalyst occurs through the detachment of one of the PPh$_3$ ligands, whereas in that of the complex **2** the rupture of one coordinative bond in the cyclooctadiene group occurs. Thereby, the steric hindrance in the complex **2** is greater than that in the complex **1** which results in a decrease in the selectivity and the formation of a large amount of the *trans* isomer.

Table 1. Conversion of alkynes and selectivity for isomers in the hydrosilylation of RC≡CH with HMTS.

Catalyst	R					
	–C$_5$H$_{11}$		–C$_6$H$_{13}$		–Ph	
	Conv. [%]	α/β(Z)/β(E) [%]	Conv. [%]	α/β(Z)/β(E) [%]	Conv. [%]	α/β(Z)/β(E) [%]
[Rh(PPh$_3$)$_3$Cl]	100	5/80/15	94	0/95/5	99	0/94/6
1	100	0/91/9	100	0/96/4	70	0/100/0
[{Rh(μ-Cl)(cod)}$_2$]	100	0/54/46	95	12/50/38	0	0
2	100	0/88/12	100	0/89/11	87	0/86/14

[RC≡CH]:[HSi≡]:[cat] = 1:1.3:2 × 10^{-3}; T = 90 °C, t = 1 h.

The catalytic study on the reaction between phenylacetylene and HMTS has shown that the catalyst **1** (where Wilkinson's catalyst was the metal precursor) was characterized by a higher selectivity towards the *cis* product (100%) compared to that of its precursor (in the case of the precursor the *trans* product was formed as well). Hydrosilylation of phenylacetylene catalyzed by the complex **2** also results in the formation of the *cis* product; however, in that case, the selectivity was 86%. The formation of predominant amounts of the *cis* isomer is frequently observed for hydrosilylation reactions catalyzed by neutral rhodium complexes [6,7].

Analogous measurements of catalytic activity were carried out for hydrosilylation of alkynes with triethylsilane. In addition, in this case, the catalysts **1** and **2**, contrary to their precursors, were insoluble in the reactants; therefore, after the reaction, the catalyst was isolated from the mixture by centrifugation. Then, the composition of the post-reaction mixture was determined by using chromatographic analysis and results of the determination are presented in Table 2.

Table 2. The conversion of alkynes and selectivity to isomers in the hydrosilylation of RC≡CH with TES.

Catalyst	R					
	–C$_5$H$_{11}$		–C$_6$H$_{13}$		–Ph	
	Conv. [%]	α/β(Z)/β(E) [%]	Conv. [%]	α/β(Z)/β(E) [%]	Conv. [%]	α/β(Z)/β(E) [%]
[Rh(PPh$_3$)$_3$Cl]	87	0/95/5	79	0/97/3	88	0/25/75
1	100	0/93/7	95	0/95/5	89	0/17/83
[{Rh(μ-Cl)(cod)}$_2$]	85	20/76/4	97	21/79/0	0	0
2	100	0/89/11	100	0/92/8	96	0/53/47

[RC≡CH]:[HSi≡]:[cat] = 1:1.3:2 × 10^{-3}; T = 90 °C, t = 1 h.

Based on the performed measurements, one can say that in all the cases the reactions catalyzed by the complexes **1** and **2** resulted in higher yields than those catalyzed by the precursors. In the case of hydrosilylation of aliphatic alkynes, an analogous tendency is observed as that occurring for the reactions with HMTS, i.e., that the selectivity towards the formation of *cis* isomers increases with the growth of chain lengths. In addition, a bit higher effectiveness of the complex **1**, compared to the complex **2**, was established. A relatively low activity was also confirmed and, in particular, low selectivity in the reactions catalyzed by the complex [{Rh(μ-Cl)(cod)}$_2$]. However, the most interesting results were observed in the case of the reaction between TES and phenylacetylene. Apart from the total lack of activity of the complex [{Rh(μ-Cl)(cod)}$_2$], in other cases, a change in the selectivity to the predominant formation of the *trans* isomer was found. As it was mentioned earlier, the preference for the formation of a particular isomer is determined by steric and induction effects. On the one hand, the phenyl group makes some steric hindrance, but in the reaction with HMTS, which is also a bulky molecule, the *cis* isomer is preferentially formed. In this case, the steric effects are enhanced by the induction effects because triethylsilane is a strong electron-donor agent and, as a result of the interaction with phenyl group (from phenylacetylene), it favors the formation of the *trans* product. Similar effects were observed in the reaction of hydrosilylation of phenylacetylene with dimethylphenylsilane in the presence of Rh-NHC complexes [20]. The predominant formation of the *trans* product is not surprising in the case of platinum catalysts, but in that of rhodium catalysts, it is a rare case [8,16].

As was already mentioned, rhodium complexes (**1** and **2**) were insoluble in the reactants what enabled their isolation from the post-reaction mixture. However, due to the high dispersion of the catalyst which formed a fine suspension and sedimented very slowly, the separation of the catalyst from the post-reaction mixture was ineffective. For this reason, we employed a rotary centrifuge which enabled the fast separation of the catalyst from the mixture. The isolated catalyst was reused in the subsequent catalytic runs, and its activity was determined. The isolation and reusing were performed for the complex **1** which was characterized by higher activity and selectivity in all the above reactions. As a test reaction, we chose hydrosilylation of 1-octyne with HMTS. After each catalytic run, the mixture was centrifuged, followed by the collection of products (colorless and clear liquid) with a syringe. The products were subjected to chromatographic and spectral analyses and the isolated catalyst was reused in a next catalytic run carried out after a fresh portion of the reaction substrates was added. In this way, we conducted five catalytic runs and the obtained results are presented in Table 3.

Table 3. The conversion and selectivity for hydrosilylation of 1-octyne with HMTS in subsequent catalytic runs catalyzed by the same portion of the catalyst **1** isolated from the post-reaction mixture by centrifugation.

No. of Catalytic Run	Conv. of Alkyne [%]	Selectivity α/β(Z)/β(E) [%]
1	100	0/96/4
2	95	0/91/9
3	90	0/88/12
4	70	0/86/14
5	65	0/86/14

[RC≡CH]:[HSi≡]:[cat] = 1:1.3:2 × 10^{-3}; T = 90 °C, t = 1 h.

The results show that catalytic activity of the complex **1** in subsequent catalytic runs is very high, particularly in the first three runs. In the further runs, the conversion of 1-octyne decreases which can be explained by a gradual loss of catalyst as a result of incomplete centrifugation, i.e., only partial catalyst isolation. The slow sedimentation of the catalyst was the reason for which centrifugation was carried out only for a few minutes. The evaluation of the centrifugation effectiveness was based on the visual observation of the separation of two phases. This appeared to be not quite appropriate. This is why we decided to check it by using a different means of isolation. The post-reaction mixture was subjected to filtration followed by placing the filter with the precipitate in the reaction mixture and repeating the catalytic run. The conversion and selectivity in the subsequent runs are presented in Table 4. No significant reduction in the conversion was observed, contrary to the case when the catalyst was reused after centrifugation. This allows supposing that the decrease in the conversion observed previously was the effect of the incomplete isolation.

Table 4. The conversion and selectivity for hydrosilylation of 1-octyne with HMTS in subsequent catalytic runs catalyzed by the same portion of the catalyst 1 isolated from the post-reaction mixture by filtration.

No. of Catalytic Run	Conv. of Alkyne [%]	Selectivity α/β(Z)/β(E) [%]
1	100	0/96/4
2	98	0/94/6
3	98	0/94/6
4	95	0/92/8
5	93	0/90/10

[RC≡CH]:[HSi≡]:[cat] = 1:1.3:2 × 10^{-3}; T = 90 °C, t = 1 h.

A deterioration of selectivity in subsequent runs was also observed. It can be caused by catalyst modification which occurs with time. Taking into consideration that the concentration of β(E) isomer increases (which according to the literature data is observed for complexes with a bulky steric hindrance), one can suppose that, in subsequent catalytic cycles, due to weakly coordinating properties of imidazolium substituent, binding of reagents or aggregation of the complex occur which results in a larger steric hindrance. However, the possibility of at least fivefold use of the same portion of catalyst significantly reduces the necessary amount of the catalyst and thereby decreases the synthesis costs.

3. Materials and Methods

3.1. Materials

All the reagents used in the presented experiments, such as 1-octyne (97%), 1-heptyne (98%), phenylacetylene (98%), 1,1,1,3,5,5,5-heptamethyltrisiloxane (97%), triethylsilane (99%), and n-decane (99%), were supplied by Sigma Aldrich (Poznań, Poland) and used as received. Wilkinson's catalyst and chloro(1,5-cyclooctadiene)rhodium(I) dimer were also purchased from Sigma Aldrich (Poznań, Poland).

3.2. Analytical Techniques

The yield of the product of hydrosilylation of alkynes was determined by using a Clarus 680 gas chromatograph (Perkin Elmer, Shelton, CT, USA) equipped with a 30 m capillary column Agilent VF-5ms (Santa Clara, CA, USA) and TCD detector, employing the following temperature program: 60 °C (3 min.), 10 °C min^{-1}, and 290 °C (5 min.). Characteristic retention times were used for partial identification of the obtained products. The GC-MS analysis was conducted using a Varian 3300 chromatograph (Mundelein, IL, USA) equipped with a 30 m DB-1 capillary column connected to the Finnigan Mat 700 mass detector (Mundelein, IL, USA). For the products obtained, NMR spectra were made with Bruker BioSpin (400 MHz) spectrometer (Ettlingen, Germany) using CDCl$_3$ and CD$_3$CN as a solvent, with tetramethylsilane as the internal standard. Proton chemical shifts are shown in parts per million (δ ppm). The FT-IR in situ measurements were performed using a Mettler Toledo ReactIR 15 instrument (Giessen, Germany). The spectra were recorded with 256 scans for 1 h at 30 s intervals with the resolution of 1 cm^{-1}. Changes in the intensity of the bands at 913 cm^{-1} and 1600 cm^{-1} were recorded using an ATR probe with a diamond window. During the experiments, the decay of the HSi≡ band originated from HMTS (913 cm^{-1}) and the increase in the band originated from the C=C bond (1600 cm^{-1}) that shows the formation of the hydrosilylation reaction product (Figure S41 and Figure S42 in Supplementary Materials) were tracked. CHN elemental analyses were performed on an elemental analyzer Vario EL III (Elementar Analysensysteme GmbH, Langenselbold, Germany).

3.3. Synthesis of Transition-Metal Based Complexes

Transition-metal based complexes were prepared using the Schlenck technique. In the Schlenck's tube equipped with a magnetic stirrer, a portion of rhodium precursor ([Rh(Cl)(PPh$_3$)$_3$] or [{Rh(μ-Cl)(cod)}$_2$]) with phosphine ligand (3-(4-(diphenylphosphanyl)butyl)-1,2-dimethylimidazolium bromide) was dissolved in toluene. The mixture was stirred for 24 h at room temperature and, after filtration, evaporation of the solvent, and drying under vacuum, the final product was obtained. The detailed synthesis procedures of rhodium ({1,2-dimethyl-3-(diphenylphosphine) butylimidazoliumbromide}bis(triphenylphosphine)chloridorhodium(I) (**1**) and {1,2-dimethyl-3-(diphenylphosphine)butylimidazoliumbromide}(η^4-cycloocta-1,5-diene)chloride-rhodium(I)) (**2**) were previously reported by our research group [37]. Spectroscopic characterization and elemental analysis of these complexes are presented in Supplementary Materials.

3.4. General Procedure for Catalytic Tests

The catalytic activity of Rh phosphine-ligated complexes was tested in the hydrosilylation of 1-heptyne, 1-octyne, and phenylacetylene with 1,1,1,3,5,5,5-heptamethyltrisiloxane (HMTS) or triethylsilane (TES). The hydrosilylation reactions were carried out in the special glass reaction vessel, which was a 5 mL reactor with a side connector for sampling. The reactor was equipped with condenser and magnetic stirrer. In addition, 2×10^{-3} mol of catalyst, 1 mmol of alkyne derivative, 1.3 mmol of HMTS or TES, and 0.5 mmol of n-decane as an internal standard were placed into the reaction vessel. The reaction was carried out at 90 °C under air with vigorous stirring for 1 h. Then, the reaction mixture was cooled, centrifuged, and subjected to GC analysis to determine the reaction yield. The obtained retention times for particular isomers were used for partial identification of hydrosilylation products. The products were isolated and subjected to NMR and GC-MS analyses. The obtained spectroscopic data were compared with the literature data [8,12,18,38–43].

1,1,1,3,5,5,5-Heptamethyl-3-[(1Z)Hept-1-Enyl)]Trisiloxane

^{1}H NMR (CDCl$_3$)ppm: β (Z): 0.10–0.14 (m, 21H, Si–CH$_3$), 0.92 (t, J = 8.2 Hz, 3H, –CH$_3$), 1.37 (m, 6H, C=C–CH$_2$–(CH$_2$)$_3$), 2.22 (m, 2H, C=C–CH$_2$), 5.34 (dd, J = 14.8, 4.8 Hz, 1H, C=CH–), 6.37 (dt, J = 14.6, 7.4 Hz, 1H, –CH=C), β(E): 5.53 (ddt, J = 21.7, 18.7, 1.5 Hz, 1H, C=CH–), 6.16 (dt, J = 18.7, 6.3 Hz, 1H, –CH=C–). **^{13}C NMR** (CDCl$_3$)ppm: β (Z): 1.81 (–Si–CH$_3$), 13.79 (–CH$_3$), 22.56 (C–CH$_3$), 29.12 (C–C–C–), 31.84 (–C–C–CH$_3$), 33.34 (C=C–C), 127.03 (Si-C=C–), 150.52 (–C=C–), β (E): 127.6 (Si–C=C–), 149.2 (–C=C–). **^{29}Si NMR** (CDCl$_3$)ppm: −35.37 (–OSiCH$_3$), 7.45 (–OSi(CH$_3$)$_3$).

GC-MS: β(Z): 305 (3%, [(–OSi(CH$_3$)$_3$)$_2$SiC$_7$H$_{14}$]$^+$), 247.5 (5%, [(–OSi(CH$_3$)$_3$)$_2$SiCH$_3$CH=CH$_2$]$^+$), 229.5 (3%, [–OSi(CH$_3$)$_3$SiCH$_3$C$_7$H$_{14}$]$^+$), 221 (100%, [(-OSi(CH$_3$)$_3$)$_2$SiCH$_3$]$^+$), 208.9 (17% [((CH$_3$)$_3$SiO)SiOH(CH$_3$)$_2$SiO]$^{2+}$), β(E): 305 (3%, [(–OSi(CH$_3$)$_3$)$_2$SiC$_7$H$_{14}$]$^+$), 247.5 (7%, [(–OSi(CH$_3$)$_3$)$_2$SiCH$_3$CH=CH$_2$]$^+$), 229.5 (3%, [–OSi(CH$_3$)$_3$SiCH$_3$C$_7$H$_{14}$]$^+$), 221 (100% [(OSi(CH$_3$)$_3$)$_2$SiCH$_3$]$^+$), 208.9 (23% [((CH$_3$)$_3$SiO)SiOH(CH$_3$)$_2$SiO]$^{2+}$) α: 305 (2.3%, [(–OSi(CH$_3$)$_3$)$_2$SiCH$_3$C$_6$H$_{12}$]$^+$), 247.5 (5% [(–OSi(CH$_3$)$_3$)$_2$ SiCH$_3$CH=CH$_2$]$^+$), 221 (100%, [(–OSi(CH$_3$)$_3$)$_2$SiCH$_3$]$^+$), 208.9 (31%, [((CH$_3$)$_3$SiO)SiOH(CH$_3$)$_2$SiO]$^+$).

1,1,1,3,5,5,5-Heptamethyl-3-[(1Z)Oct-1-Enyl)]Trisiloxane

^{1}H NMR (CDCl$_3$) ppm: β (Z): 0.10–0.15 (m, 21H, Si–CH$_3$), 0.91 (t, J = 6.9 Hz, 3H, –CH$_3$), 1.29–1.45 (m, 8H, C=C–CH$_2$–(CH$_2$)$_4$), 1.95–2.13 (m, 2H, C=C–CH$_2$), 5.37 (dt, J = 14.2, 1.2 Hz, 1H, C=CH–), 6.31 (dt, J = 14.5, 7.4 Hz, 1H, –CH=C), β(E): 5.49 (dt, J = 18.7, 1.5 Hz, 1H, C=CH–), 6.16 (dt, J = 18.7, 6.3 Hz, 1H, –CH=C). **^{13}C NMR** (CDCl$_3$)ppm: 1.71–2.01 (–Si–CH$_3$), 14.35 (–CH$_3$), 22.67 (C–CH$_3$), 29.11, 29.62 (C–C–C–), 31.86 (–C–C–CH$_3$), 36.60 (C=C–C), 126.93 (Si–C=C–), 150.47 (–C=C–). **^{29}Si NMR** (CDCl$_3$)ppm: −35.39 (–OSiCH$_3$), 8.07 (–OSi(CH$_3$)$_3$).

GC-MS: β(Z): 317 (25%, [(–OSi(CH$_3$)$_3$)$_2$SiCH=CH(CH$_2$)$_5$CH$_3$)]$^+$), 221 (100%, [(–OSi(CH$_3$)$_3$)$_2$SiCH$_3$]$^+$), 208 (19% [((CH$_3$)$_3$SiO)SiOH(CH$_3$)$_2$SiO]$^+$), 134 (4%, [–OSi(CH$_3$)$_3$-SiCH$_3$]$^{2+}$), β(E): 317 (30%, [(–OSi(CH$_3$)$_3$)$_2$ SiCH=CH(CH$_2$)$_5$CH$_3$)]$^+$), 221 (100%, [(–OSi(CH$_3$)$_3$)$_2$SiCH$_3$]$^+$), 208 (20% [((CH$_3$)$_3$SiO)SiOH(CH$_3$)$_2$SiO]$^+$), 134 (4%, [–OSi(CH$_3$)$_3$-SiCH$_3$]$^{2+}$), α: 317 (10%, [(–OSi(CH$_3$)$_3$)$_2$SiCH=CH(CH$_2$)$_5$CH$_3$)]$^+$), 221 (100%, [(–OSi(CH$_3$)$_3$)$_2$SiCH$_3$]$^+$), 208 (18% [(CH$_3$)$_3$SiO)SiOH(CH$_3$)$_2$SiO]$^+$), 134 (3.7%, [-OSi(CH$_3$)$_3$-SiCH$_3$]$^{2+}$).

1,1,1,3,5,5,5-Heptamethyl-3-[(1Z)-2-Phenylethenyl]Trisiloxane

^{1}H NMR (CD$_3$CN) ppm: β (Z): 0.10–0.24 (m, 21H, –OSiCH$_3$), 5.68 (d, J = 15.5 Hz, 1H, C=CH–Si), 7.31 (d, J = 15.7 Hz, 1H, HC=CH), 7.23–7.39 (m, 3H, C–C=C), 7.49–7.56 (m, 2H, C=C–C), β (E): 6.33 (d, J = 19.3 Hz, 1H, C=CH–Si), 7.04 (d, J = 19.3 Hz, 1H, SiC=CH). **^{13}C NMR** (CD$_3$CN) ppm: −0.5–1.05 (–OSi(CH$_3$)$_3$, –OSiCH$_3$), 127.86 (C=C–C), 128.02, 128.37, 129.37, 139.16, 147.38. **^{29}Si NMR** (CD$_3$CN)ppm: −37.04 (-OSiCH$_3$), 8.07 (–OSi(CH$_3$)$_3$).

GC-MS: β(Z): 324 (5%, [(–OSi(CH$_3$)$_3$)$_2$SiCH$_3$–CH=CH$_2$Ph]$^+$), 311 (6.5%, [(–OSi(CH$_3$)$_3$)$_2$SiCH=CH$_2$Ph]$^+$ 221 (18.7% [(–OSi(CH$_3$)$_3$-SiCH$_3$]$^+$), 208 (30%, [(–OSi(CH$_3$)$_3$)$_2$Si–]$^{2+}$), 161 (100%, [–OSi(CH$_3$) –Si–CH=CH$_2$–]$^{2+}$), 149 (31%, [–SiCH$_3$–CH=CH$_2$–Ph]$^{2+}$), 104 (5.5%, [CH=CH$_2$–Ph]$^+$), 77.1 (6.3%, [–C$_6$H$_5$]$^+$), β(E): 324 (2%, [(–OSi(CH$_3$)$_3$)$_2$SiCH$_3$–CH=CH$_2$Ph]$^+$), 311 (5%, [(–OSi(CH$_3$)$_3$)$_2$SiCH=CH$_2$Ph]$^+$ 221 (100% [(–OSi(CH$_3$)$_3$-SiCH$_3$]$^+$), 208 (12%, [(–OSi(CH$_3$)$_3$)$_2$Si–]$^{2+}$), 161 (17%, [–OSi(CH$_3$) –Si–CH=CH$_2$–]$^{2+}$), 149 (15%, [–SiCH$_3$–CH=CH$_2$–Ph]$^{2+}$), 104 (6%, [CH=CH$_2$–Ph]$^+$), 77.1 (5%, [–C$_6$H$_5$]$^+$).

(Z)-Triethyl(Hept-1-Enyl)Silane

^{1}H NMR (CDCl$_3$) ppm: β(Z): 0.63 (m, 4H, –CH$_2$–CH$_3$), 0.94 (m, 11H, –CH$_3$), 1.32 (m, 6H, –CH$_2$–CH$_2$–CH$_2$–), 2.15 (m, 2H, –C=C-CH$_2$–, J=4.8 Hz), 5.42 (dt, J = 14.1, 1.3 Hz 1H, Si–CH=C–), 6.41 (dt, J = 14.4, 7.3 Hz 1H, C=CH–), β(E): 5.59 (dt, J = 18.5, 1.4 Hz, 1H, Si–CH=C–), 6.08 (dt, J = 18.7, 6.3 Hz, 1H, C=CH–). **^{13}C NMR** (CDCl$_3$)ppm: β(Z): 4.73 (Si–CH$_2$–), 7.54 (Si–CH$_2$–CH$_3$), 14.04 (–CH$_3$), 22.61 (C–CH$_3$), 29.49 (–C–C–C–), 31.66 (–C–C–CH$_3$), 34.08 (C=C–C–), 124.89 (Si–C=C–), 150.38 (C=C–C–); **β (E)**: 125.52 (Si–C=C–), 148.83 (C=C–C). **^{29}Si NMR** (CDCl$_3$)ppm: −2.83 ((C$_2$H$_5$)$_3$SiH).

GC-MS: β(Z): 183 (100%, [–Si(C$_2$H$_5$)$_2$HC=CH$_2$C$_5$H$_{11}$]$^+$), 156 (5%, [SiC$_2$H$_5$HC=CH$_2$C$_5$H$_{11}$]$^+$), 115.1 (16%, [–Si(C$_2$H$_5$)$_3$]$^+$), 99 (37%, [HC=CH$_2$C$_5$H$_{11}$]$^+$), 89.1 (16%, [–Si(C$_2$H$_5$)$_2$]$^{2+}$); β(E): 183 (100%, [–Si(C$_2$H$_5$)$_2$HC=CH$_2$C$_5$H$_{11}$]$^+$), 156 (6%, [SiC$_2$H$_5$HC=CH$_2$C$_5$H$_{11}$]$^+$), 115.1 (14%, [–Si(C$_2$H$_5$)$_3$]$^+$), 99 (50%, [HC=CH$_2$C$_5$H$_{11}$]$^+$), 89.1 (28%, [–Si(C$_2$H$_5$)$_2$]$^{2+}$); α: 183 (100%, [–Si(C$_2$H$_5$)$_2$HC=CH$_2$C$_5$H$_{11}$]$^+$), 156 (5%, [SiC$_2$H$_5$HC=CH$_2$C$_5$H$_{11}$]$^+$), 115.1 (30%, [–Si(C$_2$H$_5$)$_3$]$^+$), 89.1 (70%, [–Si(C$_2$H$_5$)$_2$]$^{2+}$).

(Z)-Triethyl(Oct-1-Enyl)Silane

^{1}H NMR (CDCl$_3$)ppm: β(Z): 0.55 (m, 4H, –CH$_2$–CH$_3$), 0.96 (t, 3H, –CH$_3$), 0.98 (m, 11H, –CH$_2$–CH$_3$), 1.31–1.46 (m, 8H, –(CH$_2$)$_4$–), 1.96–2.17 (m, 2H, –C=C–CH$_2$–), 5.39 (dd, J = 13.8, 6.1 Hz, 1H, Si–CH=C–), 6.39 (dt, J = 14.4, 7.3 Hz, 1H, C=CH–), β (E): 5.59 (dt, J = 18.7, 1.5 Hz, 1H, –CH=C–), 6.02 (dt, J = 18.7, 6.3 Hz, 1H, C=CH–), α: 5.29 (ddd, J = 12.2, 9.9, 3.8 Hz,1H, –CH=C–). **^{13}C NMR** (CDCl$_3$)ppm: β(Z): 3.55 (Si–CH$_2$–), 7.41 (Si–CH$_2$–CH$_3$), 14.07 (–CH$_3$), 22.63 (C–CH$_3$), 28.80 (–C–C–C–), 29.66 (–C–C–C–),), 31.38 (–C–C–CH$_3$), 32.78 (C=C–C–), 126.00 (Si–C=C–), 150.36 (C=C–C–), β(E): 125.48 (Si–C=C–), 148.82 (C=C–C–). **^{29}Si NMR** (CDCl$_3$)ppm: −5.78 ((C$_2$H$_5$)$_3$SiH).

GC-MS: β(Z): 197 (100%, [SiEt$_2$C$_8$H$_{15}$]$^+$), 141.3 (2% [SiEt$_3$CH=CH]$^+$), 115.1 (72.5%, [SiEt$_3$]$^+$), 85.1 (6%, [C$_6$H$_{11}$]$^+$), β(E): 197 (100%, [SiEt$_2$C$_8$H$_{15}$]$^+$), 141.3 (15% [SiEt$_3$CH=CH]$^+$), 115.1 (15%, [SiEt$_3$]$^+$), 85.1 (26%, [C$_6$H$_{11}$]$^+$), α: 197 (100%, [SiEt$_2$C$_8$H$_{15}$]$^+$), 141.3 (12% [SiEt$_3$CH=CH]$^+$), 115.1 (30%, [SiEt$_3$]$^+$), 85.1 (25%, [C$_6$H$_{11}$]$^+$).

(E)-Triethyl(Phenyl-1-Ethene)Silane

^{1}H NMR (CD$_3$CN) ppm: β(E): (CD$_3$CN) ppm: 0.72–1.04 (m, 15H, –Si(Et)$_3$), 6.50 (d, J = 19.3 Hz, 1H, C=CH–Si), 6.97 (d, J = 19.3 Hz, 1H, HC=C–Si), 7.29 (m, 1H, C=C–C), 7.37 (m, 2H, C=C–C), 7.50 (m, 2H, C–C=C), β(Z): 5.79 (d, J=15.2 Hz, 1H, C=CH–Si) **^{13}C NMR** (CD$_3$CN) ppm: 3.33 (–SiCH$_2$CH$_3$), 7.12 (–SiCH$_2$–), 125.95 (C=C–C–), 126.32, 127.89, 128.48, 138.53, 145.19 **^{29}Si NMR** (CD$_3$CN)ppm: 0.55 (–Si(Et)$_3$).

GC-MS: β(E): 189 (100%, [Si(C$_2$H$_5$)$_2$–HC=CH$_2$-Ph], 134.1 (25%, [Si–HC=CH$_2$Ph]$^{3+}$), 115.1 (5%, [–Si(C$_2$H$_5$)$_3$]$^+$), 104 (25%, [HC=CH$_2$–Ph]$^+$), β(Z): 189 (100%, [Si(C$_2$H$_5$)$_2$–HC=CH$_2$-Ph], 134.1 (52%, [Si–HC=CH$_2$Ph]$^{3+}$), 115.1 (10%, [–Si(C$_2$H$_5$)$_3$]$^+$), 104 (30%, [HC=CH$_2$-Ph]$^+$).

3.5. General Procedure for Catalyst Isolation and Tests with Subsequent Catalytic Runs

The catalytic activity of the complex **1** in subsequent catalytic cycles was tested in the hydrosilylation of 1-octyne with 1,1,1,3,5,5,5-heptamethyltrisiloxane (HMTS). The hydrosilylation reaction was conducted in a glass reaction vessel equipped with a condenser and magnetic stirrer. In addition, 2×10^{-3} mol of catalyst, 1 mmol of 1-octyne, 1.3 mmol of HMTS, and 0.5 mmol of n-decane as an internal standard were placed into the reaction vessel. The reaction was carried out at 90 °C under air with vigorous stirring for 1 h. Then, the reaction mixture was cooled and centrifuged for a few minutes or filtered. Next, the whole amount of product was collected with a syringe equipped with a needle, followed by subjecting to GC analysis to determine the reaction yield. For the catalyst isolated by centrifugation, that remained in the reaction vessel, another portion of the same substrates was added, whereas, for filtered post-reaction mixture, the filter with isolated catalyst was added to the new portion of reagents. In both cases, the isolated catalyst was not subjected to any washing or regeneration. The aforementioned operation was repeated five times, in order to perform five catalytic runs.

4. Conclusions

The modification of the starting rhodium complexes [Rh(PPh$_3$)$_3$Cl] and [{Rh(μ-Cl)(cod)}$_2$] with the phosphine ligand containing imidazolium ionic liquid as a substituent resulted in obtaining complexes **1** and **2** which are insoluble in the reactants used in the hydrosilylation react on. The above complexes are very durable and stable in air which facilitates their use. Both complexes showed a very high catalytic activity for alkyne hydrosilylation. A complete conversion of alkynes was achieved already after 1 h from the beginning of the reaction. However, the FT-IR in situ analysis of the reactions catalyzed by the complex **1** and the Wilkinson's catalyst has shown that, in the case of the former catalyst, full conversion can be achieved in a considerably shorter time (15 min), whereas, in that of the Wilkinson's catalyst, it requires 55 min. The above studies are a preliminary step to the determination of the kinetics of hydrosilylation of alkynes in the presence of the new rhodium catalysts which we are going to carry out in the immediate future. Three alkynes (1-heptyne, 1-octyne and phenylacetylene) were subjected to hydrosilylation with 1,1,1,3,5,5,5-heptamethyltrisiloxane (HMTS) or triethylsilane (TES). In all the cases, the complexes **1** and **2** were more active than their precursors. In the reactions of hydrosilylation of aliphatic alkynes catalyzed by both complexes, mainly β(Z) isomers are formed, which is typical of the major part of rhodium catalysts applied hitherto. In addition, selectivities achieved in this case are higher than those observed for reactions catalyzed by their precursors and range from 89 to 96%. Additionally, a small increase in the selectivity is observed with alkyne chain growth. When comparing selectivities obtained for both catalysts, one can notice that complex **1** is characterized by a higher selectivity. In the reactions with aliphatic alkynes, the effect of the kind of hydrosilylating agent was not observed; both in the reactions with HMTS and TES, *cis* isomer is mainly formed. A somewhat different situation occurs in the hydrosilylation of phenylacetylene, where, depending on the kind of hydrosilylating agent, either *cis* or *trans* isomer predominates. In the reaction with HMTS, the isomer β(Z) prevails, whereas, in that with TES, the β(E) product is formed. This is caused, besides the steric effect of the phenyl group, also by strongly electronegative nature of triethylsilane whose interaction with phenyl group results in the predominant formation of the *trans* isomer. It is also worth mentioning that the catalyst **2** was active both in reactions with HMTS and TES, whereas its precursor was inactive (maybe because of too mild reaction conditions or too short reaction time).

The most important result of our study is proving the possibility of multiple uses of the same portion of the catalyst. Due to the heterogeneous nature of the complexes **1** and **2**, their isolation and reusing are feasible. The results of conducting five catalytic runs with the use of the same portion of catalyst show that the activity and particularly selectivity are high. It has been proved that the most efficient way of the catalyst isolation from the post-reaction mixture was filtration, which made it possible to reuse the catalyst five times with the preservation of its high activity. Generally, the complexes **1** and **2** that contain the imidazolium- phosphine ligand are characterized by a higher catalytic activity than their precursors. According to the literature data, the rhodium complexes containing ligands with such heteroatoms as P–, N– or NHC– group (which are electron donors) show an increase in the selectivity and yield of the reaction due to a higher stabilization of the Rh-phosphine bond [44–46]. On the other hand, phosphine ligand with imidazolium ionic liquid has a significant effect on the heterogenization of the catalytic system. The obtained catalysts enable their easy isolation and reusability which is of substantial economic and ecological importance.

Supplementary Materials: The following are available online at http://www.mdpi.com/2073-4344/10/6/608/s1, Figure S1: ^{1}H NMR spectrum of 1,1,1,3,5,5,5-Heptamethyl-3-[(1Z)hept-1-enyl)]trisiloxane. Figure S2: ^{13}C NMR spectrum of 1,1,1,3,5,5,5-Heptamethyl-3-[(1Z)hept-1-enyl)]trisiloxane. Figure S3: ^{29}Si NMR spectrum of 1,1,1,3,5,5,5-Heptamethyl-3-[(1Z)hept-1-enyl)]trisiloxane. Figure S4: ^{1}H NMR spectrum of 1,1,1,3,5,5,5-Heptamethyl-3-[(1Z)oct-1-enyl)]trisiloxane, Figure S5: ^{13}CNMR spectrum of 1,1,1,3,5,5,5-Heptamethyl-3-[(1Z)oct-1- enyl)]trisiloxane. Figure S6: ^{29}Si NMR spectrum of 1,1,1,3,5,5,5-Heptamethyl-3-[(1Z)oct-1-enyl)] trisiloxane. Figure S7: ^{1}H NMR spectrum of 1,1,1,3,5,5,5-Heptamethyl-3-[(1Z)-2-phenylethenyl]trisiloxane. Figure S8: ^{13}CNMR spectrum of 1,1,1,3,5,5,5-Heptamethyl-3-[(1Z)-2-phenylethenyl]trisiloxane. Figure S9: ^{29}SiNMR spectrum of 1,1,1,3,5,5,5-Heptamethyl-3-[(1Z)-2-phenylethenyl]-trisiloxane, Figure S10: ^{1}H NMR spectrum of (Z)- triethyl(hept-1-enyl)silane. Figure S11: ^{13}C NMR spectrum of (Z)-triethyl(hept-1-enyl)silane, Figure S12: ^{29}SiNMR spectrum of (Z)-triethyl(hept-1-enyl)silane. Figure S13: ^{1}H NMR spectrum of (Z)-triethyl(oct-1-enyl)silane. Figure S14: ^{13}C NMR spectrum of (Z)-triethyl(oct-1-enyl)silane. Figure S15: ^{29}Si NMR spectrum of (Z)-triethyl(oct -1-enyl)silane. Figure S16: ^{1}H NMR spectrum of (E)-triethyl(phenyl-1-ethene)silane. Figure S17: ^{13}C NMR spectrum of (E)-triethyl(phenyl-1-ethene)silane. Figure S18: ^{29}Si NMR spectrum of (E)-triethyl(phenyl-1-ethene)silane. Figure S19: GC chromatogram of 1,1,1,3,5,5,5-Heptamethyl-3-[(1Z)hept-1-enyl)]-trisiloxane, 1,1,1,3,5,5,5-Heptamethyl-3-[(1E)hept -1-enyl)]trisiloxane and 1,1,1,3,5,5,5-Heptamethyl-3-[(α)hept-1-enyl)]trisiloxane. Figure S20: MS spectrum of 1,1,1,3,5,5,5-Heptamethyl-3-[(1Z)hept-1-enyl)]trisiloxane. Figure S21: MS spectrum of 1,1,1,3,5,5,5-Heptamethyl-3-[(1E)hept-1-enyl)]trisiloxane. Figure S22: MS spectrum of 1,1,1,3,5,5,5-Heptamethyl-3-[(1α)hept-1-enyl)]trisiloxane. Figure S23: GC chromatogram of 1,1,1,3,5,5,5-Heptamethyl-3-[(1Z)oct-1-enyl)]trisiloxane1,1,1,3,5,5,5-Heptamethyl -3-[(1E)oct-1-enyl)] trisiloxane and 1,1,1,3,5,5,5-Heptamethyl-3-[(1α)oct-1-enyl)]trisiloxane. Figure S24: MS spectrum of 1,1,1,3,5, 5-Heptamethyl-3-[(1Z)oct-1-enyl)]trisiloxane. Figure S25: MS spectrum of 1,1,1,3,5,5,5-Heptamethyl-3- [(1E)oct-1-enyl)]trisiloxane. Figure S26: MS spectrum of 1,1,1,3,5,5,5-Heptamethyl-3-[(1α)oct-1-enyl)]trisiloxane. Figure S27: GC chromatogram of 1,1,1,3,5,5,5-Heptamethyl-3-[(1Z)-2-phenylethenyl]trisiloxane and 1,1,1,3,5,5,5-Heptamethyl-3-[(1E)-2-phenylethenyl]trisiloxane. Figure S28: MS spectrum of 1,1,1,3,5,5,5-Heptamethyl-3-[(1Z)-2-phenylethenyl]trisiloxane. Figure S29: MS spectrum of 1,1,1,3,5,5,5-Heptamethyl-3-[(1E)-2-phenylethenyl]trisiloxane. Figure S30: GC chromatogram of (Z)-triethyl(hept-1-enyl)silane, (E)-triethyl(hept-1-enyl)silane and (α)-triethyl(hept-1-enyl)silane. Figure S31: MS spectrum of (Z)-triethyl(hept-1-enyl)silane. Figure S32: MS spectrum of (E)-triethyl(hept-1-enyl)silane. Figure S33: MS spectrum of (α)-triethyl(hept-1-enyl)silane. Figure S34: GC chromatogram of (Z)-triethyl(oct-1-enyl)silane, (E)-triethyl(oct-1-enyl)silane and (α)-triethyl(oct-1-enyl)silane. Figure S35: MS spectrum of (Z)-triethyl(oct-1-enyl)silane. Figure S36: MS spectrum of (E)-triethyl(oct-1-enyl)silane. Figure S37: MS spectrum of (α)-triethyl(oct-1-enyl)silane. Figure S38: GC chromatogram of (E)-triethyl(phenyl-1-ethene)silane and (Z)-triethyl(phenyl-1-ethene)silane. Figure S39: MS spectrum of (E)-triethyl(phenyl-1-ethene)silane. Figure S40: MS spectrum of (Z)-triethyl(phenyl-1-ethene)silane. Figure S41: FT-IR spectra with characteristic peaks at 1600 cm^{-1} and 913 cm^{-1} which change with time of the hydrosilylation reaction between 1-octyne and HMTS, carried out in the presence of the Wilkinson's catalyst. Figure S42: FT-IR spectra with characteristic peaks at 1600 cm^{-1} and 913 cm^{-1} which change with time of the hydrosilylation reaction between 1-octyne and HMTS, carried out in the presence of catalyst **1**.

Author Contributions: Catalytic tests, methodology—O.B.; conceptualization—O.B. and H.M.; synthesis of phosphine ligated Rh complexes—M.J.-W.; writing—original draft preparation—O.B. and H.M., writing—review and editing—O.B., M.J.-W. and H.M.; supervision—H.M.; funding acquisition—H.M. and O.B. All authors have read and agreed to the published version of the manuscript.

Funding: This research was supported by grant No. POWR.03.02.00-00-I023/17 co-financed by the European Union through the European Social Fund under the Operational Program Knowledge Education Development and grant OPUS UMO-2014/15/B/ST5/04257 funded by National Science Center (Poland).

Conflicts of Interest: The authors declare no conflict of interest.

References

1. Luh, T.-Y.; Liu, S.-T. Synthetic Applications of Allylsilanes and Vinylsilanes. In *The Chemistry in Organic Silicon Compounds*; Rappoport, Z., Apeloig, Y., Eds.; Willey: New York, NY, USA, 1998.

2. Fleming, I.; Barbero, A.; Walter, D. Stereochemical Control in Organic Synthesis Using Silicon-Containing Compounds. *Chem. Rev.* **1997**, *97*, 2063–2192. [CrossRef]
3. Nakao, Y.; Hiyama, T. Silicon-based cross-coupling reaction: An environmentally benign version. *Chem. Soc. Rev.* **2011**, *40*, 4893–4901. [CrossRef]
4. Hiyama, T.; Diederich, F.; Stang, P.J. *Metal-Catalyzed Cross-Coupling Reactions*; Willey-VCH: New York, NY, USA, 1998; pp. 421–453.
5. Lim, D.S.W.; Anderson, E.A. Synthesis of Vinylsilanes. *Synthesis* **2012**, *44*, 983–1010.
6. Marciniec, B.; Maciejewski, H.; Pawluć, P. Hydrosilylation of Carbon-Carbon Multiple Bonds—Applications in Synthesis and Materials Science. In *Organosilicon Compounds*; Lee, V.Y., Ed.; Academic Press: Cambridge, MA, USA, 2017; Volume 2, Chapter 5.
7. Marciniec, B.; Maciejewski, H.; Pietraszuk, C.; Pawluć, P. *Hydrosilylation. A Comprehensive Review on Recent Advances*; Marciniec, B., Ed.; Springer: Dordrecht, The Netherlands, 2009.
8. Takeuchi, R.; Tanouchi, N. Solvent-controlled Stereoselectivity in the Hydrosilylation of Alk-1-ynes Catalysed by Rhodium Complexes. *J. Chem. Soc. Perkin Trans.* **1994**, *1*, 2909–2913. [CrossRef]
9. Wu, W.; Zhang, X.Y.; Kang, S.X.; Gao, Y.M. Tri(*t*-butyl)phosphine-assisted selective hydrosilylation of terminal alkynes. *Chin. Chem. Lett.* **2010**, *21*, 312–316. [CrossRef]
10. Dierick, S.; Vercruysse, E.; Berthon-Gelloz, G.; Marko, I.E. User-Friendly Platinum Catalysts for the Highly Stereoselective Hydrosilylation of Alkynes and Alkenes. *Chem. Eur. J.* **2015**, *21*, 17073–17078. [CrossRef] [PubMed]
11. Mutoh, Y.; Mohara, Y.; Saito, S. (Z)-Selective Hydrosilylation of Terminal Alkynes with HSiMe(OSiMe$_3$)$_2$ Catalyzed by Ruthenium Complex Containing an N-Heterocyclic Carbene. *Org. Lett.* **2017**, *19*, 5204–5207. [CrossRef] [PubMed]
12. Chaulagain, M.R.; Mahandru, G.M.; Montgomery, J. Alkyne hydrosilylation catalysed by nickel complexes of N-heterocyclic carbenes. *Tetrahedron* **2006**, *62*, 7560–7566. [CrossRef]
13. Field, L.D.; Ward, A.J. Catalytic hydrosilylation of acetylenes mediated by phosphine complexes of cobalt (I), rhodium(I) and iridium(I). *J. Organomet. Chem.* **2003**, *681*, 91–97. [CrossRef]
14. Lewis, N.L.; Sy, K.G.; Bryant, G.L.; Donahue, P.E. Platinum-catalyzed hydrosilylation of alkynes. *Organometallics* **1991**, *10*, 3750–3759. [CrossRef]
15. Takeuchi, R.; Nitta, S.; Watanabe, D. Cationic Rhodium Complex-catalysed Highly Selective Hydrosilylation of Propynylic Alcohols: A Convenient Synthesis of (a-y-Silyl Allylic Alcohols. *J. Chem. Soc. Chem. Commun.* **1994**, 1777–1778. [CrossRef]
16. Takeuchi, R.; Tanouchi, N. Complete reversal of stereoselectivity in rhodium complex-catalysed hydrosilylation of alk-1-yne. *J. Chem. Soc. Chem. Commun.* **1993**, 1319–1320. [CrossRef]
17. Sato, A.; Kinoshita, H.; Shinokubo, H.; Oshima, K. Hydrosilylation of Alkynes with a Cationic Rhodium Species Formed in an Anionic Micellar System. *Org. Lett.* **2004**, *6*, 2217–2220. [CrossRef] [PubMed]
18. Hamze, A.; Provot, O.; Brion, J.D.; Alami, M. Xphos ligand and platinum catalysts: A versatile catalyst for synthesis of functionalized β(E)-vinylsilanes for terminal alkynes. *J. Org. Chem.* **2008**, *693*, 2789–2797. [CrossRef]
19. Mori, A.; Takahisa, E.; Kajiro, H.; Nishihara, Y.; Hiyama, T. Stereodivergent hydrosilylation of 1-alkynes catalysed by RhI(PPh$_3$)$_3$ leading to (E)- and (Z)-alkenylsilanes and the application to polymer. *Polyhedron* **2000**, *19*, 567–568. [CrossRef]
20. Ojima, I.; Clos, N.; Donovan, R.J.; Ingallina, P. Hydrosilylation of 1-hexyne catalyzed by rhodium and cobalt-rhodium mixed-metal complexes. Mechanism of apparent trans addition. *Organometallics* **1990**, *9*, 3127–3133. [CrossRef]
21. Ojima, I.; Kumagai, M.; Nagai, Y. The stereochemistry of the addition of hydrosilanes to alkyl acetylenes catalyzed by tris(triphenylphosphine)-chlororhodium. *J. Organomet. Chem.* **1974**, *66*, C14–C16. [CrossRef]
22. Jun, C.H.; Crabtree, R.H. Dehydrogenative Silation, Isomerization and the Control of Syn- vs. Antiaddition in the Hydrosilation of Alkynes. *J. Organomet. Chem.* **1993**, *447*, 177–187. [CrossRef]
23. Jimenes, M.V.; Perez-Torrente, J.J.; Bartolome, M.I.; Gierz, V.; Lahoz, F.J.; Oro, L.A. Rhodium(I) Complexes with Hemilabile N-heterocyclic Carbenes: Efficient Alkyne Hydrosilylation Catalysts. *Organometallics* **2008**, *27*, 224–234. [CrossRef]

24. Andavan, G.T.S.; Bauer, E.B.; Letko, C.S.; Hollis, T.K.; Tham, F.S. Synthesis and Characterization of a Free Phenylene Bis(N-heterocyclic Carbene) and Its Di-Rh Complex: Catalytic Activity of the Di-Rh and CCC_NHC Rh Pincer Complexes in Intermolecular Hydrosilylation of Alkynes. *J. Organomet. Chem.* **2005**, *690*, 5938–5947. [CrossRef]
25. Iglesias, M.; Sanz Miguel, P.J.; Polo, V.; Fernandez-Alvarez, F.J.; Perez-Torrente, J.J.; Oro, L.A. An Alternative Mechanistic Paradigm for the β-Z Hydrosilylation of Terminal Alkynes: The Role of Acetone as a Silane Shuttle. *Chem. Eur. J.* **2013**, *19*, 17559–17566. [CrossRef] [PubMed]
26. Busetto, L.; Cassani, M.C.; Femoni, C.; Mancinelli, M.; Mazzanti, A.; Mazzoni, R.; Solinas, G. N-Heterocyclic Carbene-Amide Rhodium(I) Complexes: Structures, Dynamics, and Catalysis. *Organometallics* **2011**, *30*, 5258–5272. [CrossRef]
27. De Bo, G.; Berthon-Gelloz, G.; Tinant, B.; István, E.; Markó, I.E. Hydrosilylation of Alkynes Mediated by N-Heterocyclic Carbene Platinum(0) Complexes. *Organometallics* **2006**, *25*, 1881–1890. [CrossRef]
28. Maciejewski, H.; Szubert, K.; Marciniec, B.; Pernak, J. Hydrosilylation of functionalised olefins catalysed by rhodium siloxide complexes in ionic liquids. *Green Chem.* **2009**, *11*, 1045–1051. [CrossRef]
29. Maciejewski, H.; Wawrzynczak, A.; Dutkiewicz, M.; Fiedorow, R. Silicone waxes—synthesis via hydrosilylation in homo- and heterogeneous systems. *J. Mol. Catal. Chem.* **2006**, *257*, 141–148. [CrossRef]
30. Zielinski, W.; Kukawka, R.; Maciejewski, H.; Smiglak, M. Ionic Liquids as Solvents for Rhodium and Platinum Catalysts Used in Hydrosilylation Reaction. *Molecules* **2016**, *21*, 1115. [CrossRef]
31. Jankowska-Wajda, M.; Kukawka, R.; Smiglak, M.; Maciejewski, H. The effect of the catalyst and the type of ionic liquid on the hydrosilylation process under batch and continuous reaction conditions. *New J. Chem.* **2018**, *42*, 5229–5236. [CrossRef]
32. Maciejewski, H.; Jankowska-Wajda, M.; Dabek, I.; Fiedorow, R. The effect of the morpholinium ionic liquid anion on the catalytic activity of Rh (or Pt) complex–ionic liquid systems in hydrosilylation processes. *RSC Adv.* **2018**, *8*, 26922–26927.
33. Pernak, J.; Swierczynska, A.; Kot, M.; Walkiewicz, F.; Maciejewski, H. Pyrylium sulfonate based ionic liquids. *Tetrahedron Lett.* **2011**, *52*, 4342–4345. [CrossRef]
34. Maciejewski, H.; Szubert, K.; Fiedorow, R.; Giszter, R.; Niemczak, M.; Pernak, J.; Klimas, W. Diallyldimethylammonium and trimethylvinylammonium ionic liquids—Synthesis and application to catalysis. *Appl. Catal.* **2013**, *451*, 168–175. [CrossRef]
35. Luska, K.L.; Demmans, K.Z.; Stratton, S.A.; Moores, A. Rhodium complexes stabilized by phosphine-functionalized phosphonium ionic liquids used as higher alkene hydroformylation catalysts: Influence of the phosphonium headgroup on catalytic activity. *Dalton Trans.* **2012**, *41*, 13533–13540. [CrossRef] [PubMed]
36. Jin, X.; Xu, X.-f.; Zhao, K. Amino acid- and imidazolium-tagged chiral pyrrolidinodiphosphine ligands and their applications in catalytic asymmetric hydrogenations in ionic liquid systems. *Tetrahedron Asymm.* **2012**, *23*, 1058–1067. [CrossRef]
37. Jankowska-Wajda, M.; Bartlewicz, O.; Szpecht, A.; Zając, A.; Śmiglak, M.; Maciejewski, H. Platinum and rhodium complexes ligated by imidazolium- substituted phosphine as the efficient and recyclable catalysts for hydrosilylation. *RSC Adv.* **2019**, *9*, 29396–29404. [CrossRef]
38. Stefanowska, K.; Franczyk, A.; Szyling, J.; Salamon, K.; Marciniec, B.; Walkowiak, J. An effective hydrosilylation of alkenyles in supracritical CO_2—A green approach to alkenyl silanes. *J. Catal.* **2017**, *356*, 206–213. [CrossRef]
39. Berthon-Gelloz, G.; Schumers, J.M.; De Bo, G.; Marko, I.E. Highly β-(E)-Selective Hydrosilylation of Terminal and Internal Alkynes Catalyzed by a (IPr)Pt(diene) Complex. *J. Org. Chem.* **2008**, *73*, 4190–4197. [CrossRef] [PubMed]
40. Cheng, C.; Simmons, E.M.; Hartwig, J.F. Iridium-Catalyzed, Diastereoselective Dehydrogenative Silylation of Terminal Alkenes with $(TMSO)_2MeSiH$. *Angew. Chem.* **2013**, *125*, 9154–9159. [CrossRef]
41. Bokka, A.; Jeon, J. Regio- and Stereoselective Dehydrogenative Silylation and Hydrosilylation of Vinylarenes Catalyzed by Ruthenium Alkylidenes. *Org. Lett.* **2016**, *18*, 5324–5327. [CrossRef]
42. Zhao, X.; Yang, D.; Zhang, Y.; Wang, B.; Qu, J. Highly β(Z)-Selective Hydrosilylation of Terminal Alkynes Catalyzed by Thiolate-Bridged Dirhodium Complexes. *Org. Lett.* **2018**, *20*, 5357–5361. [CrossRef]

43. Faller, J.W.; D'Alliessi, D.G. Tunable Stereoselective Hydrosilylation of PhC≡CH Catalyzed by Cp*Rh Complexes. *Organometallics* **2002**, *21*, 1743–1746. [CrossRef]
44. Li, J.; Peng, J.; Bai, Y.; Zhang, G.; Lai, G.; Li, X. Phosphines with 2-imidazolium ligands enhance the catalytic activity and selectivity of rhodium complexes for hydrosilylation reactions. *J. Org. Chem.* **2010**, *695*, 431–436. [CrossRef]
45. Chen, S.J.; Wang, Y.Y.; Yao, W.M.; Zhao, X.L.; Thanh, G.V. An ionic phosphine-ligated rhodium (III) complexes as the efficient and recyclable catalyst for biphasic hydroformylation of 1-octene. *J. Mol. Catal.* **2013**, *378*, 293–298. [CrossRef]
46. Consorti, C.S.; Aydos, G.L.P.; Ebeling, G.; Dupont, J. Ionophilic phosphines: Versatile ligands for ionic liquid biphasic catalysis. *Org. Lett.* **2008**, *10*, 237–240. [CrossRef] [PubMed]

© 2020 by the authors. Licensee MDPI, Basel, Switzerland. This article is an open access article distributed under the terms and conditions of the Creative Commons Attribution (CC BY) license (http://creativecommons.org/licenses/by/4.0/).

Article

Piperidinium and Pyrrolidinium Ionic Liquids as Precursors in the Synthesis of New Platinum Catalysts for Hydrosilylation

Magdalena Jankowska-Wajda [1,*], Olga Bartlewicz [1], Przemysław Pietras [2] and Hieronim Maciejewski [1,2]

1. Faculty of Chemistry, Adam Mickiewicz University, Uniwersytetu Poznańskiego 8, 61-614 Poznań, Poland; olga.bartlewicz@amu.edu.pl (O.B.); hieronim.maciejewski@amu.edu.pl (H.M.)
2. Adam Mickiewicz University Foundation, Poznań Science and Technology Park, Rubież 46, 61-612 Poznań, Poland; przemyslaw.pietras@ppnt.poznan.pl
* Correspondence: magdajw@amu.edu.pl

Received: 11 July 2020; Accepted: 7 August 2020; Published: 10 August 2020

Abstract: Six new air-stable anionic platinum complexes were synthesized in simple reactions of piperidinium [BMPip]Cl or pyrrolidinium [BMPyrr]Cl ionic liquids with platinum compounds ([Pt(cod)Cl$_2$] or K$_2$[PtCl$_6$]). All these compounds were subjected to isolation and spectrometric characterization using NMR and ESI-MS techniques. Furthermore, the determination of melting points and thermal stability of the above derivatives was performed with the use of thermogravimetric analysis. The catalytic performance of the synthesized complexes was tested in hydrosilylation of 1-octene and allyl glycidyl ether with 1,1,1,3,5,5,5-heptamethyltrisiloxane. The study has shown that they have high catalytic activity and are insoluble in the reaction medium which enabled them to isolate and reuse them in consecutive catalytic cycles. The most active complex [BMPip]$_2$[PtCl$_6$] makes it possible to conduct at least 10 catalytic runs without losing activity which makes it an attractive alternative not only to commonly used homogeneous catalysts, but also to heterogeneous catalysts for hydrosilylation processes. The activity of the studied catalysts is also affected by the kind of anion and, to some extent, the kind of cation.

Keywords: ionic liquids; biphasic catalysis; platinum complexes; hydrosilylation

1. Introduction

Environmental, economic and technological reasons prompt researchers to pay more and more attention to planning new paths of synthesis of chemical compounds and continuing work on already known reactions and processes successfully employed in the industry, which can be optimized in the aspects of the improvement in yield, reduction in produced waste and possibility of reusing catalysts. Attempts are made at reaching the equilibrium between the costs of conducting a process, its yield, and the environmental impact. All these aspects are in line with the premises of "green chemistry" [1]. One of the processes commonly used on the industrial scale and being the main way of synthesis of organosilicon compounds is hydrosilylation [2–4]. Catalysts for this process are most often transition-metal-based systems of which the most important are platinum complexes, particularly Karstedt ([Pt$_2${H$_2$C=CHSiMe$_2$}$_2$O]$_3$) and Speier (H$_2$PtCl$_6$/i-PrOH) catalysts [5]. Taking into consideration high price of platinum and the necessity to separate the catalyst from the postreaction mixture (because the presence of heavy metals, even in trace amounts, is impermissible in many applications of the reaction product), efforts are made to employ catalytic systems based on other metals or to heterogenize the most active platinum complexes [5,6].

In recent years, a significant role in the catalytic processes has been played by ionic liquids which can serve as solvents, immobilizing agents for metal complexes, cocatalysts, and catalysts [7–12].

The ionic liquids have been employed in many reactions, one of which is hydrosilylation. In most cases of the latter reaction, ionic liquids dissolved and immobilized metal complexes (mainly those of platinum and rhodium), and formed biphasic systems with the reagents. This role of ionic liquids enabled easy isolation of a catalyst dissolved in them and its reuse. The most often applied ionic liquids were imidazolium [13–17], phosphonium [18,19], ammonium [20], pyrylium [21], and morpholinium [22] ones.

The ionic liquid can also be the element of a complex compound structure which is exemplified by the employment of ionic liquids as substituents in ligands (most often phosphine ones) which at a further stage serve for metal complexation [23–25]. Platinum complexes of this type were applied in reactions of hydrosilylation of alkenes [26] and alkynes [27]. Ionic liquids that contain metal atoms in their structure are another possibility of this kind of application. This group of derivatives is exemplified by halometallate ionic liquids which can be prepared easily by reacting metal halide with organic halide [28–33]. The first and prevalent derivatives have been chloroaluminate ionic liquids [34], albeit currently ionic liquids containing Co, Ir, Au, Ni, Pd, and Pt are also known [28–33]. The platinates known hitherto were solely imidazole derivatives: $[EMIM]_2[PtCl_4]$, $[EMIM]_2[PtCl_6]$, and $[BMIM]_2[PtCl_4]$, $[BMIM]_2[PtCl_6]$ [35,36]. Recently, our research group obtained next platinates with imidazole and pyridine derivatives and applied them as hydrosilylation catalysts [37,38]. It was the first report on the employment of platinum anionic complexes of this type in the hydrosilylation process. All the complexes have shown high catalytic activity and insolubility in the reagents which enabled them to separate and recycle them.

The simple method of synthesis, high stability, and high activity of the catalysts, as well as the possibility to recycle the latter, which translates into economic and ecological effects, have inspired us to continue studies of this subject and obtain new derivatives. This work was aimed at obtaining new platinates by reactions of platinum salts and platinum chloride complexes with pyrrolidine and piperidine derivatives, their isolation and spectroscopic characterization, as well as the determination of their catalytic activity for hydrosilylation. For the synthesis of the platinates, we have chosen pyrrolidine and piperidine derivatives to compare their properties with those of imidazole and pyridine derivatives obtained earlier [37].

2. Results and Discussion

To synthesize platinum-containing complexes, two platinum precursors, $[Pt(cod)Cl_2]$ and $K_2[PtCl_6]$, as well as derivatives of piperidine (1-butyl-1-methyl piperidinium chloride, [BMPip]Cl) and pyrrolidine (1-butyl-1-methylpyrrolidinium chloride, [BMPyrr]Cl) have been used. In the starting precursors, platinum was present at different oxidation states which made it possible to obtain tetrachloroplatinates and hexachloroplatinates. In the case of the synthesis of tetrachloroplatinates, depending on the amount of the precursor and ionic liquid, complexes were formed with a different form of the anion. When equimolar amounts of the precursor and ionic liquid were used, the complex with anion in the form of the dimer was obtained, whereas in the case of using two-fold excess of ionic liquid the complex with anion in the monomeric form was created. The syntheses of six new platinum complexes were conducted according to Scheme 1.

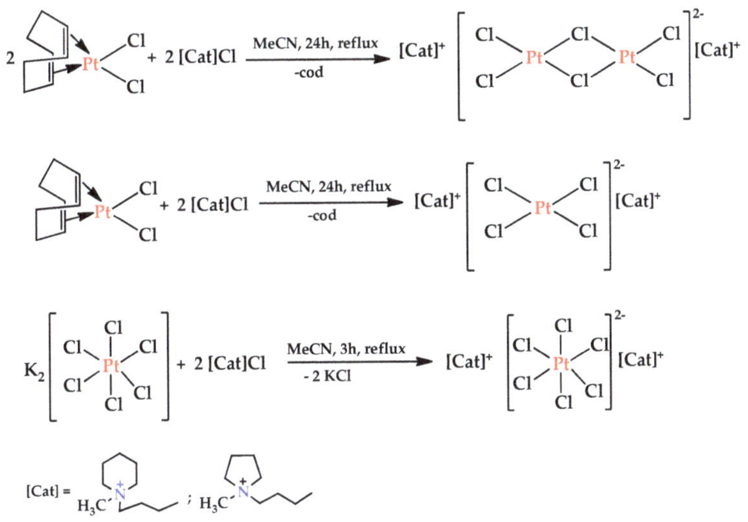

Scheme 1. The methods of the synthesis of tetrachloroplatinate, hexachlorodiplatinate, and hexachloroplatinate complexes applied in the study.

The synthesis of the above complexes is very simple and consists of the dissolution of platinum precursor and ionic liquid in acetonitrile followed by stirring under reflux for several hours. In the case of the reaction with [Pt(cod)Cl$_2$], after cooling down, the solvent was evaporated together with cyclooctadiene, whereas in the case of the reaction with K$_2$[PtCl$_6$], the filtration of the precipitated KCl preceded the solvent evaporation. All the complexes were obtained with very high yields ranging from 89 to 96%. These compounds are stable in air and no special storage conditions are required.

The obtained complexes were subjected to characterization by ^{1}H and ^{13}C NMR spectroscopy and ESI-MS mass spectrometry. The spectra of the complexes were compared with the spectra of starting reagents. Noteworthy differences were found in the values of chemical shifts of the signals originated from methyl group and CH$_2$ groups bound directly to the nitrogen present in piperidinium and pyrrolidinium cations in the starting ionic liquids compared to tetrachloroplatinate, hexachlorodiplatinate, and hexachloroplatinate complexes with the same cations. The differences in chemical shifts observed for tetrachloroplatinate and hexachloroplatinate complexes range from 0.4 to 0.6 ppm, whereas for hexachlorodiplatinate ones from 0.4 to 0.8 ppm and depend on the kind of cation. The presence of a hydrogen bond had no significant effect on chemical shift values in the ^{13}C NMR spectra.

In the ESI-MS spectra, MS(+) signals from cation and MS(−) ones from chloroplatinate anion were observed. In the MS(+) spectra of all complexes, very intense signals corresponding to molecular peaks of cations: m/z 156.17 [BMPip]$^+$ and 142.13 [BMPyrr]$^+$ were present. The MS(−) spectra of tetrachloroplatinate compounds contain multiplets resulting from the presence of three platinum isotopes. The most intense signal was at m/z 335.84 originating from [PtCl$_4$]$^{2-}$. Signals indicating the presence of [PtCl$_2$]$^{2-}$, m/z 265.15 and m/z 300.89, corresponding to [PtCl$_3$]$^-$/[Pt$_2$Cl$_6$]$^{2-}$, were also observed. On the other hand, in the case of hexachloroplatinate compounds, signals originating from cations: 156.17 [BMPip]$^+$ and 142.13 [BMPyrr]$^+$ were seen. In the MS(−) spectra, characteristic multiplets were visible which originated from platinum isotopes: the most intense of them at m/z 265.15 corresponded to [PtCl$_2$]$^{2-}$, another one at m/z 300.98 was ascribed to [PtCl$_3$]$^-$/[Pt$_2$Cl$_6$]$^{2-}$, and the signal at m/z 603.47, which is characteristic of the dimeric complex, originated from [Pt$_2$Cl$_6$]$^{2-}$.

For all obtained complexes, melting points were measured and the results are shown in Table 1.

Table 1. Melting points of the synthesized chloroplatinate complexes.

Catalyst	Melting Point [°C]
[BMPip]$_2$ [PtCl$_4$]	151
[BMPyrr]$_2$ [PtCl$_4$]	134
[BMPip]$_2$ [PtCl$_6$]	178
[BMPyrr]$_2$ [PtCl$_6$]	169
[BMPip]$_2$ [Pt$_2$Cl$_6$]	189
[BMPyrr]$_2$ [Pt$_2$Cl$_6$]	165

The obtained results permit us to conclude that from among complexes with the same cation, hexachloroplatinate complexes have higher melting points than tetrachloroplatinate ones.

Melting points of complexes with anion in the dimeric form are close to or higher than those of hexachloroplatinate complexes. According to the literature, complexes of higher anion symmetry have higher melting points [36] and hexachloroplatinates are characterized by a higher symmetry (O_h) compared to tetrachloroplatinates (D_{4h} symmetry). In the case when complexes with the same anion are compared, one can note that complexes with the [BMPip] cation have higher melting points than complexes with the [BMPyrr] cation.

The thermal stability of the studied compounds was determined by conducting thermogravimetric analysis (TGA) and the results are presented in Figure 1. The temperature at which 10% weight loss occurred was taken as the decomposition temperature (Table 2). The above weight loss value has been chosen to distinguish the decomposition temperature from water desorption temperature. The results listed in Table 2 made it possible to establish that hexachloroplatinate and hexachlorodiplatinate complexes are characterized by higher decomposition temperatures than tetrachloroplatinate complexes. From among the studied complexes, the highest decomposition temperature was found for hexachloroplatinate complex with pyrrolidinium cation, albeit all the temperatures were fairly close one to another and exceeded 200 °C which permits to classify all the complexes as thermally stable. This is crucial from the viewpoint of their application in the catalytic processes conducted at elevated temperatures.

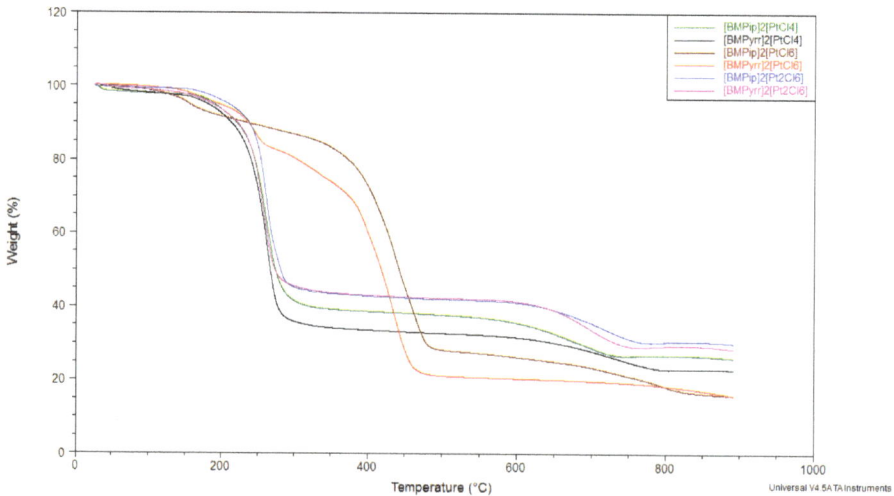

Figure 1. Thermogravimetric curves of chloroplatinate complexes.

Table 2. Decomposition temperatures of the studied compounds at 10% of weight loss.

Catalyst	Decomposition Temperature [°C]
[BMPip]$_2$ [PtCl$_4$]	219.84
[BMPyrr]$_2$ [PtCl$_4$]	212.95
[BMPip]$_2$ [PtCl$_6$]	225.88
[BMPyrr]$_2$ [PtCl$_6$]	233.07
[BMPip]$_2$ [Pt$_2$Cl$_6$]	220.69
[BMPyrr]$_2$ [Pt$_2$Cl$_6$]	237.40

The essential stage of the study was the application of the prepared complexes as catalysts for hydrosilylation. The activity and selectivity of the developed catalysts were evaluated in the model reactions of hydrosilylation of 1-octene and allyl glycidyl ether with 1,1,1,3,5,5,5-heptamethyltrisiloxane (HMTS). Based on the results of our earlier studies, we conducted the catalytic measurements in analogous conditions, i.e., at 110 °C for 1 h [37]. After the reaction completion, the composition of the postreaction mixture was analyzed using GC techniques. The analyses have shown the formation of β-addition products only (in compliance with Scheme 2). No presence of other products, e.g., α-addition or the competitive reaction of olefin isomerization, was observed.

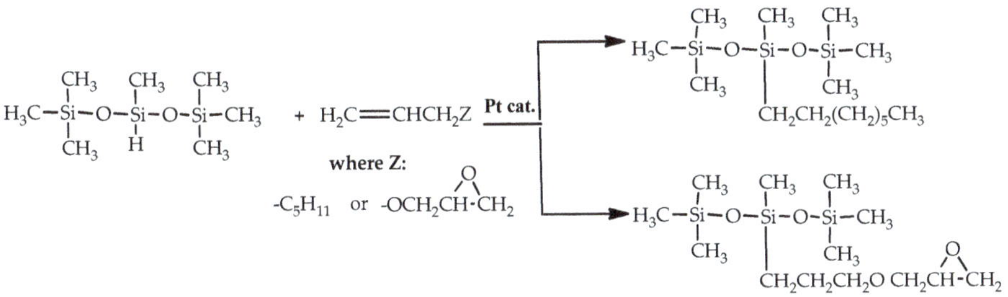

Scheme 2. The model reaction of hydrosilylation of 1-octene/allyl glycidyl ether with 1,1,1,3,5,5,5-heptamethyltrisiloxane.

The product yields in the reactions of hydrosilylation of 1-octene and allyl glycidyl ether, catalyzed by the studied platinum complexes are presented in Table 3.

Table 3. The product yields in the reactions of hydrosilylation of 1-octene and allyl glycidyl ether with 1,1,1,3,5,5,5-heptamethyltrisiloxane, catalyzed by platinum complexes.

Catalyst	Product Yield in the Reaction with	
	1-octene [1] [%]	Allyl Glycidyl Ether [2] [%]
[BMPip]$_2$[PtCl$_4$]	94	96
[BMPip]$_2$[PtCl$_6$]	99	94
[BMPip]$_2$[Pt$_2$Cl$_6$]	94	99
[BMPyrr]$_2$[PtCl$_4$]	93	85
[BMPyrr]$_2$[PtCl$_6$]	95	89
[BMPyrr]$_2$[Pt$_2$Cl$_6$]	93	98

[1] [HSi≡]:[-CH=CH$_2$]:[cat] = 1:1:10^{-4}; T = 110 °C; t = 1 h, [2] [HSi≡]:[-CH = CH$_2$]:[cat] = 1:1.2:10^{-4}; T = 110 °C; t = 1 h.

Based on the obtained results one can say that all the complexes have shown high catalytic activity and enabled them to obtain a product with high yield. In the case of the reaction with 1-octene, yields obtained in the presence of respective complexes were very similar, whereas, in that of the reaction with allyl glycidyl ether catalyzed by complexes with piperidinium cation, the yield was a bit higher. However, also in the latter case, the differences were small.

All employed complexes are insoluble in reagents which allows their isolation from postreaction mixtures and their reuse in subsequent reaction cycles after adding a new portion of reactants.

The yields obtained in 10 subsequent cycles of the reaction of 1-octene hydrosilylation, catalyzed by the same portion of catalyst, are shown in Figure 2 and Table 4.

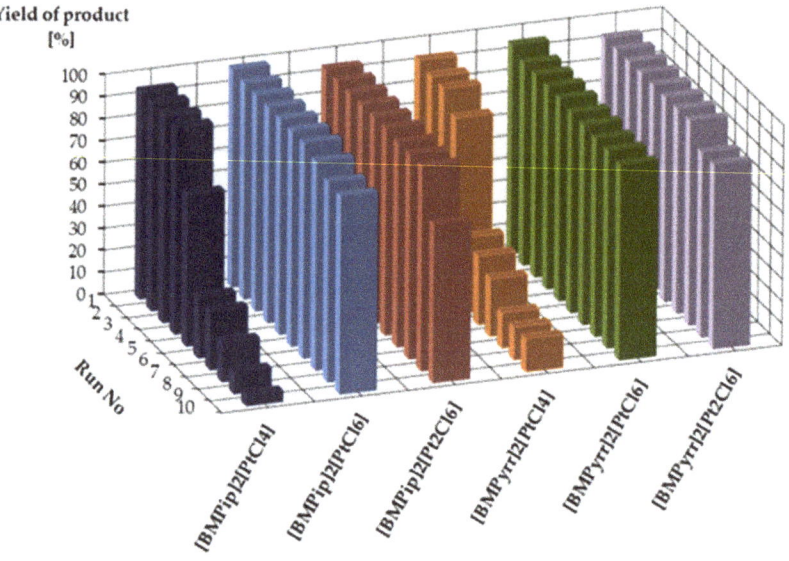

Figure 2. Yields of the product of hydrosilylation of 1-octene with 1,1,1,3,5,5,5-heptamethyltrisiloxane as determined for 10 subsequent reaction cycles catalyzed by the same catalyst portion.

Table 4. Product yields and TON values for hydrosilylation of 1-octene with heptamethyltrisiloxane (HMTS) catalyzed by anionic platinum complexes.

Catalyst	Yield of Product in Subsequent Cycle [%]	Total TON
[BMPip]$_2$[PtCl$_4$]	94 (94, 94, 94, 69, 28, 27, 18, 10, 5)	53,300
[BMPip]$_2$[PtCl$_6$]	99 (98, 98, 98, 98, 98, 98, 95, 91, 90)	96,300
[BMPip]$_2$[Pt$_2$Cl$_6$]	94 (94, 94, 94, 94, 94, 93, 93, 93, 72)	91,500
[BMPyrr]$_2$[PtCl$_4$]	93 (91, 91, 83, 34, 31, 26, 16, 15, 15)	49,500
[BMPyrr]$_2$[PtCl$_6$]	95 (92, 92, 92, 92, 91, 91, 90, 90, 89)	91,400
[BMPyrr]$_2$[Pt$_2$Cl$_6$]	93 (93, 93, 93, 93, 93, 93, 93, 84, 84)	91,200

[HSi≡]:[CH=CH]:[cat] = 1:1:10^{-4}; T = 110 °C; t = 1 h.

The obtained results point to significant differences in the activity of the studied platinum-containing catalysts and to a significant influence of the kind of anion on the catalytic activity observed in subsequent reaction cycles. Figure 2 clearly shows that complexes with [PtCl$_6$]$^{2-}$ and [Pt$_2$Cl$_6$]$^{2-}$ anions (particularly with the former one) show the highest stability and reproducibility. The mentioned complexes are permitted to obtain the product with very high yields in all 10 cycles. The catalytic activity can be easily compared by calculating TON values which are presented in Table 4. It is worth mentioning that the TON values were calculated (for the sake of comparison) for 10 conducted reaction cycles only, although the activity of some complexes was still very high, hence they could be employed in further cycles. Based on the obtained results, one can say that in this case, the effect of the cation is small, albeit the complexes with piperidinium cation have slightly higher activity. From among all complexes studied, the complex [BMPip]$_2$[PtCl$_6$] was the most active. To confirm the reaction course in subsequent cycles, we studied the reaction using an in situ FTIR probe that made it possible to follow the reaction course in real-time. In the above study, we tracked the decline in the band characteristic of the ≡SiH group. Due to the selective formation of only one product, as shown by chromatograhic analysis of the postreaction mixture, the measured conversion well correlates with the values of product yield determined by chromatographic methods. The obtained ≡SiH conversions in the hydrosilylation reaction catalyzed by the same portion of the [BMPip]$_2$[PtCl$_6$] complex in subsequent cycles are presented in Figure 3.

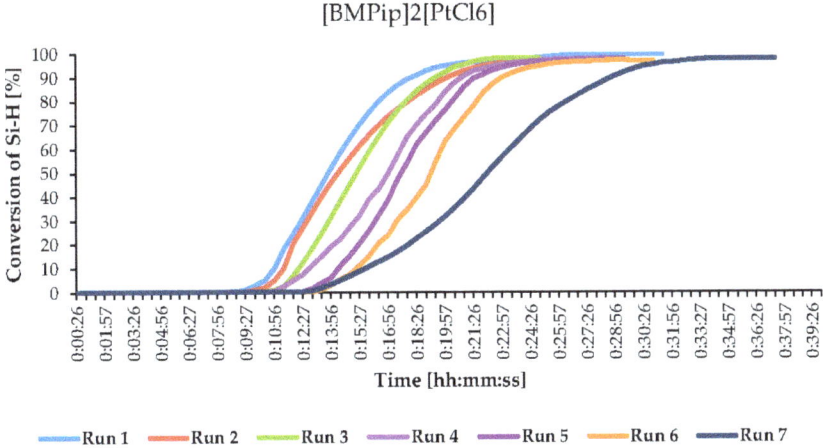

Figure 3. The change in the conversion of ≡SiH as a function of time for the reaction catalyzed by [BMPip]$_2$[PtCl$_6$].

The measurements were conducted for seven cycles and it was found that in each cycle (from the time of the reaction initiation to that at which the final conversion was reached), the reaction course was very fast and lasted from 8 to 10 min. It was also noticed that the inductive period became longer in subsequent cycles, but the reaction profiles were similar and the conversions were on the same level. Analogous measurements were carried out for the reaction catalyzed by the complex [BMPyrr]$_2$[PtCl$_4$] whose activity was the lowest. The obtained results are presented in Figure 4.

In this case, the catalytic activity decreased in subsequent cycles. Although in the first cycle the reaction proceeds very fast (about 10 min), in the further cycles (until reaching the final conversion) the reaction time becomes longer and longer. Moreover, the activity considerably decreases after the fourth cycle (this was also noticed in the results of chromatographic analysis) and the conversion reaches the value of about 30%.

The second reaction studied was hydrosilylation of allyl glycidyl ether in which the activity of all complexes in the first cycle was high (Table 3). This is why we isolated the catalysts and used them in subsequent catalytic cycles. The obtained results are presented in Figure 5 and Table 5. In this case, the results were even more diversified and the highest stability and reproducibility were found for the complexes with a hexachloroplatinic anion.

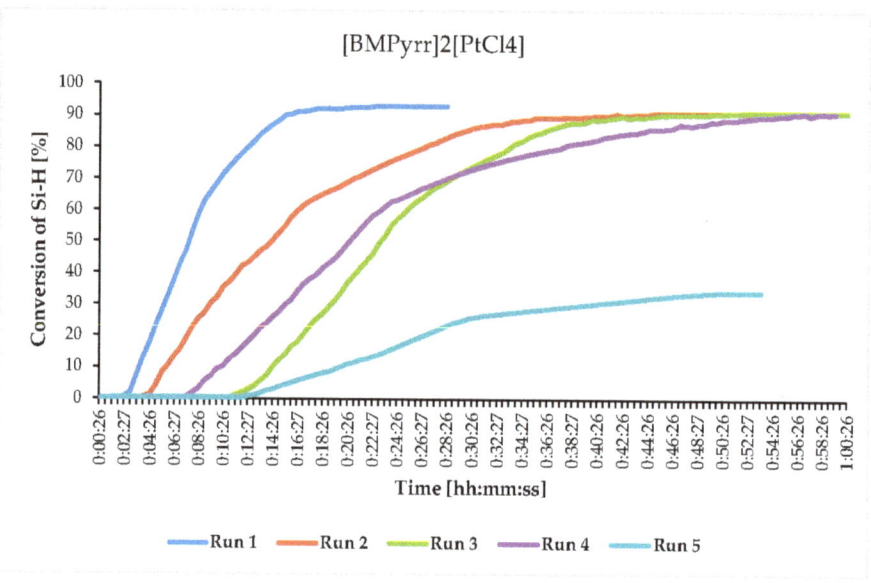

Figure 4. The change in the conversion of ≡SiH as a function of time for the reaction catalyzed by [BMPyrr]$_2$[PtCl$_4$].

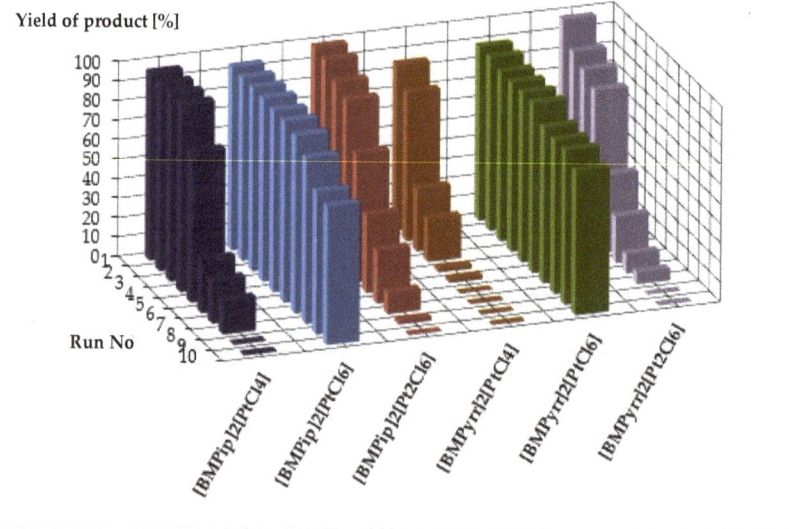

Figure 5. Yields of the product of hydrosilylation of allyl glycidyl ether with 1,1,1,3,5,5,5-heptamethyltrisiloxane as determined for 10 subsequent reaction cycles catalyzed by the same catalyst portion.

Table 5. Product yields and TON values for hydrosilylation of allyl glycidyl ether with HMTS, catalyzed by anionic platinum complexes.

Catalyst	Yield of Product in Subsequent Cycle [%]	Total TON
[BMPip]$_2$[PtCl$_4$]	94 (94, 94, 94, 69, 28, 27, 18, 10, 5)	53,300
[BMPip]$_2$[PtCl$_6$]	99 (98, 98, 98, 98, 98, 98, 95, 91, 90)	96,300
[BMPip]$_2$[Pt$_2$Cl$_6$]	94 (94, 94, 94, 94, 94, 93, 93, 93, 72)	91,500
[BMPyrr]$_2$[PtCl$_4$]	93 (91, 91, 83, 34, 31, 26, 16, 15, 15)	49,500
[BMPyrr]$_2$[PtCl$_6$]	95 (92, 92, 92, 92, 91, 91, 90, 90, 89)	91,400
[BMPyrr]$_2$[Pt$_2$Cl$_6$]	93 (93, 93, 93, 93, 93, 93, 93, 84, 84)	91,200

[HSi≡]:[CH=CH]:[cat] = 1:1.2·10^{-4}; T = 110 °C; t = 1 h.

Additionally, the effect of the cation is more visible because the activity of complexes with piperidinium cation are higher than those with pyrrolidinium cation. This is confirmed by TON values shown in Table 5.

The poorest catalyst turned out to be, also in this case, the complex [BMPyrr]$_2$[PtCl$_4$] whose activity began to decline already in the second cycle thus making it impossible to carry out 10 reaction cycles.

The decline in the activity of tetrachloroplatinic complexes in subsequent cycles of both reactions studied can be explained by gradual leaching of these complexes with new portions of reagents. To confirm this explanation we performed the ICP analysis of postreaction mixtures (obtained after the reaction with 1-octene) for the most stable complex [BMPip]$_2$[PtCl$_6$] and the least stable one [BMPyrr]$_2$[PtCl$_4$]. It has been found that in the case of the former complex, 0.22% of the initial platinum content was leached after the first cycle, whereas after further cycles the amount of leached platinum was below the detection limit. In the case of the latter complex, 2.7% of the initial platinum content was leached after the first cycle and in subsequent cycles the amounts of leached platinum were comparable. The complexes with hexachloroplatinic anions have stronger ionic character than those with tetrachloroplatinic anion, hence they are less susceptible to leaching with reagents. This is more visible in the case of the reaction with allyl glycidyl ether which is a polar compound, hence the decline in the catalytic activity is faster due to a reduction in catalyst concentration caused by stronger leaching. Taking into consideration the effect of the kind of cation, it is worth mentioning that both cations are heterocyclic derivatives with a six-membered ring in the case of piperidine and five-membered ring in that of pyrrolidine. The structure of the six-membered ring additionally stabilizes the complex and due to this, the performance of the complex with pyrrolidinium cation is more reproducible in the reaction with allyl glycidyl ether. In our earlier studies, we determined the activity of chloroplatinic complexes with cations being derivatives of imidazole and pyridine [37]. Considering analogous complexes with piperidinium [BMPip]$_2$[PtCl$_6$] and pyridinium [BMPy]$_2$[PtCl$_6$] cations and comparing TON values for the reaction with octene 96,300 and 95,200, respectively [37] and that with allyl glycidyl ether 88,200 and 60,300, respectively [37], one can additionally confirm that saturated six-membered ring better stabilizes the complex than its aromatic counterpart.

3. Materials and Methods

3.1. Materials

All reagents applied in catalytic measurements, i.e., 1-octene, allyl glycidyl ether, n-decane, and 1,1,1,3,5,5,5-heptamethyltrisiloxane were purchased from Sigma Aldrich (Poznań, Poland) and used as received. Additionally, metal precursors: [Pt(cod)Cl$_2$], K$_2$[PtCl$_6$] was supplied by Sigma Aldrich (Poznań, Poland). The ionic liquids: 1-Butyl-1-methylpyrrolidinium chloride [BMPyrr]Cl, 1-Butyl-1-methylpiperidinium chloride [BMPip]Cl, were purchased from Iolitec GmbH (Heilbronn, Germany).

3.2. Analytical Techniques

^{1}H NMR and ^{13}C NMR spectra were recorded in acetonitrile-d3 and chloroform-d, as a solvent, on a Varian 400 operating at 402.6 and 101.2 MHz, respectively. GC analyses were carried out on a Clarus 680 gas chromatograph (Perkin Elmer, Shelton, CT, USA) equipped with a 30 m capillary column Agilent VF-5ms (Santa Clara, CA, USA) and TCD detector, using the temperature program: 60 °C (3 min), 10 °C min^{-1}, 290 °C (5 min). ESI-MS spectra were recorded using a QTOF-type mass spectrometer (Impact HD, Bruker, Ettlingen, Germany). Fourier transform infrared FTIR spectra were recorded on a Bruker Tensor 27 Fourier (Billerica, MA, USA) transform spectrophotometer equipped with a SPECAC Golden Gate, diamond ATR unit with a resolution of 2 cm^{-1}. Thermogravimetric analysis (TGA) was carried out using a TA Instruments TG Q50 analyzer (New Castle, DE, USA) at a linear heating rate of 10 °C/min under synthetic air (50 mL/min). The tested samples were placed in a platinum pan and the weight of the samples was kept within 9–10 mg. The experimental error was 0.5% for weight and 1 °C for temperature. Melting points were measured on the Melting Point M-565 instrument (Buchi, Essen, Germany) equipped with a video camera. Temperature gradient: 10 °C/min. FTIR in situ measurements were performed using a Mettler Toledo ReactIR 15 instrument (Giessen, Germany). For selected samples, spectra were recorded with 256 scans for 1 h at 30 s intervals with the resolution of 1 cm^{-1}. Intensity change of the band at 913 cm^{-1}, characteristic of ≡Si–H bond, was recorded using an ATR probe with a diamond window. The ICP-MS analysis of postreaction samples was carried out on a Perkin Elmer Nexion 300D (Waltham, MA, USA) inductively coupled mass spectrometer.

3.3. Synthesis of Transition-Metal-Based Complexes

Synthesis of Bis(1-butyl-1-methylpiperidinium) tetrachloroplatinate(II), [BMPip]$_2$[PtCl$_4$]

A 25 mL high-pressure Schlenk vessel equipped with a magnetic stirring bar was charged 0.50 g (2.6 mmol) of [BMPip]Cl and 0.50 g (1.3 mmol) of [Pt(cod)Cl$_2$] in hot CH$_3$CN (2 mL). The reaction mixture was stirred for 24 h under reflux. After this time, the solution was cooled to room temperature and the solvent was evaporated. The white product was washed with diethyl ether (3 × 5 mL) and dried under vacuum. Product yield: 93% anal. calcd.
^{1}H NMR (ACN-d_6) δ (ppm): 3.21–3.16 (m, 12H, J = 7.24 Hz, N–CH$_2$), 2.88 (s, J = 7.14, 6H, N–CH$_3$), 1.90–1.85 (m, 12H, J = 7.04, –CH$_2$–), 1.60–1.56 (m, 4H, J = 7.08, –CH$_2$–), 1.32–1.26 (m, 4H, J = 7.01, –CH$_2$–), 0.91–0.88 (t, 6H, J = 7.13, –CH$_3$)
^{13}C NMR (ACN-d_6) δ (ppm): 64.79 (N–CH$_2$), 64.49 (N–CH$_2$), 48.88 (N–CH$_3$), 31.01 (–CH$_2$–), 25.97, 22.12, 19.72 (–CH$_2$–), 13.12 (–CH$_3$)
ESI-MS(+): 156.17 [BMPip]$^+$
ESI-MS(−): 265.15 [PtCl$_2$]$^{2-}$, 300.86 [PtCl$_3$]$^-$/[Pt$_2$Cl$_6$]$^{2-}$, 335.84 [PtCl$_4$]$^{2-}$

Synthesis of Bis(1-butyl-1-methylpyrrolidinium) tetrachloroplatinate(II), [BMPyrr]$_2$[PtCl$_4$]

A 25 mL high-pressure Schlenk vessel equipped with a magnetic stirring bar was charged 0.47 g (2.6 mmol) of [BMPyrr]Cl and 0.50 g (1.3 mmol) of [Pt(cod)Cl$_2$] in hot CH$_3$CN (2 mL). The reaction mixture was stirred for 24 h under reflux. After this time, the solution was cooled to room temperature and the solvent was evaporated. The white product was washed with diethyl ether (3 × 5 mL) and dried under vacuum. Product yield: 92% anal. calcd.
^{1}H NMR (ACN-d_6) δ (ppm): 3.49–3.44 (m, 8H J = 7.3, –N–CH$_2$), 3.31–3.27 (m, 4H, J = 7.4, –N–CH$_2$), 2.99 (s, 6H, J = 7.25, –N–CH$_3$), 1.99–1.96 (m, 8H, J = 7.21, –CH$_2$–), 1.76–1.73 (m, 4H, J = 7.18, –CH$_2$–), 1.42–1.38 (m, 4H, J = 7.15, –CH$_2$–), 1.01–0.98 (t, 6H, J = 7.13, –CH$_3$)
^{13}C NMR (ACN-d_6) δ (ppm): 64.79 (N–CH$_2$–); 48.63 (–N–CH$_3$); 31.09, 25.95, 21.87, 20.05 (–CH$_2$–) 13.50 (–CH$_3$)
ESI-MS(+): 142.13 [BMPyrr]$^+$
ESI-MS(−): 265.15 [PtCl$_2$]$^{2-}$, 300.98 [PtCl$_3$]$^-$/[Pt$_2$Cl$_6$]$^{2-}$, 335.86 [PtCl$_4$]$^{2-}$

Synthesis of Bis(1-butyl-1-methylpiperidinium) hexachloroplatinate(IV), [BMPip]$_2$[PtCl$_6$]

A 25 mL high-pressure Schlenk vessel equipped with a magnetic stirring bar was charged 0.37 g (2 mmol) of [BMPip]Cl and 0.53 g (1 mmol) of K$_2$[PtCl$_6$] in hot CH$_3$CN (2 mL) were added. The mixture was stirred for 3 h under reflux followed by filtration by a cannula system. The solvent was evaporated and dried under vacuum to yield orange product. Product yield: 91% anal. calcd.

^{1}H NMR (ACN-d_6) δ (ppm): 3.39–3.29 (m, 12H, J = 7.25 Hz, –N–CH$_2$), 3.02 (s, J = 7.11, 6H, –N–CH$_3$), 2.06–2.035 (m, 12H, J = 7.01, –CH$_2$–), 1.77–1.73 (m, 4H, J = 7.06, –CH$_2$–), 1.49–1.43 (m, 4H, J = 7.04, –CH$_2$–), 1.08–1.05 (t, 6H, J = 7.10, –CH$_3$)

^{13}C NMR (ACN-d_6) δ (ppm): 61.63 (N–CH$_2$), 48.18 (N–CH$_3$), 30.03 (–CH$_2$–), 23.93, 21.19, 20.14, 20.00 (–CH$_2$–), 13.26 (–CH$_3$)

Synthesis of Bis(1-butyl-1-methylpyrrolidinium) hexachloroplatinate(IV), [BMPyrr]$_2$[PtCl$_6$]

A 25 mL high-pressure Schlenk vessel equipped with a magnetic stirring bar 0.39 g (2 mmol) of [BMPyrr]Cl and 0.56 g (1.1 mmol) of K$_2$[PtCl$_6$] in hot CH$_3$CN (2 mL) were added. The mixture was stirred for 3 h under reflux followed by filtration by a cannula system. The solvent was evaporated and dried under vacuum to yield orange product. Product yield: 89% anal. calcd.

^{1}H NMR (ACN-d_6) δ (ppm): 3.44–3.43 (m, 8H J = 7.11, –N–CH$_2$), 3.28–3.24 (m, 4H, J = 7.38, –N–CH$_2$), 2.97 (s, 6H, J = 7.21, –N–CH$_3$), 1.80–1.70 (m, 8H, J = 7.19, –CH$_2$–), 1.44–1.30 (m, 4H, J = 7.17, –CH$_2$–), 1.17–1.13 (m, 4H, J = 7.12, –CH$_2$–), 1.01–0.97 (t, 6H, J = 7.09, –CH$_3$)

^{13}C NMR (ACN-d_6) δ (ppm): 64.79, 64.53 (N–CH$_2$–); 48.95 (–N–CH$_3$); 31.26, 25.87, 22.07, 19.78 (–CH$_2$–) 13.54 (–CH$_3$)

Synthesis of Bis(1-butyl-1-methylpiperidinium) hexachlorodiplatinate(II), [BMPip]$_2$[Pt$_2$Cl$_6$]

A 25 mL high-pressure Schlenk vessel equipped with a magnetic stirring bar was added 0.25 g (1.3 mmol) of [BMPip]Cl and 0.50 g (1.3 mmol) of [Pt(cod)Cl$_2$] in hot CH$_3$CN (2 mL). The mixture was stirred for 24 h under reflux. After this time, the solution was cooled to room temperature and the solvent was evaporated. The white product was washed with diethyl ether (3 × 5 mL) and dried under vacuum. Product yield: 93% anal. calcd.

^{1}H NMR (CDCl$_3$) δ (ppm): 3.87–3.81 (m, 4H, J = 7.21 Hz, –N–CH$_2$), 3.66–3.60 (m, 8H, J = 7.23 Hz, –N–CH$_2$), 3.37 (s, J = 7.14, 6H, –N–CH$_3$), 1.91–1.82 (m, 12H, J = 7.03, –CH$_2$–), 1.84–1.66 (m, 4H, J = 7.06, –CH$_2$–), 1.51–1.41 (m, 4H, J = 7.01, –CH$_2$–), 1.03–0.99 (t, 6H, J = 7.11, –CH$_3$)

^{13}C NMR (CDCl$_3$) δ (ppm): 60.58 (N–CH$_2$), 48.39 (N–CH$_3$), 30.67 (–CH$_2$–), 23.73, 20.52, 19.99, 19.56 (–CH$_2$–), 13.46 (–CH$_3$)

ESI-MS(+): 156.17 [BMPip]$^+$

ESI-MS(−): 265.15 [PtCl$_2$]$^{2-}$, 300.98 [PtCl$_3$]$^-$/[Pt$_2$Cl$_6$]$^{2-}$, 603.47 [Pt$_2$Cl$_6$]^2

Synthesis of Bis(1-butyl-1-methylpyrrolidinium) hexachlorodiplatinate(II), [BMPyrr]$_2$[Pt$_2$Cl$_6$]

A 25 mL high-pressure Schlenk vessel equipped with a magnetic stirring bar was added 0.24 g (1.3 mmol) of [BMPyrr]Cl and 0.50 g (1.3 mmol) of [Pt(cod)Cl$_2$] in hot CH$_3$CN (2 mL). The mixture was stirred for 24 h under reflux. After this time, the solution was cooled to room temperature and the solvent was evaporated. The white product was washed with diethyl ether (3 × 5 mL) and dried under vacuum. Product yield: 90% anal. calcd.

^{1}H NMR (CDCl$_3$) δ (ppm): 3.89–3.75 (m, 8H, J = 7.42, –N–CH$_2$), 3.63–3.59 (m, 4H, J = 7.24, –N–CH$_2$), 3.30 (s, 6H, J = 7.20, –N–CH$_3$), 1.81–1.53 (m, 12H, J = 7.18, –CH$_2$–), 1.51–1.44 (m, 4H, J = 7.16, –CH$_2$–), 1.05–1.01 (t, 6H, J = 7.09, –CH$_3$)

^{13}C NMR (CDCl$_3$) δ (ppm): 64.51 (N–CH$_2$–), 64.05 (–N–CH$_3$), 30.09, 26.09, 21.63, 19.74 (–CH$_2$–) 13.74 (–CH$_3$)

ESI-MS(+): 142.13 [BMPyrr]$^+$

ESI-MS(−): 265.15 [PtCl$_2$]$^{2-}$, 300.98 [PtCl$_3$]$^-$/[Pt$_2$Cl$_6$]$^{2-}$, 602.89 [Pt$_2$Cl$_6$]$^{2-}$

NMR and ESI-MS spectra of these complexes are presented in Supplementary Materials.

3.4. General Procedure for Catalytic Tests

To investigate the catalytic activity of platinum anionic complexes the hydrosilylation reactions of 1-octene or allyl glycidyl ether with 1,1,1,3,5,5,5-heptamethyltrisiloxane (HMTS) were carried out. The 5 mL glass reactor equipped with a reflux condenser was charged with 3.68 mmol of 1-octene or 4.41 mmol of allyl glycidyl ether and 3.68 mmol of HMTS. Then, 10^{-4} mol of Pt per 1 mol of Si–H was applied. As an internal standard 1 mmol of n-decane was added. The reaction was carried out in the presence of air at 110 °C for 1 h, without stirring. After each catalytic cycle, the reaction mixture was cooled down and subjected to GC analysis to determine the reaction yield. The product was isolated and subjected to NMR analyses. Due to a very small amount of catalyst, the products were entirely taken with a needle-equipped syringe after the reaction completion and a new portion of the reaction substrates was added to the reaction vessel, followed by conducting the reaction in the same way as described above. The catalyst remaining in the flask was not washed or regenerated in any way. The above operation was repeated 10 times.

3-octyl-1,1,1,3,5,5,5-heptamethyltrisiloxane:

^{1}H NMR (CDCl$_3$) ppm: 1.36–1.27 (m; 12H; CH$_2$–CH$_2$–CH$_2$); 0.9 (t; 3H; CH$_2$–CH$_3$), 0.48 (t; 2H; Si–CH$_2$), 0.11 (m, 18H, Si–(CH$_3$)$_3$), 0.02 (s, 3H Si–CH$_3$).
^{13}C NMR (CDCl$_3$)ppm: 33.25 (C–C–C); 31.95 (C–C–C), 29.35, 29.27 (C–C–C), 23.07 (C–CH$_3$), 22.70 (Si–C–C), 17.63 (C–Si), 14.10 (C–CH$_3$), 1.89 (Si–CH$_3$), 0.28 (O–Si–CH$_3$).
^{29}Si NMR (CDCl$_3$)ppm: −2.19 (–O–Si–O–), 6.75 (OSi(CH$_3$)$_3$).

3-(3-glycidyloxypropyl)-1,1,1,3,5,5,5-heptamethyltrisiloxane:

^{1}H NMR (CD$_3$CN)ppm: 3.69 (m, J = 17.1 Hz, 1H, –O–CH$_2$–); 3.43 (m, 2H, –CH$_2$–O–CH$_2$–); 3.27 (dd, J = 11.5 Hz, 1H, –O–CH$_2$–); 3.08 (m, J = 6.8 Hz, 1H, HC–O–CH$_2$–); 2.74 (dd, J = 5.1 Hz, 1H, HC–CH$_2$–O); 2.54 (m, J = 5.1 Hz, 1H, HC–CH$_2$–O); 0.05 (m, 3H, –SiCH$_3$); 1.59 (m, J = 11.3 Hz, 2H, –Si–CH$_2$–CH$_2$–); 0.5 (m, 2H, –Si–CH$_2$–); 0.13 (m, 18H, –Si(CH$_3$)$_3$);
^{13}C NMR (CD$_3$CN, δ, ppm): 73.67 (–C–C–O–); 71.41 (–O–C–C–); 50.71 (–C–O–C–); 43.56 (–C–O–C–); 23.16 (–Si–C–C–); 12.77 (–Si–C–); 0.37–1.39 (–Si(CH$_3$)$_3$); −1.0 (–Si–CH$_3$).
^{29}Si NMR (CD$_3$CN, δ, ppm): −20.52 (–O–Si–O–), 8.07 (OSi(CH$_3$)$_3$).

NMR spectra of these products are presented in Supplementary Materials.

4. Conclusions

Six new air-stable anionic platinum complexes were synthesized with high yields in the reaction of a suitable ionic liquid and platinum compound and the complexes were fully characterized. Derivatives of piperidine and pyrrolidine, which were precursors of the formed complexes, have been chosen for the syntheses. Due to high melting points (above 100 °C), the newly formed complexes cannot be formally classified into ionic liquids, but the ionic structure significantly influences their catalytic activity and stability. All the complexes proved to be highly active in the reactions of hydrosilylation of 1-octene and allyl glycidyl ether with 1,1,1,3,5,5,5-heptamethyltrisiloxane. Their insolubility (or limited solubility) in the reagents enabled easy isolation from postreaction mixtures and multiple uses of them in subsequent reaction cycles. The performed studies have shown that the kind of anion influences to a large extent the catalytic activity and, first and foremost, the stability of the complexes. The most stable complexes turned to be those with [PtCl$_6$]$^{2-}$ anion due to their stronger ionic character compared to analogous complexes with [PtCl$_4$]$^{2-}$ anion. Moreover, the complexes containing the six-membered heterocyclic ring (piperidinium) are more stable than those containing the five-membered ring (pyrrolidinium). Hexachloroplatinic complexes, particularly the complex [BMPip]$_2$[PtCl$_6$], make an attractive alternative not only to known homogeneous complexes but also to heterogeneous catalysts applied in hydrosilylation processes. High catalytic activity and stability of the latter complex make possible its multiple uses which is of high importance both from an ecological and economic viewpoint.

Supplementary Materials: The following are available online at http://www.mdpi.com/2073-4344/10/8/919/s1, Figure S1–Figure S12: NMR Spectra of complexes, Figure S13–Figure S20: ESI-MS spectra of complexes, Figure S21–Figure S26: NMR spectra of isolated products

Author Contributions: Synthesis of platinum complexes, methodology, M.J.-W.; catalytic tests, O.B.; FTIR in situ analyzes, O.B. and P.P.; conceptualization, M.J.-W. and H.M.; writing—original draft preparation, M.J.-W. and H.M.; writing—review and editing, M.J.-W., H.M., and O.W.; supervision, H.M.; funding acquisition, H.M. All authors have read and agreed to the published version of the manuscript.

Funding: This research was supported by grant OPUS UMO-2014/15/B/ST5/04257, funded by National Science Center (Poland)

Conflicts of Interest: The authors declare no conflicts of interest.

References

1. Anastas, P.T.; Warner, J.C. *Green Chemistry: Theory and Practice*; Oxford University Press: New York, NY, USA, 1998; p. 30.
2. Marciniec, B.; Maciejewski, H.; Pietraszuk, C.; Pawluć, P. *Hydrosilylation: A Comprehensive Review on Recent Advances*; Marciniec, B., Ed.; Springer: Dordrecht, The Netherlands, 2009.
3. Trogel, D.; Strohrer, J. Recent advances and actual challenges in late transition metal catalyzed hydrosilylation of olefins from an industrial point of view. *Coord. Chem. Rev.* **2011**, *255*, 1440–1459. [CrossRef]
4. Marciniec, B.; Maciejewski, H.; Pawluć, P. *Organosilicon Compounds*; Lee, V.Y., Ed.; Academic Press: Cambridge, MA, USA, 2017; Chapter 5; pp. 169–218.
5. Marciniec, B.; Maciejewski, H.; Pietraszuk, C.; Pawluć, P. *Applied Homogeneous Catalysis with Organometallic Compounds*; Cornils, B., Hermann, W.A., Belier, M., Pawelo, R., Eds.; Wiley: New York, NY, USA, 2017; Chapter 8; pp. 569–620.
6. Nakajima, Y.; Shimada, S. Hydrosilylation reaction of olefins: Recent advances and perspectives. *RSC Adv.* **2015**, *5*, 20603–20616. [CrossRef]
7. Roger, R.D.; Seddon, K.R. (Eds.) *Ionic Liquids–Industrial Applications to Green Chemistry*; ACS: Washington, DC, USA, 2002.
8. Dyson, P.J.; Geldbach, T.J. *Metal Catalysed Reactions in Ionic Liquids*; Springer: Dordrecht, The Netherlands, 2005.
9. Hardacre, C.; Parvulescu, V. *Catalysis in Ionic Liquids. From Catalyst Synthesis to Application*; RS Chemistry: London, UK, 2014.
10. Vekariya, R.L. A review of ionic liquids: Applications towards catalytic organic transformations. *J. Mol. Liq.* **2017**, *227*, 44–60. [CrossRef]
11. Ozokwelu, D.; Zhang, S.; Okafor, O.C.; Cheng, W.; Litombe, N. *Novel Catalytic and Separation Processes Based on Ionic Liquids*; Elsevier: Amsterdam, The Netherlands, 2017.
12. Lozano, P. (Ed.) *Sustainable Catalysis in Ionic Liquids*; CRC Press: Boca Raton, FL, USA, 2019.
13. Behr, A.; Toslu, N. Hydrosilylation reactions in single and two phases. *Chem. Eng. Technol.* **2000**, *23*, 122–125. [CrossRef]
14. Hofmann, N.; Bauer, A.; Auer, T.; Stanjek, V.; Schulz, P.; Taccardi, N.; Wasserscheid, P. Liquid-liquid biphasic, platinum-catalyzed hydrosilylation of allyl chloride with trichlorosilane using an ionic liquid catalyst phase in a continuous loop reactor. *Adv. Synth. Catal.* **2008**, *350*, 2599–2609. [CrossRef]
15. Taccardi, N.; Fekete, M.; Berger, M.E.; Stanjek, V.; Schulz, P.; Wasserscheid, P. Catalyst recycling in monophasic Pt-catalyzed hydrosilylation reactions using ionic liquids. *Appl. Catal. A* **2011**, *399*, 69–74. [CrossRef]
16. Weyershausen, B.; Hell, K.; Hesse, U. Industrial application of ionic liquids as process aid. *Green Chem.* **2005**, *7*, 283–287. [CrossRef]
17. Schulz, T.; Strassner, T. Biphasic platinum catalyzed hydrosilylation of terminal alkenes in TAAILs. *J. Organometal. Chem.* **2013**, *744*, 113–118. [CrossRef]
18. Maciejewski, H.; Szubert, K.; Marciniec, B.; Pernak, J. Hydrosilylation of functionalised olefins catalysed by rhodium siloxide complexes in ionic liquids. *Green Chem.* **2009**, *11*, 1045–1051. [CrossRef]
19. Zielinski, W.; Kukawka, R.; Maciejewski, H.; Smiglak, M. Ionic liquids as solvents for rhodium and platinum catalysts used in hydrosilylation reaction. *Molecules* **2016**, *21*, 1115. [CrossRef]
20. Maciejewski, H.; Szubert, K.; Fiedorow, R.; Giszter, R.; Niemczak, M.; Pernak, J. Diallyldimethylammonium and trimethylvinylammonium ionic liquids—Synthesis and application to catalysis. *Appl. Catal. A* **2013**, *451*, 168–175. [CrossRef]

21. Maciejewski, H.; Jankowska-Wajda, M.; Dabek, I.; Fiedorow, R. The effect of the morpholinium ionic liquid anion on the catalytic activity of Rh (or Pt) complex–ionic liquid systems in hydrosilylation processes. *RSC Adv.* **2018**, *8*, 26922–26927.
22. Jankowska-Wajda, M.; Kukawaka, R.; Smiglak, M.; Maciejewski, H. The effect of the catalyst and the type of ionic liquid on the hydrosilylation process under batch and continuous reaction conditions. *New J. Chem.* **2018**, *42*, 5229–5236. [CrossRef]
23. Luska, K.L.; Demmans, K.Z.; Stratton, S.A.; Moores, A. Rhodium complexes stabilized by phosphine-functionalized phosphonium ionic liquids used as higher alkene hydroformylation catalysts: Influence of the phosphonium headgroup on catalytic activity. *Dalton Trans.* **2012**, *41*, 13533–13540. [CrossRef]
24. Jin, X.; Xu, X.-F.; Zhao, K. Amino acid- and imidazolium-tagged chiral pyrrolidinodiphosphine ligands and their applications in catalytic asymmetric hydrogenations in ionic liquid systems. *Tetrahedron Asymmetry* **2012**, *23*, 1058–1067. [CrossRef]
25. Chen, S.J.; Wang, Y.Y.; Yao, W.M.; Zhao, X.L.; Vo-Thanh, G.; Liu, Y. An ionic phosphine-ligated rhodium (III) complex as the efficient and recyclable catalyst for biphasic hydroformylation of 1-octene. *J. Mol. Catal. A Chem.* **2013**, *378*, 293–298. [CrossRef]
26. Jankowska-Wajda, M.; Bartlewicz, O.; Szpecht, A.; Zając, A.; Smiglak, M.; Maciejewski, H. Platinum and rhodium complexes ligated by imidazolium-substituted phosphine as efficient and recyclable catalysts for hydrosilylation. *RSC Adv.* **2019**, *9*, 29396–29404. [CrossRef]
27. Bartlewicz, O.; Jankowska-Wajda, M.; Maciejwski, H. Highly efficient and reusable alkyne hydrosilylation catalysts based on rhodium complexes ligated by imidazolium- substituted phosphine. *Catalysts* **2020**, *10*, 608. [CrossRef]
28. Parvulescu, V.I.; Hardacre, C.h. Catalysis in ionic liquids. *Chem. Rev.* **2007**, *107*, 2615–2665. [CrossRef]
29. Lee, J.W.; Chan, Y.S.; Jang, H.B.; Song, C.E.; Lee, S.-G. Toward understanding the origin of positive effects of ionic liquids on catalysis: Formation of more reactive catalysts and stabilization of reactive intermediates and transition states in ionic liquids. *Acc. Chem. Res.* **2010**, *43*, 985–994. [CrossRef]
30. Wang, S.; Hui, S.; Shengjie, C.h.; Honqxing, Y.H.; Ye, L. Applications of transition metallates in catalysis. *Prog. Chem.* **2012**, *24*, 2287–2298.
31. Estager, J.; Holbrey, J.D.; Swadzba-Kwasny, M. Halometallate ionic liquids–revisited. *Chem. Soc. Rev.* **2014**, *43*, 985–994. [CrossRef] [PubMed]
32. Chiappe, C.; Ghilardi, T.; Pomelli, C.S. Structural features and properties of metal complexes in ionic liquids: Application in alkylation reactions. *Top. Organomet. Chem.* **2015**, *51*, 79–94.
33. Brown, L.C.; Hogg, J.M.; Swadzba-Kwasny, M. Lewis Acidic Ionic Liquids. *Top. Curr. Chem.* **2017**, *5*, 78–117. [CrossRef]
34. Wilkes, J.S. A short history of ionic liquids—From molten salts to neoteric solvents. *Green Chem.* **2002**, *4*, 73–80. [CrossRef]
35. Hasan, M.; Kozhevnikov, I.V.; Siddiqui, M.R.H.; Femoni, C.; Steiner, A.; Winterton, N. N,N'-dialkylimidazolium chloroplatinate(II), chloroplatinate(IV), and chloroiridate(IV) salts and an N-heterocyclic carbene complex of platinum(II): Synthesis in ionic liquids and crystal structures. *Inorg. Chem.* **2001**, *40*, 795–800. [CrossRef]
36. Zhong, C.; Sasaki, T.; Jimbo-Kobayashi, A.; Fujiwara, E.; Kobayashi, A.; Tada, M.; Iwasawa, Y. Syntheses, structures, and properties of a series of metal ion-containing dialkylimidazolium ionic liquids. *Bull. Chem. Soc. Jpn.* **2007**, *12*, 2365–2374. [CrossRef]
37. Jankowska-Wajda, M.; Bartlewicz, O.; Walczak, A.; Stefankiewicz, A.R.; Maciejewski, H. Highly efficient hydrosilylation catalysts based on chloroplatinate ionic liquids. *J. Catal.* **2019**, *374*, 266–275. [CrossRef]
38. Maciejewski, H.; Jankowska-Wajda, M.; Bartlewicz, O. New Anionic Platinum Complexes, Method for Obtaining Them and Application, Preferably for Hydrosilylation Processes. Polish Patent PL 233547, 11 March 2019.

© 2020 by the authors. Licensee MDPI, Basel, Switzerland. This article is an open access article distributed under the terms and conditions of the Creative Commons Attribution (CC BY) license (http://creativecommons.org/licenses/by/4.0/).

Article

Ru-Catalyzed Repetitive Batch Borylative Coupling of Olefins in Ionic Liquids or Ionic Liquids/scCO$_2$ Systems

Jakub Szyling [1,2,*], Tomasz Sokolnicki [1,2], Adrian Franczyk [1] and Jędrzej Walkowiak [1,*]

[1] Center for Advanced Technology, Adam Mickiewicz University in Poznań, Uniwersytetu Poznańskiego 10, 61-614 Poznań, Poland; tomasz.sokolnicki@amu.edu.pl (T.S.); adrian.franczyk@amu.edu.pl (A.F.)
[2] Faculty of Chemistry, Adam Mickiewicz University in Poznań, Uniwersytetu Poznańskiego 8, 61-614 Poznań, Poland
* Correspondence: j.szyling@amu.edu.pl (J.S.); jedrzej.walkowiak@amu.edu.pl (J.W.)

Received: 8 June 2020; Accepted: 5 July 2020; Published: 8 July 2020

Abstract: The first, recyclable protocol for the selective synthesis of (*E*)-alkenyl boronates via borylative coupling of olefins with vinylboronic acid pinacol ester in monophasic (cat@IL) or biphasic (cat@IL/scCO$_2$) systems is reported in this article. The efficient immobilization of [Ru(CO)Cl(H)(PCy$_3$)$_2$] (1 mol%) in [EMPyr][NTf$_2$] and [BMIm][OTf] with the subsequent extraction of products with *n*-heptane permitted multiple reuses of the catalyst without a significant decrease in its activity and stability (up to 7 runs). Utilization of scCO$_2$ as an extractant enabled a significant reduction in the amount of catalyst leaching during the separation process, compared to extraction with *n*-heptane. Such efficient catalyst immobilization allowed an intensification of the processes in terms of its productivity, which was indicated by high cumulative TON values (up to 956) in contrast to the traditional approach of applying volatile organic solvents (TON = ~50–100). The reaction was versatile to styrenes with electron-donating and withdrawing substituents and vinylcyclohexane, generating unsaturated organoboron compounds, of which synthetic utility was shown by the direct transformation of extracted products in iododeborylation and Suzuki coupling processes. All synthesized compounds were characterized using ^{1}H, ^{13}C NMR and GC-MS, while leaching of the catalyst was detected with ICP-MS.

Keywords: homogeneous catalysis; ionic liquids; supercritical CO$_2$; borylative coupling; catalyst recycling; green chemistry; ruthenium catalyst; vinyl boronates; organoboron compounds

1. Introduction

Homogeneous catalysis has remained a key part of chemistry for several decades and is a powerful tool in the synthesis of valuable compounds. The high activity and selectivity of molecular catalysts under mild reaction conditions, the lack of diffusion barriers in comparison to heterogeneous systems, and the variability of their electronic and steric properties tuned by the proper design and choice of ligands and metal centers lead to their application in the chemical industry in the production of advanced polymers and fine chemicals [1,2]. On the other hand, homogeneous conditions generate notable problems in recovery and reuse of catalysts, which are mostly based on expensive noble transition metals (TM), for example, rhodium, iridium, platinum, palladium, or ruthenium. To obtain high TON values and the proper process selectivity, this precious catalyst is often sacrificed within the separation process. Moreover, homogeneous conditions require a considerable amount of volatile organic solvent to dissolve all reaction components: reagents and catalysts. Such an approach generates problems with the process economy resulting from the high solvent consumption within the process and separation but also influences toxicological and environmental aspects. As a consequence of these drawbacks, new more sustainable methods for the synthesis of advanced organic and organometallic molecules by the application of environmentally benign solvents and/or catalyst immobilization and recycling are continuously being developed in academia and industry [3–8].

Ionic liquids (ILs) constitute an interesting green alternative to typical organic solvents because of their unique properties, such as incombustibility, negligible vapor pressure and abilities for the dissolution of many inorganic, organic and organometallic compounds. As a result of their physicochemical properties, ILs have immense structural variability which results in a wide range of their applications, for example, lubricants [9,10], solar cells [11,12], biomass processing [13,14] and solvents in chemical syntheses [15–17]. Their affinity to TM-complexes, as well as their high polarity, mean that ILs are frequently used in catalysis as a typical solvent, ligand source or even the catalyst itself. Moreover, ILs can also be used for catalyst immobilization, very often without any interference in its structure, and therefore without affecting its initial activity.

An efficient immobilization of the catalyst in IL and a properly designed product separation strategy significantly reduce metal leaching into the final product and allow for the recycling of the expensive TM-complex. This can be achieved by the extraction of products from the ionic phase through organic solvents or more preferably, supercritical CO_2 (scCO_2) [18]. Combining ILs and scCO_2 permits the development of an approach benefitting from the advantages of both homogeneous and heterogeneous catalysis, i.e., (i) high catalyst activity, selectivity, and stability, (ii) solvent (IL) and catalyst recyclability (iii) enhanced productivity and (iv) product separation simplicity. Moreover, the presence of scCO_2 significantly reduces the viscosity of ILs facilitating mass transfer, increasing the reaction rate and allows for effective extraction of the products by compressed CO_2 from ILs [19].

There are a lot of examples of ILs and scCO_2 being used as benign, eco-friendly alternative solvents in many catalytic processes such as hydrogenations [20,21], oxidations [22], hydroformylations [23], Pd-catalyzed C–C bond-forming couplings [24], hydrometallations [25–27] or silylative coupling [28]. These solvents permit catalyst immobilization in liquid or solid-state, and processes to be carried out under repetitive batch and continuous flow methods. Our group also has experience in the application of green solvents (ILs, scCO_2, PEGs) in designing sustainable protocols for the synthesis of organoboron and organosilicon compounds by the immobilization of the molecular catalyst and their recycling [29–35].

Due to their low toxicity, high reactivity under specified reaction conditions, and ease of handling, these organometallic compounds are valuable synthons in many chemical transformations, for example, Suzuki or Hiyama couplings [36,37] or halodemetalations [34,38]. Most of the organosilicon and organoboron compounds are produced under homogeneous conditions, with the use of TM-complexes as catalysts, thus searching for efficient and more sustainable protocols for their preparation is of great importance.

The utilization of ILs and/or scCO_2 for the synthesis of unsaturated organoboron compounds is poorly explored. Up to date, only three reports focusing on the catalytic hydroboration of alkynes in these solvents have been described [30,31,39]. Because of the challenging regio- and stereoselectivity control of the catalytic hydroboration of alkynes, the exclusive formation of one isomer was difficult to achieve, despite the good or very good recyclability and stability of catalytic systems. In contrast to the catalytic hydroboration of alkynes, borylative coupling (trans-borylation) of vinyl boronates with olefins, especially styrene derivatives, in the presence of ruthenium hydride catalysts, smoothly leads to the exclusive formation of (E)-alkenyl boronates [40].

In continuation of our studies on the application of green solvents in the synthesis of unsaturated organoboron compounds, herein, we would like to report the first application of ILs and the Ru-H catalyst [Ru(CO)Cl(H)(PCy$_3$)$_2$] in the monophasic ([Ru(CO)Cl(H)(PCy$_3$)$_2$]@ILs) or biphasic ([Ru(CO)Cl(H)(PCy$_3$)$_2$]@ILs/scCO_2) recyclable borylative coupling of vinylboronic acid pinacol ester with olefins (Scheme 1). The presented method is an interesting variant of traditional protocols through (i) the ability of catalysts and ILs to recycle, (ii) product separation simplicity, (iii) the high productivity and stability of the systems.

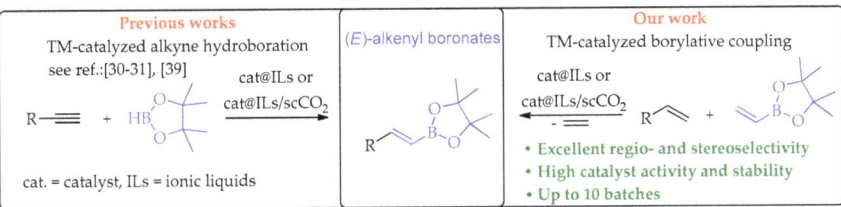

Scheme 1. Ionic liquids-based recyclable protocols for the synthesis of unsaturated organoboron compounds.

2. Results and Discussion

2.1. Reaction and Extraction Conditions Screening

The studies were initiated by an investigation of the borylative coupling of vinylboronic acid pinacol ester (**1**) with styrene (**2a**) as a model reaction performed in several ILs based on pyrrolidinium ([EMPyr]$^+$) or 1-butyl-3-methylimidazolium ([BMIm]$^+$) cations and various inorganic and organic anions (Figure 1). Most of them were successfully applied in TM-catalyzed transformations [16]. Borylative coupling is a catalytic reaction which occurs in the presence of Ru-H catalysts with the activation of the C–B bond in vinyl boronates and the C–H bond in olefins or O–H bond in alcohols, silanols or boronic acids furnishing boryl-substituted olefins, boronic esters, borasiloxanes or boroxanes respectively, with the simultaneous evolution of ethylene [29,40–43].

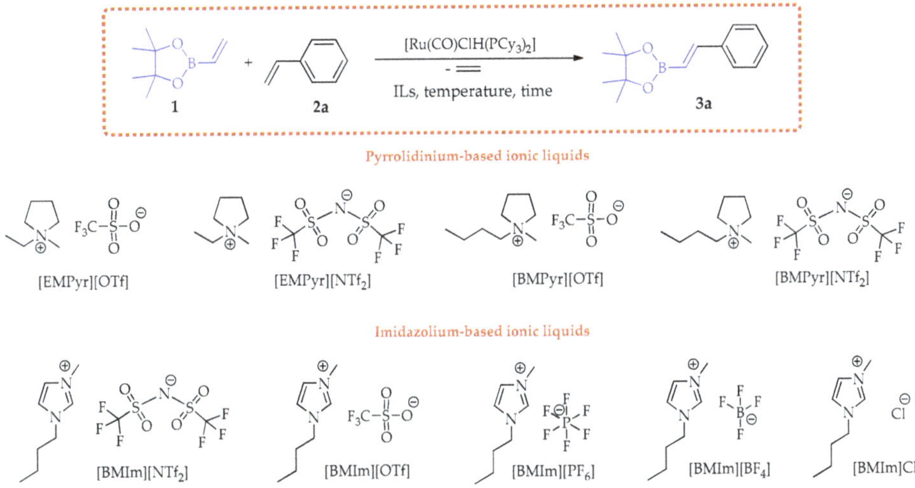

Figure 1. Model reaction of **1** with **2a** used for the conditions screening. Pyrrolidinium and imidazolium-based ionic liquids used as solvents and catalyst immobilization media.

For the initial ionic liquids screening, we applied slightly more rigorous reaction conditions if compared to the traditional approach of utilizing volatile organic solvents (a higher process temperature to ensure good homogeneity of the reaction mixture). [Ru(CO)Cl(H)(PCy$_3$)$_2$] was used as a catalyst, which was previously described as the most effective in this transformation. In all pyrrolidinium-based ionic liquids, a conversion of **1** was over 90% (Table 1, entry 1–4). The complete conversion of **1** was observed only for [EMPyr][NTf$_2$]. In general, a higher conversion of **1** was observed for pyrrolidinium cations with ethyl substituents. When butyl groups were attached to the pyrrolidinium cation, the product yield was slightly lower. Simultaneously, the application of trifluoromethanesulfonate anion ([OTf]$^-$) resulted in the lower conversion of substrates than for bis(trifluoromethylsulfonyl)imide anion ([NTf$_2$]$^-$).

Table 1. Borylative coupling of vinylboronic acid pinacol ester (**1**) with styrene (**2a**) in the presence of [Ru(CO)Cl(H)(PCy$_3$)$_2$]@IL—IL screening. IL—Ionic liquids.

Entry	Ionic Liquid	Conversion of 1 [%] [a]	Yield of 3a [%] [b]
1	[EMPyr][OTf]	95	95
2	**[EMPyr][NTf$_2$]**	**100**	**100**
3	[BMPyr][OTf]	91	91
4	[BMPyr][NTf$_2$]	93	93
5	[BMIm][NTf$_2$]	97	97
6	**[BMIm][OTf]**	**100**	**100**
7	[BMIm][PF$_6$]	88	88
8	[BMIm][BF$_4$]	54	54
9	[BMIm]Cl	trace	n.d

Reaction conditions: [Ru-H]:**1**:**2a** = 0.03:1:4, 130 °C, ionic liquid (1 g), inert atmosphere. [a] Determined by GC-MS. [b] Determined by GC-MS and ^{1}H NMR.

Borylative coupling of **1** with **2a** in 1-butyl-3-methylimidazolium-based ionic liquids with [OTf]$^-$ or [NTf$_2$]$^-$ anions occurred smoothly with very high yields with the formation of the desired product **3a** and excellent selectivity (Table 1, entry 5–6). In contrast to pyrrolidinium cation, the quantitative conversion of **1** was observed for [OTf]$^-$ anion. Lower yields were observed when moisture-sensitive anions such as hexafluorophosphate ([PF$_6$])$^-$ or tetrafluoroborate ([BF$_4$])$^-$ were used (Table 1, entry 7–8). The presence of traces of water in those ILs probably caused partial deactivation of the moisture-sensitive catalyst. When [BMIm]Cl was used as a solvent and immobilization medium, no conversion of **1** was noticed. This can be explained by the strong coordination ability of chloride anion to the metal center of the catalyst, and inhibition of its activity. This observation is in agreement with previous reports [28,30].

[EMPyr][NTf$_2$] and [BMIm][OTf], which ensured the complete conversion of **1** (Table 1, entry 2 and 6), were chosen for further reaction conditions screening. We optimized parameters crucial for obtaining the highest reaction yield, i.e., temperature, time, the molar ratio of substrates, as well as catalyst loading (Table 2). Initially, for [EMPyr][NTf$_2$], the high catalyst loading of [Ru(CO)Cl(H)(PCy$_3$)$_2$] (3 mol%) was maintained while the other parameters were tuned. The quantitative yields of **3a** were observed above 110 °C. The same results were noticed when the time of the reaction was reduced from 24 to 6 h. Moreover, only a 1.5-fold excess of **2a** towards **1** was necessary to get a full conversion of **1**. In the final stage of optimization studies in [EMPyr][NTf$_2$], the influence of catalyst loading on reaction efficiency was determined. It was found that 1 mol% of [Ru(CO)Cl(H)(PCy$_3$)$_2$] is essential for achieving a very high reaction yield (Table 2, entry 12). Although the application of the equimolar ratio of substrates resulted in a complete conversion of **1**, the presence of a homocoupling product of **1** was observed (Table 2, entry 14). Therefore, a small excess of olefin is essential for the elimination of the side-homocoupling reaction of vinyl boronate. Replacement of ruthenium hydride catalyst with coordinated PCy$_3$ ligand to [Ru(CO)Cl(H)(PPh$_3$)$_3$] leads to a lower conversion of **1** and the formation of undefined by-products (Table 2, entry 15).

Based on the most effective reaction conditions in [EMPyr][NTf$_2$], borylative coupling of **1** with **2a** in [BMIm][OTf] was determined. Similar to pyrrolidinium-based IL, the reaction in [BMIm][OTf] should be performed at least at 110 °C for 6 h with a slight excess of **2a** towards **1** and 1 mol% of [Ru(CO)Cl(H)(PCy$_3$)$_2$], to achieve the best process efficiency (Table 2, entry 16). It is worth emphasizing that the application of ILs as a solvent enhances reaction rate compared to an analogous reaction performed in poly (ethylene glycol), another green solvent with similar properties to ILs [29] The same phenomenon has already been observed by us during studies on the catalytic hydroboration of alkynes in Ru(CO)Cl(H)(PPh$_3$)$_3$@ILs systems [30].

Table 2. Reaction conditions screening for borylative coupling of **1** with **2a** in the presence of immobilized [Ru(CO)Cl(H)(PCy$_3$)$_2$]@[EMPyr][NTf$_2$] or @[BMIm][OTf].

Entry	Ionic Liquid	Temperature [°C]	Time [h]	1:2a [mol]	Catalyst Loading [mol%]	Conversion of 1 [%] [a]	Yield of 3a [%] [b]
1	[EMPyr][NTf$_2$]	120		1:4	3	100	100
2	[EMPyr][NTf$_2$]	110	24	1:4	3	100	100
3	[EMPyr][NTf$_2$]	100	24	1:4	3	96	96
4	[EMPyr][NTf$_2$]	80	24	1:4	3	88	88
5	[EMPyr][NTf$_2$]	110	18	1:4	3	100	100
6	[EMPyr][NTf$_2$]	110	6	1:4	3	100	100
7	[EMPyr][NTf$_2$]	110	3	1:4	3	94	94
8	[EMPyr][NTf$_2$]	110	6	1:3	3	100	100
9	[EMPyr][NTf$_2$]	110	6	1:2	3	100	100
10	[EMPyr][NTf$_2$]	110	6	1:2	2	100	100
11	[EMPyr][NTf$_2$]	110	6	1:1.5	2	100	100
12	**[EMPyr][NTf$_2$]**	**110**	**6**	**1:1.5**	**1**	**100**	**100**
13	[EMPyr][NTf$_2$]	110	6	1:1.5	0.5	93	93
14	[EMPyr][NTf$_2$]	110	6	1:1	1	100	83 [c]
15 [d]	[EMPyr][NTf$_2$]	110	6	1:1.5	1	46	37 [e]
16	**[BMIm][OTf]**	**110**	**6**	**1:1.5**	**1**	**100**	**100**
17	[BMIm][OTf]	80	6	1:1.5	1	53	53
18	[BMIm][OTf]	110	3	1:1.5	1	91	91
19	[BMIm][OTf]	110	6	1:1	1	100	88 [c]
20 [d]	[BMIm][OTf]	110	6	1:1.5	1	41	33 [e]
21	[BMIm][OTf]	110	6	1:1.5	0.5	89	89

[a] Determined by GC-MS analysis. [b] Determined by GC-MS and ^{1}H NMR analyses. [c] Homocoupling of **1** was observed. [d] 1 mol% of [Ru(CO)Cl(H)(PPh$_3$)$_3$] was used as a catalyst. [e] Undefined side products were observed.

Among many possibilities for the separation of products from the homogeneous reaction mixture, extraction seems to be a fast and efficient method. The extractant should dissolve the reaction products well, not mix with an ionic liquid to prevent leaching of the catalyst, as well as not inhibiting or deactivating the catalyst. The right choice is crucial for the ability to reuse the catalytic system. To find the most appropriate medium for product extraction, we examined several organic solvents and scCO$_2$ (Table 3). Non-polar aliphatic hydrocarbons (*n*-hexane or *n*-heptane) are completely immiscible with [EMPyr][NTf$_2$] and [BMIm][OTf], and during the extraction, the two phases were observed. The extractant (upper) phase was transparent or pale yellow. Application of more polar solvents such as toluene, dichloromethane or tetrahydrofuran resulted in partial or complete dissolution of the catalytic systems regardless of the ionic liquid used.

Table 3. Extractant screening.

Entry	Solvent	[EMPyr][NTf$_2$]	[BMIm][OTf]
1 [a]	*n*-Hexane	−	−
2 [a]	*n*-Heptane	−	−
3 [a]	Toluene	+	+
4 [b]	Dichloromethane	+	+
5 [a]	Tetrahydrofuran	+	+
6 [c]	Supercritical CO$_2$	−	−

Extraction conditions: 3 × 5 mL of the solvent at [a] 60 or [b] 40 °C. [c] 160–180 bar of CO$_2$ at 40 °C, 8 mL/min, 45 min). (−) Biphasic system. (+) Monophasic system or catalyst leaching from IL.

Similar to aliphatic hydrocarbons, ScCO$_2$ creates a biphasic solvent system with [EMPyr][NTf$_2$] and [BMIm][OTf]. Additionally, it is known that scCO$_2$ is partially soluble in ILs, lowering their viscosity, while ILs are not soluble in this supercritical fluid. Therefore, such a biphasic system is suitable for carrying out the reaction/extraction process [44]. Based on extractant screening for further experiments, *n*-heptane (lower neurotoxicity than *n*-hexane) and scCO$_2$ were chosen as solvents for product extraction from the post-reaction mixtures.

2.2. Scope of Substrates Investigations

Having optimized the reaction and extraction conditions in hand, we have studied a scope of olefins, mainly electron-deficient, neutral or donating styrenes, in borylative coupling with **1** in [BMIm][OTf] (Table 4). The reaction of 1 with bromo-(**2b**) or fluoro-substitued (**2c**) styrenes in *para* position led to (*E*)-alkenyl boronates with high yields (Table 4, entry 2–3). Application of electron-withdrawing groups such as –CF$_3$ (**2d–e**) also gave an excellent conversion of **1**. Trans-borylation of **1** with a weakly donating –CH$_3$ groups in *meta* (**2f**) or *para* (**2g**) positions resulted in the desired products **3f** and **3g** with very good isolation yields (Table 4, entry 6–7). However, the utilization of the sterically hindered alkyl group –C(CH$_3$)$_3$ (**2h**) gave a lower conversion of **1**. A similar result was noticeable when the electron-donating –OCH$_3$ group (**2i**) was applied. We also examined the reactivity of vinylcyclohexane (**2j**) in borylative coupling with **1**. After 6 h, **3j** was observed with a very high yield (Table 4, entry 10).

Table 4. Borylative coupling of **1** with a various olefins **2a–j** in Ru(CO)Cl(H)(PCy$_3$)$_2$@[BMIm][OTf] in optimized reaction and extraction conditions.

Entry	Olefin	2	Product	3	Conversion of 1 [%] [a]	Extraction Yield [%] [b]	Isolation Yield [%] [c]
1	Ph-CH=CH$_2$	a	Ph-CH=CH-Bpin	a	100	94 [d] / 90 [e]	81
2	4-Br-C$_6$H$_4$-CH=CH$_2$	b	4-Br-C$_6$H$_4$-CH=CH-Bpin	b	99	91 [d] / 88 [e]	77
3	4-F-C$_6$H$_4$-CH=CH$_2$	c	4-F-C$_6$H$_4$-CH=CH-Bpin	c	99	90 [d] / 91 [e]	78
4	4-F$_3$C-C$_6$H$_4$-CH=CH$_2$	d	4-F$_3$C-C$_6$H$_4$-CH=CH-Bpin	d	100	93 [d] / 94 [e]	83
5	3,5-(F$_3$C)$_2$-C$_6$H$_3$-CH=CH$_2$	e	3,5-(F$_3$C)$_2$-C$_6$H$_3$-CH=CH-Bpin	e	99	90 [d] / 92 [e]	81
6	4-CH$_3$-C$_6$H$_4$-CH=CH$_2$	f	4-CH$_3$-C$_6$H$_4$-CH=CH-Bpin	f	100	93 [d] / 91 [e]	79
7	2-CH$_3$-C$_6$H$_4$-CH=CH$_2$	g	2-CH$_3$-C$_6$H$_4$-CH=CH-Bpin	g	98	84 [d] / 81 [e]	69
8	4-*t*Bu-C$_6$H$_4$-CH=CH$_2$	h	4-*t*Bu-C$_6$H$_4$-CH=CH-Bpin	h	93	81 [d] / 77 [e]	73
9	4-MeO-C$_6$H$_4$-CH=CH$_2$	i	4-MeO-C$_6$H$_4$-CH=CH-Bpin	i	89	74 [d] / 75 [e]	71
10	Cy-CH=CH$_2$	j	Cy-CH=CH-Bpin	j	100	95 [d] / 96 [e]	87

Reaction conditions: [Ru-H]:**1**:**2a–j** = 0.01:1:1.5, 110 °C, 6 h, 1g [BMIm][OTf], inert atmosphere. Extraction conditions: 3 × 5 mL of the *n*-heptane at 60 °C or 160–180 bar of CO$_2$ at 40 °C, 8 mL/min, 45 min. [a] Determined by GC-MS. [b] Calculated on total amount of reagents. [c] After *n*-heptane extraction; calculated on a theoretical mass of the product. [d] *n*-Heptane extraction. [e] Supercritical CO$_2$ extraction.

Simultaneously, the efficiency of extraction for different products of **3** was investigated. *n*-Heptane (3 × 5 mL at 60 °C) or scCO$_2$ (160–180 bar of CO$_2$ at 40 °C, 8 mL/min, 45 min) were applied as extractants. Very high extraction yields (over 90%) using *n*-heptane were obtained for most of the products (**3a–f, 3j**). Similar results were achieved for the product extraction with scCO$_2$. It is worth noticing that when CO$_2$-philic groups such as –F or –CF$_3$ were attached to the phenyl rings (**3c–e**), the extraction yields in scCO$_2$ were slightly higher (Table 4, entry 3–5). The extraction of product **3j** with a cyclohexyl ring was almost quantitative due to its good affinity to both extractants.

2.3. Repetitive Batch Borylative Coupling in the Monophasic Solvent System—[Ru(CO)Cl(H)(PCy$_3$)$_2$]@ILs

A key objective of our research was to develop an efficient and more sustainable protocol for the synthesis of (*E*)-alkenyl boronates via borylative coupling of vinyl boronates with olefins in ILs. Having optimized the process (reaction and extraction) conditions, as well as the information on the reactivity of olefins (**2a–j**), we verified the possibility of catalytic system recycling after each run using [BMIm][OTf] and [EMPyr][NTf$_2$] as solvents. For repetitive batch borylative coupling of **1** with olefins, **2a**, and **2d** were chosen as reagents due to their excellent reaction and extraction yields. Trans-borylation of **1** with **2a** in [BMIm][OTf] and [EMPyr][NTf$_2$] (Figure 2a,b) gave a final product **3a** with very high yields up to the sixth run. For both ionic liquids, a gradual decrease in the reaction yield in the seventh up to the 10th cycle was observed. A higher extraction yield for extractant-soluble components (**3a** and excess or unreacted substrate) was noticed when *n*-heptane was used. Moreover, the cumulative turnover number (cTON) was several times higher than in the classical approach, utilizing volatile organic solvents (930–948 vs. approx. 50–100).

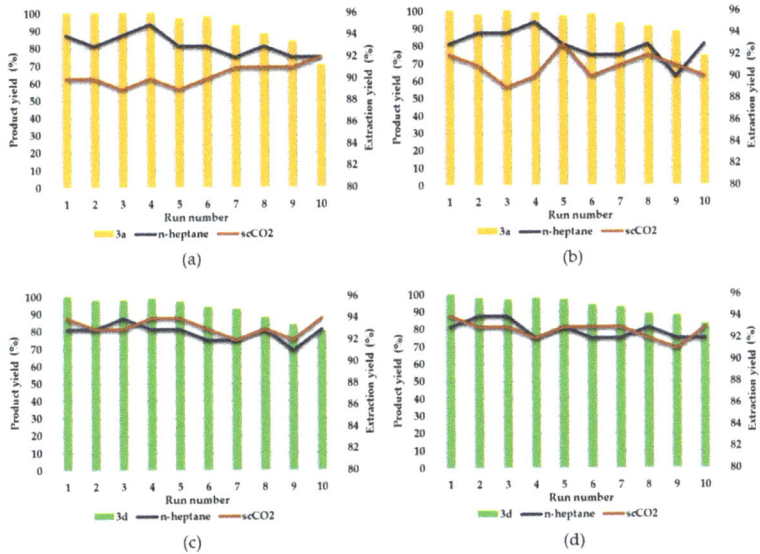

Figure 2. Repetitive batch borylative coupling of **1** with **2a** in [BMIm][OTf] (**a**) and [EMPyr][NTf$_2$] (**b**) or **1** with **2d** in [BMIm][OTf] (**c**) and [EMPyr][NTf$_2$] (**d**) under optimized reaction ([Ru-H]:**1**:**2a** or **2d** = 0.01:1:1.5, 110 °C, 6 h, 1 g of IL, inert atmosphere conditions) and extraction (3 × 5 mL of *n*-heptane at 60 °C or 160–180 bar of CO$_2$ at 40 °C, 8 mL/min, 45 min) conditions. The product yields were determined by GC-MS. The extraction yields refer to the mass of products and reactants.

Repetitive batch borylative coupling of **1** with **2d** containing a CO$_2$-phlilic trifluoromethyl group gave a very high product yield (over 90%) up to the seventh cycle for [BMIm][OTf] and [EMPyr][NTf$_2$] (Figure 2c,d), respectively). The last three cycles resulted in a slight decrease in the reaction efficiency, but it was still at an acceptable level. Nevertheless, high values of cTON were obtained (941–956), showing good system activity.

To investigate the possible cause of the gradual decrease in catalyst activity during the subsequent runs, extracts after the first, second, fifth and eighth runs were analyzed by inductively coupled plasma–mass spectrometry (ICP-MS). The highest ruthenium content was observed for extracts after the first and second runs (7.1–6.6 ppm) if n-heptane was applied regardless of the IL used. A pale yellow color was visible in both extracts. Although extracts analyzed after the fifth and eighth cycles were colorless, the ruthenium content was only a bit lower (6.4–4.8 ppm). Clearly better results were observed when scCO$_2$ was used as an extractant. For all ILs, ruthenium content in the analyzed extracts was comparable and lower than < 1 ppm. It should be pointed out that, despite significantly lower catalyst leaching in scCO$_2$, the activity of catalytic systems was the same as for processes that used n-heptane for extraction. Thus, a gradual decrease in catalyst activity is rather caused by the formation of inactive catalyst species or non-coordinative cyclohexylphosphine oxide than catalyst leaching. Similar observations were made by us during recyclable hydroboration of alkynes in [Ru(CO)Cl(H)(PPh$_3$)$_3$]@[EMPyr][NTf$_2$] or [Ru(CO)Cl(H)(PPh$_3$)$_3$]@[EMPyr][OTf]/scCO$_2$ systems [30].

2.4. Repetitive Batch Borylative Coupling in the Biphasic Solvent System—[Ru(CO)Cl(H)(PCy$_3$)$_2$]@ILs/scCO$_2$

In the final stage of our study we performed a repetitive borylative coupling of **1** with **2a**, **2d** and **2j** in the biphasic catalytic system [Ru(CO)Cl(H)(PCy$_3$)$_2$]@[BMIm][OTf]/scCO$_2$ (Figure 3a–c), respectively). The reactions were performed in 10-mL autoclaves equipped with sapphire windows under optimized reaction and extraction conditions at 180–190 bar of scCO$_2$ (Figure 3d). Similar to the repetitive trans-borylation of **1** with **2a** in [Ru(CO)Cl(H)(PCy$_3$)$_2$]@[BMIm][OTf], very high reaction and extraction yields of **3a** in the biphasic solvent system, were observed. Cumulative TON values were comparable to those obtained for the monophasic system (930 vs. 924).

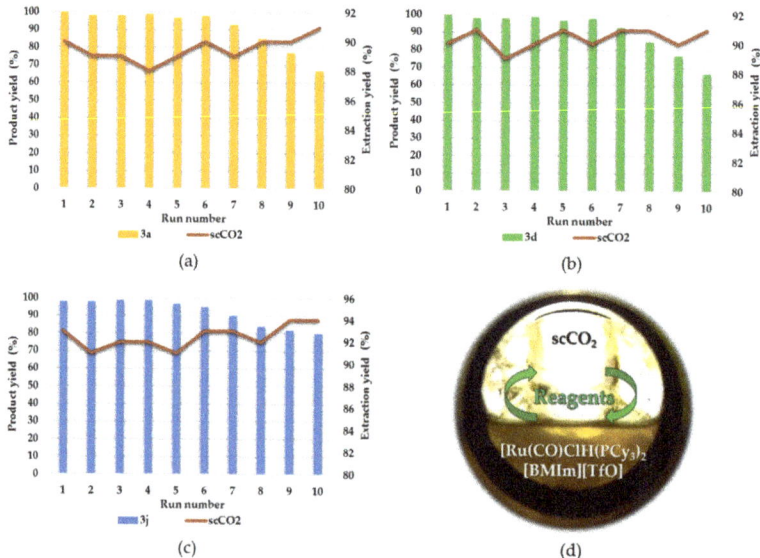

Figure 3. Repetitive batch borylative coupling of **1** with **2a** (**a**), **2d** (**b**) and **2j** (**c**) in [Ru(CO)Cl(H)(PCy$_3$)$_2$]@[BMIm][OTf]/scCO$_2$ biphasic solvent systems under optimized reaction ([Ru-H]:**2a**, **2d** or **2j** = 0.01:1:1.5, 110 °C, 6 h, 1 g of [BMIm][OTf], inert atmosphere conditions) and extraction (160–180 bar of CO$_2$ at 40 °C, 8 mL/min, 45 min) conditions. Visualization of the reaction mixture through the sapphire window of high-pressure autoclave's (**d**). The product yields were determined by GC-MS. The extraction yields refer to the mass of products and reactants.

For 4-(trifluoromethyl)styrene (**2d**) and vinylcyclohexane (**2j**), very high reaction and extraction yields of products were observed up to the sixth run, whereupon a gradual decrease in reaction yields was noticed. Extraction yields remained at a very high level regardless of the run. Additionally, the negligible amount of metal in all analyzed extracts (0.88–0.54 ppm) clearly suggested that [Ru(CO)Cl(H)(PCy$_3$)$_2$] was strongly immobilized in [BMIm][OTf] under applied reaction conditions. The satisfactory stability and activity of the [Ru(CO)Cl(H)(PCy$_3$)$_2$]@[BMIm][OTf]/scCO$_2$ system creates an opportunity for its application in continuous-flow borylative coupling reactions in the future.

At the end of our investigation on repetitive batch borylative coupling in the biphasic system, we verified the possibility of direct use of extracts in deborylation protocols, i.e., Suzuki coupling and iododeborylation (Scheme 2).

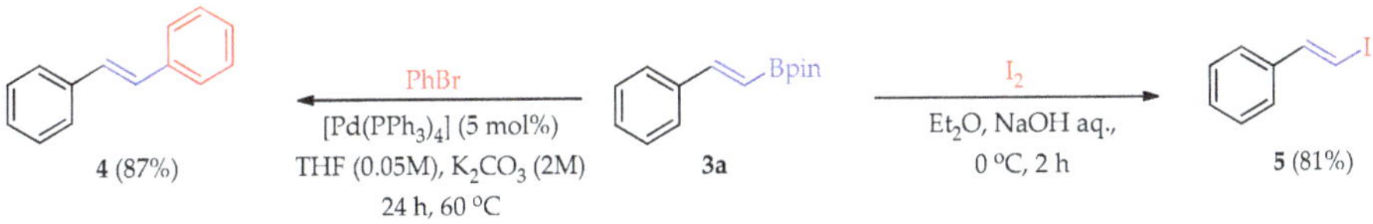

Scheme 2. Transformations of **3a** via Suzuki coupling and iododeborylation reactions.

Suzuki coupling of **3a** with bromobenzene in the presence of [Pd(PPh$_3$)$_4$] occurred with a complete conversion of **3a** and with a very high isolation yield of (*E*)-1,2-diphenylethene (**4**) (87%). Similarly, iododeborylation of **3a** with molecular iodine in the presence of sodium hydroxide resulted in a high isolation yield of (*E*)-(2-iodovinyl)benzene (**5**) (81%). The possibility of applying deborylation protocols with the use of crude extract is an attractive approach from the synthetical point of view because of the lack of necessity of carrying out time- and cost-consuming purification steps, for example, column chromatography or distillation.

3. Materials and Methods

3.1. Materials

Vinylboronic acid pinacol ester (95%, Sigma-Aldrich), styrene (97%, Sigma-Aldrich, Poznań, Poland), 4-bromostyrene (97%, Sigma-Aldrich), 4-fluorostyrene (99%, Sigma-Aldrich), 4-(trifluoromethyl)styrene (98%, Sigma-Aldrich), 3,5-bis(trifluoromethyl)styrene (98%, Alfa Aesar, Kandel, Germany), 4-methylstyrene (96%, Sigma-Aldrich), 2-methylstyrene (95%, Sigma-Aldrich), 4-*t*-butylstyrene (93%, Sigma-Aldrich), 4-methoxystyrene (97%, Sigma-Aldrich), vinylcyclohexane (99%, Sigma-Aldrich) were applied as reagents in borylative coupling reactions. Sodium hydroxide (99%, POCH Basic, Gliwice, Poland), iodine (99.8%, Sigma-Aldrich), diethyl ether (99%, POCH Basic), potassium carbonate (99%, Sigma-Aldrich), bromobenzene (99%, Sigma-Aldrich), tetrakis (triphenylphosphine)palladium (0) (99%, Sigma-Aldrich) were applied as received. 1-Ethyl-1-methylpyrrolidinium trifluoromethanesulfonate (99%, IoLiTec, Heilbronn, Germany), 1-ethyl-1-methylpyrrolidinium bis(trifluoromethylsulfonyl)imide (98%, Sigma-Aldrich), 1-butyl-1-methylpyrrolidinium trifluoromethanesulfonate (95%, Sigma-Aldrich), 1-butyl-1-methylpyrrolidinium bis(trifluoromethylsulfonyl)imide (98%, Sigma-Aldrich), 1-butyl-3-methylimidazolium bis(trifluoromethylsulfonyl)imide (98%, Sigma-Aldrich), 1-butyl-3-methylimidazolium trifluoromethanesulfonate (97%, Sigma-Aldrich), 1-butyl-3-methylimidazolium hexafluorophosphate (97%, Sigma-Aldrich), 1-butyl-3-methylimidazolium tetrafluoroborate (98%, Sigma-Aldrich), 1-butyl-3-methylimidazolium chloride (99%, Sigma-Aldrich) were dried in a vacuum before use (70 °C, 18 h). Styrene, 4-methoxystyrene, and 4-*tert*-butylstyrene were purified by the bulb to bulb distillation before use. The carbon dioxide (99.995%, Messer Polska, Chorzów, Poland) was used as a solvent and extractant in the reactions carried out in scCO$_2$. The ruthenium catalysts [Ru(CO)Cl(H)(PCy$_3$)$_2$] [45] and [Ru(CO)Cl(H)(PPh$_3$)$_3$] [46] were prepared according to literature procedures. Deuterium solvents were obtained from Dr. Glaser AG Basel. The *n*-hexane (99%, POCH

Basic), *n*-heptane (99%, Sigma-Aldrich), toluene (99%, POCH Basic), dichloromethane (99%, POCH Basic), tetrahydrofurane (99%, POCH Basic) used in the extractions were distilled and dried prior use.

3.2. General Procedures

All manipulations were performed under an argon atmosphere using standard Schlenk's techniques or in high-pressure autoclaves equipped with sapphire windows for the solubility tests of reagents and catalyst, when $scCO_2$ was used (For detailed analysis descriptions and NMR spectra of obtained products please see the Supplementary Materials).

3.2.1. Borylative Coupling in the [Ru(CO)Cl(H)(PCy$_3$)$_2$]@ILs System under Optimized Reaction/Extraction Conditions with Subsequent n-Heptane Extraction

A 50 mL Schlenk's vessel was charged with dried IL (1 g) and [Ru(CO)Cl(H)(PCy$_3$)$_2$] (0.01 mmol) in an argon atmosphere. Subsequently, vinylboronic acid pinacol ester (0.5 mmol) and olefin (0.75 mmol) were added. The reaction was carried out for 6 h at 110 °C. The reaction mixture was cooled down and the extractant soluble components were extracted at 60 °C with *n*-heptane (3 × 5 mL). After evaporation, the extracts were weighed and characterized by GC–MS and ^{1}H NMR analyses. The products (**3a–i**) were purified on silica by flash chromatography (Biotage IsoleraOne chromatograph) with a UV detector (λ_1 = 255 nm, λ_2 = 280 nm). Purification details: cartridge 10 g, flow rate: 12 mL/min, length: 12 CV (CV = column volume), phase: hexane/ethyl acetate (step 1: hexane 100% by 5 CV, step 2: gradient 10%/CV by 5 CV, step 3: hexane 50% by 2 CV). The non-aromatic **3j** was purified on silica using standard column chromatography using *n*-hexane/ethyl acetate (95/5–7/3) as eluents. The products were characterized by GC-MS, ^{1}H and ^{13}C NMR analyses.

3.2.2. Borylative Coupling in the [Ru(CO)Cl(H)(PCy$_3$)$_2$]@ILs/scCO$_2$ System with Subsequent scCO$_2$ Extraction

A high-pressure stainless steel autoclave reactor (10 mL) equipped with sapphire windows and connected to a Schlenk line, was charged with dried IL (1 g) and [Ru(CO)Cl(H)(PCy3)2] (0.01 mmol) in an argon atmosphere. Subsequently, vinylboronic acid pinacol ester (0.5 mmol) and olefin (0.75 mmol) were added and the reactor was pressurized with CO$_2$ to 55 bar, heated up to 110 °C and pressurized to the required pressure (approx. 170–190 bar). After 6 h, the reactor was cooled to 40 °C and the products were extracted in a CO$_2$ stream (160–180 bar of CO$_2$, 8 mL/min, 45 min) into a small amount (10–15 mL) of *n*-heptane (previously cooled down in dry ice/*i*-propanol bath) to avoid product loss during extraction. Then the extracts were evaporated, weighed and characterized by GC-MS and ^{1}H NMR analyses.

3.2.3. Repetitive Batch Borylative Coupling in [Ru(CO)Cl(H)(PCy$_3$)$_2$]@ILs with Subsequent Organic Solvent Extraction

After the extraction process, the Schlenk's vessel was dried under vacuum for 20 min at 60 °C. Then a new portion of substrates was added in an argon atmosphere and subsequent batches, and extractions were carried out according to the procedure described in Section 3.2.1 without the isolation step.

3.2.4. Repetitive Batch Borylative Coupling in [Ru(CO)Cl(H)(PCy$_3$)$_2$]@ILs/scCO$_2$ with Subsequent scCO$_2$ Extraction

After the extraction process, the autoclave was dried in a vacuum for 20 min at 60 °C. Then a new portion of substrates was added in an argon atmosphere and subsequent batches, and extractions were carried out according to the procedure described in Section 3.2.2.

3.2.5. Suzuki Coupling of **3a** with Bromobenzene

A 100 mL Schlenk vessel equipped with a stirring bar was charged with [Pd(PPh$_3$)$_4$] (5 mol%). Subsequently, THF (10 mL), **3a** after scCO$_2$ extraction (0.5 mmol in 10 mL of *n*-heptane), bromobenzene (0.5 mmol) and 2 M aqueous solution of K$_2$CO$_3$ (20 mL) were added. The reaction was carried out for 24 h at 50 °C. Then, the organic phase was separated, and the product was purified on silica by flash chromatography with a UV detector (λ_1 = 255 nm, λ_2 = 280 nm) using the purification conditions described in Section 3.2.1 in 87% yield. The product was characterized by GC-MS, ^{1}H and ^{13}C NMR analyses.

3.2.6. Iododeborylation of **3a** with Molecular Iodine

To a 100 mL, round-bottom flask equipped with a stirring bar **3a** after scCO$_2$ extraction, 10 mL of *n*-heptane and 5 mL of diethyl ether were added. Then 2 mL of the aqueous solution of NaOH (3 M) was dosed dropwise at 0 °C. Afterward, I$_2$ (0.85 mmol) in diethyl ether (5 mL) was slowly added. The reaction mixture was stirred for 2 h and then the excess of iodine was quenched with a saturated solution of sodium thiosulfate. The organic solution was separated, and the aqueous solution was washed with diethyl ether. The volatile organic fractions were evaporated, and the crude product was purified on silica by flash chromatography with a UV detector (λ_1 = 255 nm, λ_2 = 280 nm) using the purification conditions described in Section 3.2.1 in 81% yield. The product was characterized by GC-MS, ^{1}H and ^{13}C NMR analyses.

3.3. Product Characterization

(E)-4,4,5,5-Tetramethyl-2-styryl-1,3,2-dioxaborolane (**3a**), ^{1}H NMR (300 MHz, CDCl$_3$, 25 °C): δ = 7.49–7.10 (m, 6H, Ph, =CH(Ph)), 6.09 (d, J_{H-H} = 18.4 Hz, 1H, =CH(Bpin)), 1.22 (s, 12H, CCH$_3$) ppm. ^{13}C NMR (75 MHz, CDCl$_3$, 25 °C): δ = 149.6, 137.5, 129.0, 128.6, 127.1, 83.4 (CCH$_3$), 24.9 (CCH$_3$) ppm. C$_\alpha$ to boron atom was not observed. MS (EI) [*m/z* (%)]: 230(M$^+$, 29), 215(21), 187(11), 173(8), 157(14), 144(83), 129(100), 118(14), 105(33), 77(19). Isolated yield: 81%. NMR and GC-MS data are in agreement with the literature [29].

(E)-2-(4'-Bromostyryl)-4,4,5,5-tetramethyl-1,3,2-dioxaborolane (**3b**), ^{1}H NMR (300 MHz, acetone-*d$_6$*, 25 °C): δ = 7.66–7.45 (m, 4H, Ph), 7.31 (d, J_{H-H} = 18.4 Hz, 1H, =CH(Ph)), 6.19 (d, J_{H-H} = 18.4 Hz, 1H, =CH(Bpin)), 1.28 (s, 12H, CCH$_3$) ppm. ^{13}C NMR (75 MHz, acetone-*d$_6$*, 25 °C): δ = 148.8, 137.7, 132.8, 129.9, 123.5, 84.2 (CCH$_3$), 25.3 (CCH$_3$) ppm. C$_\alpha$ to boron atom was not observed. MS (EI) [*m/z* (%)]: 310((M + 2)$^+$, 21), 308(M$^+$, 19), 293(17), 222(6), 209(44), 208(41), 143(100), 129(81), 77(59). Isolated yield: 77%. NMR and GC-MS data are in agreement with the literature [29].

(E)-2-(4'-Fluorostyryl)-4,4,5,5-tetramethyl-1,3,2-dioxaborolane (**3c**), ^{1}H NMR (300 MHz, acetone-*d$_6$*, 25 °C): δ = 7.71–7.55 (m, 2H, Ph), 7.33 (d, J_{H-H} = 18.4 Hz, 1H, =CH(Ph)), 7.213–7.09 (m, 2H, Ph), 6.10 (d, J_{H-H} = 18.4 Hz, 1H, =CH(Bpin)), 1.27 (s, 12H, CCH$_3$) ppm. ^{13}C NMR (75 MHz, acetone-*d$_6$*, 25 °C): δ = 165.7, 162.5, 148.8, 135.0, 130.0, 129.9, 116.5, 116.3, 84.0 (CCH$_3$), 25.2 (CCH$_3$) ppm. C$_\alpha$ to boron atom was not observed. MS (EI) [*m/z* (%)]: 248(M$^+$, 23), 233(19), 191(9), 175(11), 162(73), 148(100), 136(21), 123(23), 102(13), 85(19), 57(14). Isolated yield: 78%. NMR and GC-MS data are in agreement with the literature [29].

(E)-2-(4'-(Trifluoromethyl)styryl)-4,4,5,5-tetramethyl-1,3,2-dioxaborolane (**3d**), ^{1}H NMR (300 MHz, CDCl$_3$, 25 °C): δ = 7.67–7.49 (m, 4H, Ph), 7.41 (d, J_{H-H} = 18.4 Hz, 1H, =CH(Ph)), 6.26 (d, J_{H-H} = 18.4 Hz, 1H, =CH(Bpin)), 1.32 (s, 12H, CCH$_3$) ppm. ^{13}C NMR (75 MHz, CDCl$_3$, 25 °C): δ = 147.8, 140.9, 130.8, 130.4, 127.3, 125.8, 125.7, 83.7 (CCH$_3$), 25.0 (CCH$_3$) ppm. C$_\alpha$ to boron atom was not observed. MS (EI) [*m/z* (%)]: 298(M$^+$, 33), 283(27), 241(19), 212(100), 197(77), 179(93), 151(34), 143(50), 85(41), 77(11), 57(30). Isolated yield: 83%. NMR and GC-MS data are in agreement with the literature [29].

(E)-2-(3′,5′-Bis(trifluoromethyl)styryl)-4,4,5,5-tetramethyl-1,3,2-dioxaborolane (**3e**), ^{1}H NMR (300 MHz, CDCl$_3$, 25 °C): δ = 7.88 (s, 2H, Ph), 7.78 (s, 1H, Ph), 7.41 (d, J_{H-H} = 18.4 Hz, 1H, =C$\underline{H}$(Ph)), 6.31 (d, J_{H-H} = 18.4 Hz, 1H, =C$\underline{H}$(Bpin)), 1.32 (s, 12H, CC$\underline{H}_3$) ppm. ^{13}C NMR (75 MHz, CDCl$_3$, 25 °C): δ = 146.0, 139.6, 132.2 (q, J_{C-F} = 22.3 Hz, –CF$_3$), 126.9, 124.8, 122.2, 122.0, 83.9 ($\underline{C}$CH$_3$), 25.0 (C$\underline{C}$H$_3$) ppm. C$_\alpha$ to boron atom was not observed. MS (EI) [m/z (%)]: 366(M$^+$, 44), 351(39), 347(22), 309(27), 280(100), 267(31), 247(77), 211(51), 169(19), 151(25), 97(24), 85(70), 59(77). Isolated yield: 81%. NMR and GC-MS data are in agreement with the literature [29].

(E)-4,4,5,5-Tetramethyl-2-(4′-methylstyryl)-1,3,2-dioxaborolane (**3f**), ^{1}H NMR (300 MHz, acetone-d_6, 25 °C): δ = 7.46 (d, J_{H-H} = 8.1 Hz, 2H, Ph), 7.33 (d, J_{H-H} = 18.4 Hz, 1H, =C$\underline{H}$(Ph)), 7.18 (d, J_{H-H} = 8.0 Hz, 2H, Ph), 6.09 (d, J_{H-H} = 18.4 Hz, 1H, =C$\underline{H}$(Bpin)), 2.32 (s, 3H, 4-C$\underline{H}_3$Ph), 1.27 (s, 12H, CC$\underline{H}_3$) ppm. ^{13}C NMR (75 MHz, CDCl$_3$, 25 °C): δ = 149.6, 139.0, 134.8, 129.4, 127.1, 83.3 ($\underline{C}$CH$_3$), 24.9 (C$\underline{C}$H$_3$), 21.5 (Ph$\underline{C}$H$_3$) ppm. C$_\alpha$ to boron atom was not observed. MS (EI) [m/z (%)]: 244(M$^+$, 27), 229(9), 171(7), 158(51), 143(100), 128(41), 117(51), 105(12), 91(19), 77(8), 57(7). Isolated yield: 79%. NMR and GC-MS data are in agreement with the literature [29].

(E)-4,4,5,5-Tetramethyl-2-(2′-methylstyryl)-1,3,2-dioxaborolane (**3g**), ^{1}H NMR (300 MHz, acetone-d_6, 25 °C): δ = 7.44–7.09 (m, 5H, Ph, =C$\underline{H}$(Ph)), 6.14 (d, J_{H-H} = 18.4 Hz, 1H, =C$\underline{H}$(Bpin)), 2.34 (s, 3H, 2-C$\underline{H}_3$Ph), 1.28 (s, 12H, CC$\underline{H}_3$) ppm. ^{13}C NMR (75 MHz, CDCl$_3$, 25 °C): δ = 149.8, 138.2, 137.5, 129.9, 128.5, 127.9, 124.3, 83.4 ($\underline{C}$CH$_3$), 24.9 (C$\underline{C}$H$_3$), 21.5 (Ph$\underline{C}$H$_3$) ppm. C$_\alpha$ to boron atom was not observed. MS (EI) [m/z (%)]: 244(M$^+$, 32), 229(12), 171(13), 158(45), 143(100), 128(31), 117(49), 105(11), 91(112), 77(8), 57(10). Isolated yield: 69%. NMR and GC-MS data are in agreement with the literature [29].

(E)-2-(4′-(Tert-butyl)styryl)-4,4,5,5-tetramethyl-1,3,2-dioxaborolane (**3h**), ^{1}H NMR (300 MHz, acetone-d_6, 25 °C): δ = 7.49 (d, J_{H-H} = 8.4 Hz, 2H, Ph), 7.41 (d, J_{H-H} = 8.5 Hz, 2H, Ph), 7.34 (d, J_{H-H} = 18.4 Hz, 1H, =C$\underline{H}$(Ph)), 6.11 (d, J_{H-H} = 18.4 Hz, 1H, =C$\underline{H}$(Bpin)), 1.31 (s, 9H, 4-t-$\underline{B}$uPh), 1.28 (s, 12H, CC$\underline{H}_3$) ppm. ^{13}C NMR (75 MHz, acetone-d_6, 25 °C): δ = 153.1, 150.2, 135.8, 127.8, 126.5, 83.9 ($\underline{C}$CH$_3$), 35.4, 31.6, 25.3 (C$\underline{C}$H$_3$) ppm. C$_\alpha$ to boron atom was not observed. MS (EI) [m/z (%)]: 286(M$^+$, 219), 271(100), 186(11), 171(19), 155(15), 143(23), 129(9), 83(10), 57(13). Isolated yield: 73%. NMR and GC-MS data are in agreement with the literature [29].

(E)-2-(4′-Methoxystyryl)-4,4,5,5-tetramethyl-1,3,2-dioxaborolane (**3i**), ^{1}H NMR (300 MHz, CDCl$_3$, 25 °C): δ = 7.44 (d, J_{H-H} = 8.7 Hz, 2H, Ph), 7.35 (d, J_{H-H} = 18.4 Hz, 1H, =C$\underline{H}$(Ph)), 6.86 (d, J_{H-H} = 8.7 Hz, 2H, Ph), 6.01 (d, J_{H-H} = 18.4 Hz, 1H, =C$\underline{H}$(Bpin)), 3.81 (s, 3H, 4-OC$\underline{H}_3$Ph), 1.31 (s, 12H, CC$\underline{H}_3$) ppm. ^{13}C NMR (75 MHz, CDCl$_3$, 25 °C): δ = 160.4, 149.2, 130.5, 128.6, 114.1, 83.3 ($\underline{C}$CH$_3$), 55.4, 25.0 (C$\underline{C}$H$_3$) ppm. C$_\alpha$ to boron atom was not observed. MS (EI) [m/z (%)]: 260(M$^+$, 39), 245(10), 187(9), 174(25), 160(64), 144(100), 129(17), 117(31), 91(19), 77(31), 57(16). Isolated yield: 71%. NMR and GC-MS data are in agreement with the literature [29].

(E)-2-(2-Cyclohexylvinyl)-4,4,5,5-tetramethyl-1,3,2-dioxaborolane (**3j**), ^{1}H NMR (300 MHz, CDCl$_3$, 25 °C): δ = 6.57 (dd, J_{H-H} = 18.2, 6.2 Hz, 1H, =C$\underline{H}$(c-hexyl)), 5.37 (d, J_{H-H} = 18.2 Hz, 1H, =C$\underline{H}$(Bpin)), 1.83–1.55 (m, 7H), 1.26 (s, 12H, CC$\underline{H}_3$), 1.22–1.00 (m, 4H) ppm. ^{13}C NMR (75 MHz, CDCl$_3$, 25 °C): δ = 160.1, 83.1 ($\underline{C}$CH$_3$), 43.4, 32.0, 26.3, 26.1, 24.9 (C$\underline{C}$H$_3$) ppm. C$_\alpha$ to boron atom was not observed. MS (EI) [m/z (%)]: 236(M$^+$, 4), 221(7), 178(9), 153(15), 135(21), 123(24), 108(58), 84(100), 69(16), 55(14). Isolated yield: 87%. NMR and GC-MS data are in agreement with the literature [29].

(E)-1,2-Diphenylethene (**4**), ^{1}H NMR (300 MHz, CDCl$_3$, 25 °C): δ = 7.57 (d, J_{H-H} = 8.0 Hz, 4H), 7.47–7.27 (m, 6H), 7.17 (s, 2H, =C$\underline{H}$Ph) ppm. ^{13}C NMR (75 MHz, CDCl$_3$, 25 °C): δ = 137.5 (=$\underline{C}$HPh), 128.8, 127.8, 126.6 ppm. MS (EI) [m/z (%)]: 180(M$^+$, 100), 165(22), 152(12), 102(11), 89(15), 77(11). Isolated yield: 87%. NMR and GC-MS data are in agreement with the literature [32].

(E)-(2-Iodovinyl)benzene (**5**), ^{1}H NMR (300 MHz, CDCl$_3$, 25 °C) δ = 7.42 (d, J_{H-H} = 14.9 Hz, 1H, =C$\underline{H}$), 7.28–7.08 (m, 5H), 6.77 (d, J_{H-H} = 14.9 Hz, 1H, =C$\underline{H}$). ^{13}C NMR (75 MHz, CDCl$_3$, 25 °C) δ 144.95, 138.48, 135.15, 129.51, 126.02, 75.49. MS (EI) [m/z (%)]: 230(M+, 100), 127(13), 103(85), 77(53). Isolated yield: 81%. NMR and GC-MS data are in agreement with the literature [34].

4. Conclusions

The new, effective and highly regio- and stereoselective catalytic systems based on the immobilization of [Ru(CO)Cl(H)(PCy$_3$)$_2$] in ionic liquids for the borylative coupling of vinylboronic pinacol ester with olefins was presented for the first time within these studies. The best results, characterized by excellent product yields, were obtained when [BMIm][OTf] and [EMPyr][NTf$_2$] were used as solvents and catalyst immobilization media. The system permitted a transformation of various vinyl-substituted olefins (styrene (**2a**), functionalized styrenes (**2b–i**) and vinylcyclohexane (**2j**)) into useful borylsubstituted unsaturated products, which can be used in deborylation or Suzuki coupling reactions. The application of ILs permitted catalyst reuse and the carrying out of repetitive batch borylative coupling in monophasic [Ru(CO)Cl(H)(PCy$_3$)$_2$]@ILs or biphasic [Ru(CO)Cl(H)(PCy$_3$)$_2$]@[BMIm][OTf]/scCO$_2$ systems with product extraction with non-polar *n*-heptane or scCO$_2$. These new protocols, under optimized reaction ([Ru-H]:**1**:**2** = 0.01:1:1.5, 110 °C, 6 h, 1 g IL, inert atmosphere conditions) and extraction (3 × 5 mL of the *n*-heptane at 60 °C or 160–180 bar of CO$_2$ at 40 °C, 8 mL/min, 45 min) conditions, allowed of the catalyst to be reused 5–7 times without any significant loss of activity or stability. The application of scCO$_2$ as an extractant in monophasic or biphasic solvent systems significantly reduced catalyst leaching during the separation process (<1 ppm in each batch), compared to the extraction with *n*-heptane. Moreover, the good solubility of the reagents and products in scCO$_2$ and the high catalyst stability creates future possibilities for carrying out this process under a continuous flow method. In addition, it was found that the application of ILs as reaction media has a positive impact on the reaction rate, shortening the process from 24 h to 6 h. Such an approach based on multiple catalyst use enabled an intensification of the cumulative TON values (up to 956) in comparison to the single batch (~50–100), showing the potential of the system reported within this research, which was developed according to the sustainable development rules. The catalyst reuse, simplification of the separation process, increased process productivity and possibilities to carry out the reaction in repetitive batch borylative coupling of olefins are in agreement with green chemistry paradigms.

Supplementary Materials: The following are available online at http://www.mdpi.com/2073-4344/10/7/762/s1, Procedure description of GC-MS, NMR and ICP-MS analyses, NMR spectra for all isolated products (**3a–j**, **4**, **5**).

Author Contributions: Conceptualization, J.S. and J.W.; methodology, J.S. and J.W.; investigation, J.S. and T.S.; resources, J.S., T.S., A.F. and J.W.; writing—original draft preparation, J.S.; writing—review and editing, J.W., A.F., T.S. and J.S.; supervision, J.W.; funding acquisition, T.S. All authors have read and agreed to the published version of the manuscript.

Funding: The APC was funded by grant No. POWR.03.02.00-00-I026/16 co-financed by the EU through the European Social Fund under the Operational Program Knowledge Education Development.

Acknowledgments: This work was supported by The National Centre for Research and Development in Poland, Lider Programme No. LIDER/26/527/L-5/13/NCBR/2014 and The National Science Centre No. UMO-2019/32/C/ST4/00235 and The National Centre for Research and Development in Poland, Lider Programme No. LIDER/6/0017/L-9/17/NCBR/2018 and grant No. POWR.03.02.00-00-I026/16 co-financed by the EU through the European Social Fund under the Operational Program Knowledge Education Development.

Conflicts of Interest: The authors declare no conflict of interest.

References

1. Bhaduri, S.; Mukesh, D. Chemical industry and homogeneous catalysis. In *Homogeneous Catalysis*; Bhaduri, S., Mukesh, D., Eds.; John Wiley & Sons, Inc.: Hoboken, NJ, USA, 2014; pp. 1–21.
2. Joshi, S.S.; Bhatnagar, A.; Ranade, V.V. Chapter 8—Catalysis for fine and specialty chemicals. In *Industrial Catalytic Processes for Fine and Specialty Chemicals*; Joshi, S.S., Ranade, V.V., Eds.; Elsevier: Amsterdam, The Netherlands, 2016; pp. 317–392.
3. Cole-Hamilton, D.J.; Tooze, R.P. *Catalyst Separation, Recovery and Recycling: Chemistry and Process Design*; Springer Science & Business Media: Berlin/Heidelberg, Germany, 2006; Volume 30, pp. 1–247.
4. Centi, G.; Perathoner, S. Methods and tools of sustainable industrial chemistry: Catalysis. *Sustain. Ind. Chem.* **2009**, 73–198. [CrossRef]

5. Shende, V.S.; Saptal, V.B.; Bhanage, B.M. Recent advances utilized in the recycling of homogeneous catalysis. *Chem. Rec.* **2019**, *19*, 2022–2043. [CrossRef] [PubMed]
6. Cole-Hamilton, D.J. Homogeneous catalysis—New approaches to catalyst separation, recovery, and recycling. *Science* **2003**, *299*, 1702–1706. [CrossRef] [PubMed]
7. Clarke, C.J.; Tu, W.-C.; Levers, O.; Bröhl, A.; Hallett, J.P. Green and sustainable solvents in chemical processes. *Chem. Rev.* **2018**, *118*, 747–800. [CrossRef] [PubMed]
8. Walkowiak, J.; Franciò, G.; Leitner, W. Supercritical fluids as advanced media for reaction and separation in homogeneous catalysis. In *Applied Homogeneous Catalysis with Organometallic Compounds: A Comprehensive Handbook in Four Volumes*; Wiley-VCH Verlag GmbH & Co. KGaA.: Weinheim, Germany, 2018; pp. 1221–1258.
9. Bermúdez, M.-D.; Jiménez, A.-E.; Sanes, J.; Carrión, F.-J. Ionic liquids as advanced lubricant fluids. *Molecules* **2009**, *14*, 2888–2908. [CrossRef]
10. Zhou, F.; Liang, Y.; Liu, W. Ionic liquid lubricants: Designed chemistry for engineering applications. *Chem. Soc. Rev.* **2009**, *38*, 2590–2599. [CrossRef] [PubMed]
11. Zhao, Y.; Bostrom, T. Application of ionic liquids in solar cells and batteries: A review. *Curr. Org. Chem.* **2015**, *19*, 556–566. [CrossRef]
12. MacFarlane, D.R.; Tachikawa, N.; Forsyth, M.; Pringle, J.M.; Howlett, P.C.; Elliott, G.D.; Davis, J.H.; Watanabe, M.; Simon, P.; Angell, C.A. Energy applications of ionic liquids. *Energy Environ. Sci.* **2014**, *7*, 232–250. [CrossRef]
13. Tan, S.S.Y.; MacFarlane, D.R. Ionic liquids in biomass processing. In *Ionic Liquids*; Springer: Berlin/Heidelberg, Germany, 2009; pp. 311–339.
14. Tadesse, H.; Luque, R. Advances on biomass pretreatment using ionic liquids: An overview. *Energy Environ. Sci.* **2011**, *4*, 3913–3929. [CrossRef]
15. Vekariya, R.L. A review of ionic liquids: Applications towards catalytic organic transformations. *J. Mol. Liq.* **2017**, *227*, 44–60. [CrossRef]
16. Welton, T. Ionic liquids in catalysis. *Coord. Chem. Rev.* **2004**, *248*, 2459–2477. [CrossRef]
17. Pârvulescu, V.I.; Hardacre, C. Catalysis in ionic liquids. *Chem. Rev.* **2007**, *107*, 2615–2665. [CrossRef]
18. Blanchard, L.A.; Brennecke, J.F. Recovery of organic products from ionic liquids using supercritical carbon dioxide. *Ind. Eng. Chem. Res.* **2001**, *40*, 287–292. [CrossRef]
19. Jutz, F.; Andanson, J.-M.; Baiker, A. Ionic liquids and dense carbon dioxide: A beneficial biphasic system for catalysis. *Chem. Rev.* **2011**, *111*, 322–353. [CrossRef] [PubMed]
20. Jessop, P.G.; Stanley, R.R.; Brown, R.A.; Eckert, C.A.; Liotta, C.L.; Ngo, T.T.; Pollet, P. Neoteric solvents for asymmetric hydrogenation: Supercritical fluids, ionic liquids, and expanded ionic liquids. *Green Chem.* **2003**, *5*, 123–128. [CrossRef]
21. Geier, D.; Schmitz, P.; Walkowiak, J.; Leitner, W.; Franciò, G. Continuous flow asymmetric hydrogenation with supported ionic liquid phase catalysts using modified co2 as the mobile phase: From model substrate to an active pharmaceutical ingredient. *ACS Catal.* **2018**, *8*, 3297–3303. [CrossRef]
22. Muzart, J. Ionic liquids as solvents for catalyzed oxidations of organic compounds. *Adv. Synth. Catal.* **2006**, *348*, 275–295. [CrossRef]
23. Rieger, B.; Plikhta, A.; Castillo-Molina, D.A. Ionic liquids in transition metal-catalyzed hydroformylation reactions. In *Ionic Liquids (ILs) in Organometallic Catalysis*; Dupont, J., Kollár, L., Eds.; Springer: Berlin/Heidelberg, Germany, 2015; pp. 95–144.
24. Li, J.; Yang, S.; Wu, W.; Jiang, H. Recent advances in Pd-catalyzed cross-coupling reaction in ionic liquids. *Eur. J. Org. Chem.* **2018**, *2018*, 1284–1306. [CrossRef]
25. Maciejewski, H.; Szubert, K.; Marciniec, B.; Pernak, J. Hydrosilylation of functionalised olefins catalysed by rhodium siloxide complexes in ionic liquids. *Green Chem.* **2009**, *11*, 1045–1051. [CrossRef]
26. Maciejewski, H.; Szubert, K.; Marciniec, B. New approach to synthesis of functionalised silsesquioxanes via hydrosilylation. *Catal. Commun.* **2012**, *24*, 1–4. [CrossRef]
27. Maciejewski, H.; Szubert, K.; Fiedorow, R.; Giszter, R.; Niemczak, M.; Pernak, J.; Klimas, W. Diallyldimethylammonium and trimethylvinylammonium ionic liquids—Synthesis and application to catalysis. *Appl. Catal. A* **2013**, *451*, 168–175. [CrossRef]
28. Rogalski, S.; Żak, P.; Miętkiewski, M.; Dutkiewicz, M.; Fiedorow, R.; Maciejewski, H.; Pietraszuk, C.; Śmiglak, M.; Schubert, T.J. Efficient synthesis of E-1, 2-bis (silyl) ethenes via ruthenium-catalyzed homocoupling of vinylsilanes carried out in ionic liquids. *Appl. Catal. A* **2012**, *445*, 261–268. [CrossRef]

29. Szyling, J.; Walkowiak, J.; Sokolnicki, T.; Franczyk, A.; Stefanowska, K.; Klarek, M. PEG-mediated recyclable borylative coupling of vinyl boronates with olefins. *J. Catal.* **2019**, *376*, 219–227. [CrossRef]
30. Szyling, J.; Franczyk, A.; Stefanowska, K.; Maciejewski, H.; Walkowiak, J.d. Recyclable hydroboration of alkynes using RuH@ IL and RuH@ IL/scCO2 catalytic systems. *ACS Sustain. Chem. Eng.* **2018**, *6*, 10980–10988. [CrossRef]
31. Szyling, J.; Franczyk, A.; Stefanowska, K.; Klarek, M.; Maciejewski, H.; Walkowiak, J. An effective catalytic hydroboration of alkynes in supercritical CO2 under repetitive batch mode. *ChemCatChem* **2018**, *10*, 531–539. [CrossRef]
32. Szyling, J.; Franczyk, A.; Stefanowska, K.; Walkowiak, J. A recyclable Ru (CO) Cl (H)(PPh3) 3/PEG catalytic system for regio-and stereoselective hydroboration of terminal and internal alkynes. *Adv. Synth. Catal.* **2018**, *360*, 2966–2974. [CrossRef]
33. Stefanowska, K.; Franczyk, A.; Szyling, J.; Salamon, K.; Marciniec, B.; Walkowiak, J. An effective hydrosilylation of alkynes in supercritical CO2–A green approach to alkenyl silanes. *J. Catal.* **2017**, *356*, 206–213. [CrossRef]
34. Szyling, J.; Franczyk, A.; Pawluć, P.; Marciniec, B.; Walkowiak, J. A stereoselective synthesis of (E)-or (Z)-β-arylvinyl halides via a borylative coupling/halodeborylation protocol. *Org. Biomol. Chem.* **2017**, *15*, 3207–3215. [CrossRef]
35. Szyling, J.; Walkowiak, J. Effective one-pot synthesis of (E)-poly (vinyl arylenes) via trans-borylation/Suzuki coupling protocol. *Green Proc. Synth.* **2017**, *6*, 301–310. [CrossRef]
36. Denmark, S.E.; Tymonko, S.A. Sequential cross-coupling of 1, 4-bissilylbutadienes: Synthesis of unsymmetrical 1, 4-disubstituted 1, 3-butadienes. *J. Am. Chem. Soc.* **2005**, *127*, 8004–8005. [CrossRef]
37. Carreras, J.; Caballero, A.; Pérez, P.J. Alkenyl boronates: Synthesis and applications. *Chem. Asian J.* **2019**, *14*, 329–343. [CrossRef]
38. Pawluć, P.; Franczyk, A.; Walkowiak, J.; Hreczycho, G.; Kubicki, M.; Marciniec, B. (E)-9-(2-Iodovinyl)-9H-carbazole: A new coupling reagent for the synthesis of π-conjugated carbazoles. *Org. Lett.* **2011**, *13*, 1976–1979. [CrossRef]
39. Aubin, S.; Le Floch, F.; Carrié, D.; Guegan, J.P.; Vaultier, M. Transition-metal-catalyzed hydrosilylation and hydroboration of terminal alkynes in ionic liquids. In *Ionic Liquids; ACS Symposium Series*; American Chemical Society: Washington, DC, USA, 2002; Volume 818, pp. 334–346.
40. Marciniec, B.; Jankowska, M.; Pietraszuk, C. New catalytic route to functionalized vinylboronates. *Chem. Commun.* **2005**, 663–665. [CrossRef] [PubMed]
41. Marciniec, B.; Walkowiak, J. Ruthenium (II) complex catalyzed O-borylation of alcohols with vinylboronates. *Synlett* **2009**, *2009*, 2433–2436. [CrossRef]
42. Marciniec, B.; Walkowiak, J. New catalytic route to borasiloxanes. *Chem. Commun.* **2008**, 2695–2697. [CrossRef] [PubMed]
43. Walkowiak, J.; Marciniec, B. A new catalytic method for the synthesis of boroxanes. *Tetrahedron Lett.* **2010**, *51*, 6177–6180. [CrossRef]
44. Blanchard, L.A.; Gu, Z.; Brennecke, J.F. High-pressure phase behavior of ionic liquid/CO2 systems. *J. Phys. Chem. B* **2001**, *105*, 2437–2444. [CrossRef]
45. Yi, C.S.; Lee, D.W.; Chen, Y. Hydrovinylation and [2 + 2] cycloaddition reactions of alkynes and alkenes catalyzed by a well-defined cationic ruthenium—Alkylidene complex. *Organometallics* **1999**, *18*, 2043–2045. [CrossRef]
46. Ahmad, N.; Levison, J.J.; Robinson, S.; Uttley, M.; Wonchoba, E.; Parshall, G. Complexes of ruthenium, osmium, rhodium, and iridium containing hydride carbonyl, or nitrosyl ligands. *Inorg. Synth.* **1974**, *15*, 45–64.

© 2020 by the authors. Licensee MDPI, Basel, Switzerland. This article is an open access article distributed under the terms and conditions of the Creative Commons Attribution (CC BY) license (http://creativecommons.org/licenses/by/4.0/).

MDPI
St. Alban-Anlage 66
4052 Basel
Switzerland
Tel. +41 61 683 77 34
Fax +41 61 302 89 18
www.mdpi.com

Catalysts Editorial Office
E-mail: catalysts@mdpi.com
www.mdpi.com/journal/catalysts